D0504513

L. I. H. E.
THE MARKLAND LIBRARY
STAND PARK RD., LIVERPOOL, L16 9JD

CHRIST'S COLLEGE
LIBRARY

THE STRATIGRAPHY OF THE BRITISH ISLES

BY

DOROTHY H. RAYNER, M.A., PH.D.

Senior Lecturer in the Department of Earth Sciences, University of Leeds

CAMBRIDGE
AT THE UNIVERSITY PRESS
1971

Published by the Syndics of the Cambridge University Press
Bentley House, 200 Euston Road, London NW1 2DB
American Branch: 32 East 57th Street, New York, N.Y.10022

© Cambridge University Press 1967

Library of Congress Catalogue Card Number: 67-12319

ISBN: 0 521 06047 8

First published 1967
Reprinted with corrections 1971

Printed in Great Britain
at the University Printing House, Cambridge
(Brooke Crutchley, University Printer)

CL52436

CHRIST'S COLLEGE LIBRARY
Accession No. 21480
THE MARKLAND LIBRARY
Class No. 910.517
Catal. ✓

Replacement Copy

CONTENTS

PLATES

PREFACE

British stratigraphy has been studied and discussed for some 150 years and the writer who presents yet another synthesis should set out its scope and purpose. In this book I am largely concerned with surface and near-surface processes and the rocks that resulted from them. The outcome is stratigraphy in a somewhat restricted sense and not 'historical geology'. Igneous, metamorphic and structural aspects appear only cursorily, when bound up with stratigraphical problems.

It also happens that the present time is opportune for a review of the subject. In the past two decades much has been discovered and assembled about Ireland, and Irish stratigraphy has become more closely integrated with that of Great Britain, to the benefit of understanding on both sides. Advances in allied sciences, especially sedimentology and isotopic dating, have altered subtly both the factual syntheses and deductions made from them. The permanence of ocean basins and continents has once more been called into question and the stratigraphy and structure of these islands, on the edge of a continent, have a significance that is global as well as European. Lastly by historical accident Britain is the origin of many stratigraphical terms, especially of several systems and their larger subdivisions, and definitions, re-definitions and much nomenclatural turmoil are in the air.

Thus some fresh breezes have found their way through the classic edifice of British stratigraphy, but to the writer of a textbook this also brings drawbacks and the student reader should perhaps be warned that it is precisely in the most active fields that a general work becomes dated, and that he may have to search further after a few years. For this I am not greatly concerned: any writer can only set down, as clearly as may be, the facts and interpretations of his age, make plain the areas of doubt and leave future research to confirm or reject his generalizations.

For whom is this book written? I have aimed at making the subject intelligible to those with little previous knowledge of it, but some acquaintance with the broad principles of geology. Such an acquaintance can be gained from Read and Watson's *Introduction to Geology*, vol. 1 (1962) or Holmes' *Principles of Physical Geology* (2nd edn., 1965)—to name but two recently published works. In particular I have not explained many terms available in the former, but some topics of especial importance to stratigraphy are taken a little further in the Introduction.

Selected references appear at the end of each chapter, and within the last—which is some form of concluding (but scarcely 'conclusive') epilogue, touching on a few broader or more controversial problems. All references are collected together in the list at the end. They are probably beyond the needs of the elementary student, but it has been my hope that the more advanced, or perhaps the worker in more distant fields, while familiar with the bulk of the text, may find the list useful as a first source for more detailed study. Moreover, if one principle more than another has imbued the work it is that while generalizations and oversimplified deductions are inevitably the substance of a textbook, yet they stand weakly alone, and that the evidence should be accessible as far as space allows.

At the time of writing there is no formal British Code of Stratigraphical Nomenclature, though the American Code, which is often taken as a guide, may be found in several standard works, such as Krumbein and Sloss (1963, pp. 23, 621). In these circumstances I have thought it best to retain traditional or current British terms even though a few may not be strictly correct, or there may be minor inconsistencies between one system and another. At least this retention should not introduce any new complexities.

Most of the chapters have been read in draft by one or more persons more deeply versed than I in the various geological systems and to them I am most sincerely grateful: Dr H. W. Ball, Mr M. A. Calver, Mr D. Curry, Dr W. T. Dean, Professor D. T. Donovan, Dr K. C. Dunham, Dr J. C. Harper, Professor J. E. Hemingway, Dr R. W. Hey, Professor M. R. House, Professor W. Q. Kennedy, Dr J. W. Neale, Mr L. F. Penny, Dr W. H. C. Ramsbottom, Professor R. M. Shackleton, Professor F. W. Shotton and Dr J. V. Watson. Their friendly advice has added detail and perspective to the descriptive sections and saved me from many errors and questionable deductions. They are in no way responsible for the shortcomings that remain. In addition I would like to record my thanks to very many friends, in Leeds and elsewhere, who have provided information on many topics and borne patiently with my enquiries.

The illustrations reproduce or incorporate much copyright material. The plates are reproduced by permission of the persons or organisations given in the list on page vi; the sources of the figures appear in their legends. I am glad to thank all holders of copyright for permission to reproduce this material.

D. H. R.

1966

Table 1. *The Phanerozoic time-scale*

The age of the base of each stratigraphical unit is given in millions of years. Taken from the Geological Society Phanerozoic time-scale 1964 (Harland, Gilbert Smith and Wilcock, 1964, pp. 260–2).

The stratigraphical units and zones that are well developed in the British Isles are set out in the Appendix (pp. 407–15).

CAINOZOIC	
QUATERNARY	
Pleistocene	1·5–2
TERTIARY	
PLIOCENE	*c.* 7
MIOCENE	
Upper	*c.* 12
Middle	*c.* 18–19
Lower	26
OLIGOCENE	
Upper	?
Middle	31–32
Lower	37–38
EOCENE	
Upper	*c.* 45
Middle	*c.* 49
Lower	53–54
PALAEOCENE	
Upper	58·5
Lower	65
MESOZOIC	
CRETACEOUS	
UPPER	
Maestrichtian	70
Campanian	76
Santonian	82
Coniacian	88
Turonian	94
Cenomanian	100
LOWER	
Albian	106
Aptian	112
Barremian	118
Hauterivian	124
Valanginian	130
Ryazanian	136
JURASSIC	
UPPER	
Purbeckian	141
Portlandian	146
Kimeridgian	151
Oxfordian	157
Callovian	162
MIDDLE	
Bathonian	167
Bajocian	172
LOWER	
Toarcian	178
Pliensbachian	183
Sinemurian	188
Hettangian	190–195
TRIASSIC	
UPPER	
Rhaetian	
Norian	
Karnian	[205]
MIDDLE	
Ladinian	
Anisian	[215]
LOWER	
Olenekian	
Induan	225
PALAEOZOIC	
PERMIAN	
UPPER	
Tatarian	230
Kazanian	240
LOWER	
Kungurian	255–258
Artinskian	265–268
Sakmarian	280
CARBONIFEROUS	
UPPER (SILESIAN)	
Stephanian	290–295
Westphalian	310–315
Namurian	325
LOWER (DINANTIAN)	
Viséan	335–340
Tournaisian	345
DEVONIAN	
UPPER	
Famennian	353
Frasnian	359
MIDDLE	
Givetian	
Eifelian	370
LOWER	
Emsian	374
Siegenian	390
Gedinnian	395
SILURIAN	
Ludlow	
Wenlock	
Llandovery	430–440
ORDOVICIAN	
UPPER	
Ashgill	
Caradoc	445
LOWER	
Llandeilo	
Llanvirn	
Arenig	*c.* 500
CAMBRIAN	
UPPER	515
MIDDLE	540
LOWER	570

1

INTRODUCTION

There have been many definitions of stratigraphy, differing slightly in scope and emphasis, but its core is the study of the earth's history, in so far as it can be interpreted from the outermost layers of the crust and particularly from the succession of stratified rocks. As a formal term it is a latecomer (appearing in 1865, according to the *Oxford English Dictionary*); but the description of stratified rocks had been placed on a scientific basis long before, in the late eighteenth and early nineteenth centuries—a phase of geological history that will always be associated in this country with the work of William Smith.[1] Since that time, when stratigraphy concentrated on the more accessible, fossiliferous strata, there have been some far-reaching extensions in methods and scope. More has been found out about rocks at depth, either through boring or by geophysical methods, and about those underlying seas and oceans, so that research is no longer confined to surface, terrestrial exposures. Stratigraphical and allied techniques have been applied successfully to highly metamorphosed and deformed rocks and, in particular, Pre-Cambrian history is much better known. From a greater knowledge of modern sediments, their composition and origin, our interpretations of the lithified, sedimentary rocks and their environments are now on a more secure basis. Nevertheless in all these fields much remains to be done, and they indicate some of the directions in which stratigraphy and historical geology are likely to advance in the next few decades.

This chapter is concerned with a few old-fashioned principles in stratigraphy, and particularly with the explanation and description of some of the ideas, methods and terms that appear in the following, systematic chapters. It must be emphasized that it is not a complete, definitive list of such topics, and further reading is referred to on p. 26.

Like other branches of geology, stratigraphy owes much to the physical sciences, but its basic concepts cannot really be defined in the same way as the 'laws' of those sciences. Even where the word 'law' appears

[1] References to the lives and work of such classic British geologists as William Smith, Lyell, Sedgwick, Murchison and Lapworth are given at the end of the chapter.

(The Law of the Order of Superposition of Strata, The Law of Uniformity) it is based on customary usage rather than on modern justification. These concepts are better considered as principles.

The principle of superposition summarizes the fact, so well known as to be axiomatic, that in a succession of sedimentary strata the younger beds, having been laid down on top of the older, consequently lie above them. The principle holds for horizontal or gently tilted strata, but obviously fails in those cases where the beds are vertical or overturned, when other methods must be used. In rare instances within undisturbed successions it is possible for new sediments to be found below older ones, as in vertical fissures ('Neptunean dykes') or solution hollows and their ramifications, but these minor structures are usually self-evident and cause little confusion.

The succession of life through time, and the observable fact that plant and animal species, genera or families, each have their characteristic time-span and do not recur in later ages, leads to the correlation of strata by their fossils. Bearing in mind that there are certain forms with an exceptionally long time-range, it is generally true that a few chosen short-range forms, or a whole assemblage, will distinguish a bed, or group of beds, from those above or below. There are no obvious exceptions to this rule, but again other methods must be used in unfossiliferous rocks.

The principle of uniformity or uniformitarianism—its phrasing differs slightly with different authors—is based on the idea that surface processes have not altered significantly during the earth's known history. It is summarized in slightly popular fashion as the present being the key to the past, and is a fundamental tenet in all attempts to interpret past processes and environments from a knowledge of present ones. It also arose during an historical phase in geological and biological thought when an appreciation of gradual processes and slow transitions was replacing ideas of sudden or catastrophic changes. Darwin's *Origin of Species* (1859) was part of such a change, but Darwin owed much to earlier concepts, especially to the uniformitarian principles formulated by Lyell.

In its exact form the Principle of Uniformity is probably an overstatement. There is some evidence of radical chemical and physical changes in the composition of the atmosphere and oceans early in the earth's history, and much later there must have been marked biological changes. The clothing of much of the land surface by an advanced type of vegetation is an example. Although a covering of lichens, algae and fungi is locally an effective protection against erosion, the development of a rooted, forest vegetation must have had some effect on the régimes of denudation, transportation and deposition. Similarly, economic beds of coal are known only from Devonian and later rocks. Man himself has altered the appearance and the régimes of large tracts of land, though as yet the effects have had little time to influence stratigraphical successions.

TIME IN STRATIGRAPHY

Time, enormous stretches of time which are measured in millions of years, forms a background to most aspects of geology, particularly to stratigraphical and historical geology. It is a compensation to the palaeontologist for the gross imperfections of his material so that, though he sees only the skeletal structures of a small fraction of a once-living population, he can watch the way those structures change through time. To the structural geologist it is a fundamental factor in bridging the gap between the puny tectonic and volcanic effects visible in historic times—the slow changes in land and sea, earthquakes and volcanoes (however disastrous these may be to humanity)—and the growth of great mountain ranges with their folded, faulted and overturned strata and accompanying igneous intrusions. With time the sand, silt and mud of lakes, coastal flats and seas become lithified rocks.

In the last few decades the measurement of this basic dimension has grown into a subject in its own right—geochronology. Until the early years of this century the age ascribed to geological strata or events meant stratigraphical age—that is, their position in relation to the stratigraphical systems. The assessment of age in years (earlier called 'absolute age', but the adjective is misleading and superfluous) became possible with the discovery of radioactive substances in rocks, and there are now a very large number of isotopic (or radiometric) age determinations in various parts of the world, though chiefly of igneous rocks. They are based on the analysis of one or more of the pairs of isotopes (radioactive parent and daughter) that are shown in Table 2; these occur in a number of minerals, some rare and some common.

The first of each pair is transformed into the second at a known rate and from the relative amounts of the two an estimate can be made of the time since the mineral was formed, when it contained none of the daughter isotope. Any assessment of this, the isotopic age of a rock, and its relation to the stratigraphical systems is liable to three main sources of uncertainty or error. There are possible experimental errors in the

Table 2. *Principal isotopes used in age determination*

Method	Parent	Approximate half-life*	Daughter	Minerals
Uranium-lead	U^{238}	4,500	$Pb^{206}(+8He^4)$	Uraninite, pitchblende and zircon
Uranium-lead	U^{235}	700	$Pb^{207}(+7He^4)$	Uraninite, pitchblende and zircon
Thorium-lead	Th^{232}	14,000	$Pb^{208}(+6He^4)$	Monazite, zircon
Rhenium-osmium	Re^{187}	50,000	Os^{187}	Molybdenite
Rubidium-strontium	Rb^{87}	50,000	Sr^{87}	Micas, potassium feldspars
Potassium-argon and Potassium-calcium	K^{40}	1,300	Ar^{40}	Micas, sanidine hornblende and glauconite
			Ca^{40}	Sylvine, very old micas

* The half-life (in millions of years) is that in which the quantity of the parent isotope is reduced by a half.

analysis of what may be very minute quantities of radioactive substances; there is the possibility of loss of the daughter isotope since the mineral was formed—that is, it has not remained a closed chemical system; and the position of the specimen analysed in relation to a stratigraphical stage or zone may be somewhat doubtful.

Experimental methods have been much improved and the majority of determinations now published are subject to an analytical error of 5 per cent or less. The loss of the crucial daughter isotope, however, can be very serious. This is especially true of the older (notably Pre-Cambrian) rocks, in which the minerals may have been reconstituted and have lost the daughter isotope during a later period of extensive metamorphism, so that the age determined is only that of the last metamorphism. It is thus necessary to qualify many results as minimum dates or ages; among Pre-Cambrian determinations these may be much later, even by geological estimations of time, than the original formation of the rock. A Pre-Cambrian time-scale is thus most satisfactorily established in the very old stable, shield areas of the world which have escaped much of the folding and metamorphism of later periods. From such regions dates as far back as 3,500 million years have been obtained. Moreover Pre-Cambrian 'systems' cannot, in their nature, have the same criteria as those of later, classical stratigraphy, nor the same possibilities of being (in theory at least) world-wide in application. The isotopic dates in such regions are thus the primary means of establishing their history.

The current Phanerozoic time-scale (that is, of Cambrian and later systems, Table 1, p. xi) depends on the close dating of sedimentary rocks that contain reliable stratigraphical fossils, but this is not always possible. Glauconite is a mineral found in many marine sedimentary rocks, and is thus well suited from the correlative aspect, but it loses argon easily, especially if it has been deeply buried; nevertheless it has been used extensively for Jurassic and later sediments. Lavas and tuffs containing zircon, sanidine or mica can be very useful if they are interbedded with suitable sediments, and such rocks have provided an admirable series of dates in North America, where they are interleaved with fossiliferous Cretaceous and Tertiary deposits.

To be a satisfactory gauge of stratigraphical age an intrusion should cut fossiliferous strata of one age and be overlain by (or contribute safely recognizable detritus into) a later series, with as little time gap as possible between the two. Many of the Palaeozoic dates are based on intrusions, and those from Great Britain include the Dartmoor Granite (approximately 282 million years, Hercynian) and the Shap Granite of the Lake District. The latter is a good example of an intrusion whose age, already known within fairly close stratigraphical limits, can be told more precisely by radiometric methods. The granite is intruded into Ordovician volcanic rocks and is not affected by their folding, which is the same age as that of the nearby Silurian sediments; highly characteristic detritus from the granite is found in early Carboniferous conglomerates. Thus far it might belong to any part of the Devonian period, but the average of many determinations, about 395 million years, makes a Lower Devonian age highly probable.

Another isotopic method, on a very different scale, is based on the radioactive isotope of carbon, C^{14}, and it is often called radiocarbon dating. Very small amounts of C^{14} are produced in the upper atmosphere by cosmic ray reactions; it is distributed as carbon dioxide and incorporated in the tissues of plants and animals. While these are alive their proportion of C^{14} is in equilibrium with that of the atmosphere but after the organism dies no more is absorbed and the isotope is gradually lost, the half-life period being 5,568 years. Estimations of radiocarbon allow organic substances as old as 35,000–40,000 years old to be dated, and by special techniques it can be carried back to about 60,000 years, but is unlikely to be taken much further. The method is thus valuable in late Pleistocene stratigraphy and for dating archaeological objects.

SEDIMENTARY ROCKS AND STRUCTURES

The raw materials of stratigraphy are rocks (sedimentary, metamorphic, volcanic), their structures and relationships, and any fossils they may contain. Within the scope of this book, the British Isles, sedimentary rocks are the most important, and some of the characters and structures of most importance to stratigraphy are briefly reviewed.

Clastic rocks

Clastic or detrital rocks are primarily described by their grain size, or grade (Table 3), the classification being usually based on the Wentworth scale. Arenaceous (sandy) rocks are also called psammitic, and argillaceous (clays and shales) pelitic; these two terms are also used for metamorphosed sediments (or metasediments) where 'sandy', etc., would be misleading and inappropriate. Since the term quartzite traditionally refers to a rock composed of quartz grains cemented with silica, it can be used for a metamorphosed sandstone, or for a specific type of unaltered sandstone; the latter is then distinguished as an orthoquartzite. An arkose is a sandstone with a relatively high proportion of feldspar; the grains tend to be angular and the sorting moderate.

Table 3. *Classification of sedimentary rocks according to grain size*

Limits of grades	
Boulder conglomerate	(above 256 mm.)
	256 mm.
Cobble conglomerate	
	64 mm.
Pebble conglomerate	
	4 mm.
Very fine conglomerate (granule)	
	2 mm.
Very coarse sandstone	
	1 mm.
Coarse sandstone	
	$\frac{1}{2}$ mm.
Medium sandstone	
	$\frac{1}{4}$ mm.
Fine sandstone	
	$\frac{1}{8}$ mm.
Very fine sandstone	
	$\frac{1}{16}$ mm.
Siltstone	
	$\frac{1}{256}$ mm.
Clay or shale	(below $\frac{1}{256}$ mm.)

Greywacke is an important rock-type in stratigraphy, particularly in geosynclinal successions. Definitions vary slightly, but all lay emphasis on the lack of sorting, so that a rock that is dominantly of sandstone grade also has a matrix of clay and fine silt. The coarser grains are sometimes of pre-existing rocks ('lithic fragments') and feldspar may be present as well as quartz, in additon to other relatively unstable minerals. These characters, together with the lack of sorting, point to rapid deposition. Greywackes commonly show graded bedding, or occur in units in which all bedding is obscure; fossils are usually sparse or absent.

Greywacke, arkose and orthoquartzite are the three extreme types of psammitic rocks, indicative of the maturity of the sediment and related in some degree to the stratigraphical and tectonic setting. During the weathering and erosional processes in the source area (and to a smaller extent in transport) the less resistant components of a clastic sediment are gradually eliminated, and if these processes are lengthy the mature end-product contains very little detritus besides quartz grains, usually well sorted and well rounded. Orthoquartzites, or at least highly quartzose rocks, thus commonly result from the persistent working of detritus, often in stable areas. Greywackes and arkoses, on the other hand, with their less durable components, are relatively immature sediments, characteristically resulting from uplift and erosion in unstable areas. Greywackes are especially associated with geosynclinal conditions, sometimes in great thickness; arkoses are typically formed from the rapid erosion and deposition of detritus derived from an uplifted granitic terrain. In between the three extreme types there are many intermediates, such as subarkose (feldspathic sandstone) and subgreywacke; many sandstones of the Coal Measures fall into the latter category.

'Grit' is a term found in much stratigraphical writing but it is a bad one. In this book it is avoided whenever possible, except in certain traditional names (not descriptions) such as Denbighshire Grits or Millstone Grit. Even this proviso, however, scarcely lends approval to 'Pea Grit' (Middle Jurassic of the Cotswolds) which is a pisolitic limestone. As a lithological term 'grit' cannot be defined and has often been used for a rock of 'gritty' texture that stands out by being harder than its neighbours. In practice it has been applied to anything from a siltstone to a pebble conglomerate but more commonly is a medium to coarse-grained sandstone.

Turbidites are rocks inferred to have been deposited by turbidity, or density, currents—that is currents which, being charged with sediment, are denser than the surrounding water and consequently flow down a sub-aqueous slope and spread their load out on to the sea (or occasionally lake) floor. The term is thus genetic rather than descriptive of composition. Graded bedding is typical of turbidites and many greywackes are attributed to this mechanism.

Turbidites are also characterized by sole markings, or bottom structures, on the under surface. As the dense, turbid flows pass over the soft surface of previous sediments they erode various types of hollows and grooves. These are filled by the incoming sediments and so form casts on the under bedding plane; when aligned they may show the directions of the flow. Turbidity flows thus demand a submarine, or sublacustrine, slope, and they are known to flow down the continental slope at the present day with considerable force. It is not necessary, however, to presuppose a feature of such magnitude for all turbidity flows in the past, and the slopes may have been quite gentle.

Non-clastic rocks

Among these the limestones and dolomites are the most abundant. Their principal components, calcium carbonate and magnesium carbonate, often occur intimately associated, and few thick rock sequences consist of pure limestone or pure dolomite alone; dolomites in particular are often accompanied by a small amount of ferrous carbonate. Consequently it is common to refer to a 'carbonate sequence'. Carbonates may be formed in fresh and salt waters; in more saline bodies the more soluble salts are precipitated as well and these are known as evaporites. In Great Britain evaporites in bulk are found only in Permian and Triassic rocks, and their succession and characters are described in chapter 8. The low-grade ironstones, well represented in Britain, are dominated by ferrous carbonate.

Marl is a term that often appears in stratigraphical descriptions, but it is not easy to find a definition that covers all uses. The rock is usually fine-grained, of silt or sometimes clay grade, and often contains a considerable non-clastic proportion, usually of carbonates but sometimes of evaporites. Lake marls are often highly calcareous.

Sedimentary structures

Among these there are several that, when abundant, normally indicate shallow water. They include current-bedding and aqueous current ripple marks (another type of ripple, with a lower crest compared with the wave-length, is produced by wind). Rootlet beds result from plant growth both just above and below water level. Abundant desiccation cracks are one of the clearest indications of temporary emergence. Intertidal muds and silts are often reworked by many types of marine organisms, leaving tracks and burrows, and disrupting or destroying the original bedding.

Current-bedding and graded-bedding (Fig. 1) are both used extensively to deduce the top and bottom (the orientation or 'way-up') of beds that have been much folded and lack other means for determining their

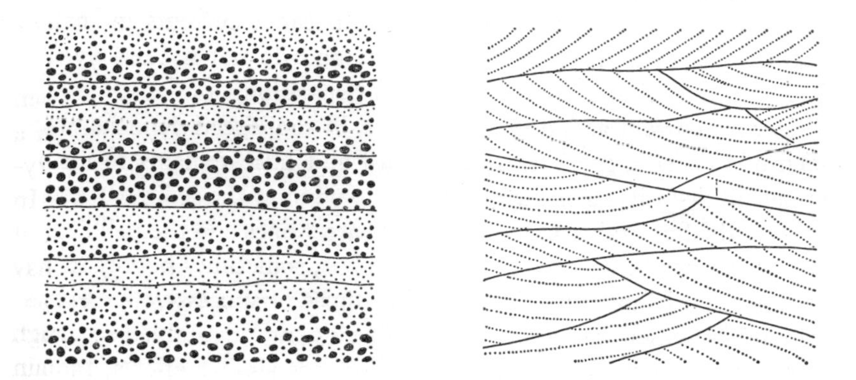

Fig. 1. Graded-bedding and current-bedding as guides to the orientation of sedimentary strata. On the left four out of the seven beds are graded and show an upward decrease in grain-size; on the right the majority of inclined foreset beds are concave upwards. Adapted from Shrock (1948) and other sources.

original superposition. In graded beds there is a gradual upward decrease in grain size, so that a unit that is dominantly arenaceous or conglomeratic at the base passes into silt or finer grade at the top. Such beds presumably represent a single influx of sediment from which the coarser particles settled more quickly than the finer. Graded-bedding in greywackes and quartzites in particular has been used to find out the orientation of local successions in the Scottish Highlands and Southern Uplands.

The use of current-bedding depends on the truncation of the inclined planes, or foreset beds. In certain types of current-bedding these tend to curve tangentially into the normal bedding planes at the base, but to be truncated at the top, where they have been eroded by the next influx of sediment. Nevertheless, if current-bedding foresets are examined in a succession whose superposition is indisputable, they commonly show a number of indeterminate examples. It is therefore imprudent to rely on isolated examples of either type of structure, but when repeated many times, and used in conjunction with structural and other types of evidence, they are valuable stratigraphical guides.

The direction of the foreset dips shows the direction of the water currents immediately responsible for them, and in certain cases, when a large number of observations can be assembled from a single stratigraphical unit, it is possible to get a picture of current directions within, say, a delta-complex or a sedimentary basin. Nevertheless, as indicators of the ultimate source of the sediments, or distribution of land and sea, they need interpreting cautiously, if only because currents in shallow irregular seas often take sinuous courses, and do not necessarily flow directly from sediment source to place of deposition.

Vigorous current action may also result in condensed deposits, when much of the sediment is carried onward and little remains to form a permanent deposit; such currents may also winnow the sediment, carrying away the fine grains and leaving the coarser ones and the pebbles. In a rock sequence this results in very thin beds, sometimes pebbly and, if the formation is normally rich in fossils, those of successive zones may be found crammed into a few inches of sediment, or jumbled together. Very slow or intermittent deposition produces similar effects, though without the signs of current action, and fossils and pebbles that remain for long periods on the sea floor, in addition to being rolled and worn, are often coated or infilled with calcium phosphate to form seams of 'phosphatic nodules'.

The term 'non-sequence' refers to the actual absence of one or more stratigraphical zones from similar causes—a smaller gap than an unconformity and lacking evidence of uplift or tilting and erosion. 'Remanié deposit' is also used for a jumbled layer of fossils and other debris. 'Derived' fossils include not only those from underlying beds, but others introduced by currents from elsewhere.

Disturbed bedding and slump structures

In most rocks that are unaffected by tectonic folding or tilting the bedding planes are horizontal or nearly so, except for those formed on an original slope of deposition, such as the foresets of current-bedding, or those laid down on a steep shore or the flanks of a reef. Occasionally, however, a normally bedded series contains units in which the bedding planes are strongly convoluted, or even disrupted altogether, and the sediment forms contorted masses or balled-up pillow-like structures. These contorted and disrupted layers are known as slump units or slump structures. They were clearly formed while the sediment was still plastic, and so are classed as contemporary rather than as post-depositional features. Slumping is most common in sandstones, but is also found in siltstones and calcareous rocks.

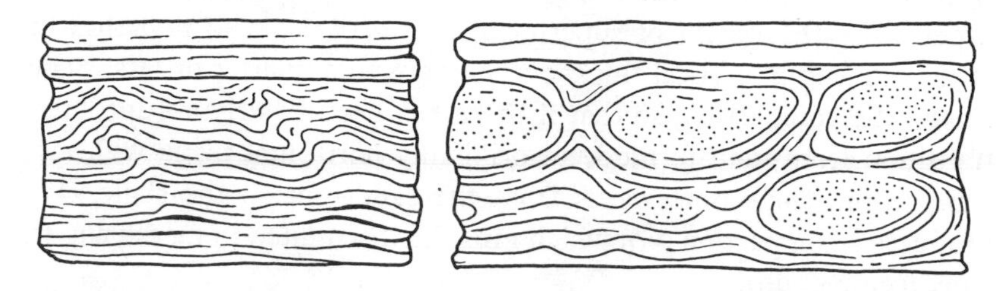

Fig. 2. Types of disturbed bedding. On the left convolute bedding planes; on the right the distortion has been carried further to form balled-up or pillow-like structures in which the bedding is largely destroyed. Both types are overlain by undisturbed beds.

A lesser form of distortion is seen where the bedding planes are wavy or convoluted but are still recognizable and continuous. Some types of disturbed bedding are associated with the curving over of the tops of foreset beds. It is probable that most slump units were formed on a sub-aqueous slope, though not necessarily a steep one, while the sediment still held enough water to be unstable, and that the dislodged mass crumpled up as it slid. Some kind of mechanism seems to be necessary to start the process and this was possibly supplied by earthquake shocks. Slump structures do not seem to be particularly related to water depth and are found associated with greywackes in geosynclinal conditions as well as in sequences that show many shallow water features.

Diagenesis

The changes that transform the incoherent sediment, whether clastic, precipitated or organic in origin, into the hard lithified rock are called diagenesis, which is still one of the more obscure processes in geology. It includes in particular the results of circulating solutions in the pore spaces of the sediment—solutions which, in addition to laying down the cement that holds the grains together, profoundly or subtly alter the character of the rock, whether locally or in bulk. Some minerals are partly or wholly dissolved, others introduced and others redistributed or recrystallized. To the stratigrapher the enigmas of diagenesis are a constant reminder that the rock he is examining has been altered physically and chemically from its original state, and that these alterations

CHRIST'S COLLEGE LIBRARY

have to be taken into account when assessing the conditions of sedimentation.

Compaction is a diagenetic process affecting clay rocks. In their earliest stages these contain a great deal of water, which is gradually expelled by the weight of superincumbent sediments so that they are much reduced in vertical thickness, perhaps to a quarter or less. Sands show little compaction, so that if there has been a fairly abrupt lateral transition from sand to mud, an originally horizontal plane will, after compaction, slope downwards from the sandstone to the shale. This produces a compaction dip, which must be discounted in measuring the true, tectonic, dip.

Facies

So far we have reviewed some characters and structures of sedimentary rocks as they might be seen individually in the field or in the laboratory. Associations of rocks with their fossils introduce the concept of facies—one essential in stratigraphy but not easy to define. 'Facies' was first used in 1838, in a description of the Swiss Jura mountains, more or less in its modern sense, to describe lateral changes in lithology within a stratigraphical unit. Bound up with this is the concept of analogous lateral changes in the environments that gave rise to the sediments.

It is thus postulated that during a certain period the conditions of deposition vary from place to place—sand banks and channels here, mud flats there, and shell banks a little way off shore; or, perhaps, a strand flat, a lagoon, a barrier reef and a steep outer slope down to the ocean floor. Each of these environments produces somewhat different bottom sediments, characterized by different faunal and floral assemblages: when 'fossilized' they become different facies, sandy, muddy or calcareous, all of the same age and passing laterally into one another. The names given to the various facies may pick out one aspect or another of the whole—the lithology (carbonate facies), fauna (shelly and graptolitic facies), tectonic or regional setting (geosynclinal and shelf facies), environment (lagoon facies). There is a greater element of deduction in the last method than in the first, but bearing in mind that any name is only a label for a complex whole, all have their uses.

The term 'reef' may conveniently be considered in this context. It is one firmly fixed in geological writing, but is not wholly satisfactory. There are the reefs of ordinary usage—very shallow or emergent ridges and mounds on the sea floor, and particularly those built up by lime-secreting organisms, of which corals are the best known, though they are often outnumbered by calcareous algae. Reefs and reef-complexes of past ages are usually mounds or ridges in carbonate (or carbonate plus evaporite) sequences. Reef limestones are often poorly bedded, and, if laid down on a steep slope, also include bands of breccia. They may or may not contain recognizable macro- or microfossils, or, though these were once present their traces may have been destroyed by diagenetic recrystallization, to which carbonate rocks are particularly susceptible. 'Reef' is thus not a very exact term and there have been attempts to replace it, for instance, by 'bioherm' for a mound primarily built by organic skeletons or sediment-binders, and 'biostrome' for flatter structures of the same type. A reef-complex is the combination of the reef itself with its neighbouring environments, particularly the back-reef, or lagoon-like area on the shoreward side, and the fore-reef of the outer slope to open sea or deeper basin. The most famous example is the Permian Reef Complex of Texas and New Mexico, but various reef elements have been recognized in carbonate formations of many ages and places, and with their accompanying, highly selective faunas form a good example of a facies.

We shall not pursue the various, intricate problems of facies any further here; it is a term best understood by example and these abound in succeeding chapters. The term has also been adopted in other branches of geology, in the concepts of metamorphic facies and mineral facies, but when unqualified it is normally the sedimentary type that is meant.

Cyclic sedimentation

This refers to the vertical repetition, several times over, of a few sedimentary types in a stratigraphical sequence, each cyclic unit being known as a cyclothem. The type of repetition in the cyclothem may be either of the following:

d	coal	cyclothem			
c	sandstone		*a*	limestone	cyclothem
b	shale		*b*	sandstone	
a	limestone		*c*	shale	
d	coal	cyclothem	*d*	coal	
c	sandstone		*c*	shale	
b	shale		*b*	sandstone	
a	limestone		*a*	limestone	

Cyclic sedimentation is generally characteristic of various types of shallow-water sediments, though some repetitive or cyclic effects have been noted among greywacke and evaporite associations. Some of the best examples result from fluctuations between the sharply distinct environments just above and below sea-level, such as those of Carboniferous sedimentation. The causes of cyclic repetitions are much debated, and probably there are several. Some seem to reflect largely sedimentary events, such as a shift in the position of major river distributaries, while for others an ultimate tectonic or climatic cause has been invoked.

Thicknesses of strata and isopachs

Stratigraphical descriptions frequently refer to the thickness (or changes in thickness) of a system or series, or some other unit. Stratal thicknesses are important in their own right, but also as some measure of the amount of sedimentation in the region. 'Thick' and 'thin' are not exact qualifications but it is convenient to know how and where the rocks of certain sedimentary basins become thicker or thinner and how these changes are related, as they often are, to facies-changes. Sedimentation is accompanied by subsidence, and if there is evidence throughout a formation that it was deposited near sea-level, then the thickness is also a measure of the subsidence. The Lower and Middle Coal Measures of England and Wales are a good example of a fairly exact relationship between sedimentation and subsidence. It is often convenient to summarize regional changes in thickness in the form of an isopach map; the isopachs, being lines denoting equal thickness, form a system of contours, based on a number of points where the total thickness of the unit can be measured. An example is shown on p. 310.

Thickness can be related in some degree to subsidence, but there is no certain way of relating it to duration, or the time taken to deposit a certain number of feet, although a commonsense assessment will suggest that 20 feet of coarse sediment are likely to be laid down in a shorter time than 20 feet of fine sediment. Rates of sedimentation are very elusive, however, and they can only be deduced at all reliably either on a small scale or on a very large one. Certain laminated beds are formed by the annual variation in the sedimentary régime, the best known being the late glacial and post-glacial varves; such beds can be counted in years. Then since isotopic age determinations give the approximate duration of the stratigraphical systems, and of most of their major subdivisions, it is possible to deduce, for instance, that 1,000 feet of Lower Jurassic clays were laid down in about 10 million years. To extrapolate, however, and deduce that a certain ten feet of those clays represent

100,000 years, is to assume a regularity of deposition for which there is little justification. Reliable methods of correlating thickness with duration on this scale are lacking.

CORRELATION

Correlation of fossiliferous strata

The basic unit in the correlation of rocks by their fossils is the palaeontological 'zone'—a term originated by the German stratigrapher and palaeontologist, Oppel, in the mid-nineteenth century, in his subdivision and classification of the Jurassic rocks of southern Germany. A zone is a small thickness of strata characterized and defined by one or a few species; or, to put it in another way, it is the sediments that were laid down during the duration (or approximate duration) of those species. Stages are larger units, also palaeontologically defined, comprising several zones. Two types of zones are sometimes distinguished, characterized either by a single index or guide species or by an assemblage of species. In practice there is some merging of the two and where one is named as the index fossil it is commonly accompanied by others which may be used in conjunction or where the index locally fails.

The chief criteria of a good zonal index are a wide geographical range, a short vertical (or time) range, and a fair abundance. The first two are best fulfilled by swimming or floating forms in a group that evolved relatively fast. An independence of bottom conditions is advantageous in that the zonal forms are much less likely to be restricted to certain conditions and hence to certain lithological facies. Complete independence, however, is uncommon. Even such widespread surface-living forms as ammonites or graptolites are rare or virtually absent from a few facies; moreover certain sediments, such as coarse sandstones, are very poor preserving media for any form of life they may have contained.

Very few zonal fossils are, in fact, world-wide in distribution, though a small number of graptolites may prove to be an exception; nevertheless many zones, and the stages into which they are grouped, are successfully applied within major geological regions such as Europe north of the Alps, or central and western North America. Since conditions on land and in fresh water vary more than those of the sea, and land deposits are often sparse in fossils, it is more difficult to apply zonal divisions and correlations to terrestrial and fresh water facies; the standard succession of zones for any system is normally based on marine rocks and faunas. Some of the most successful stratigraphical fossils belong to microfaunas and microfloras, especially the smaller, planktonic foraminifera. Although small, these have the advantage of often being very abundant, and they are the principal groups used in the palaeontological correlation of borings through Tertiary strata, from which much of the world's oil is drawn.

The succession of zones of the various systems, as they are exemplified from the British Isles or neighbouring countries, are given in the Appendix. Marker horizons, or fossil bands, characterized by a particular species are employed on a more limited scale than zones, but are useful mapping aids in a single basin of deposition. They are often only a few feet thick and sometimes include species from groups that are too facies-bound to be desirable zone fossils, provided the facies is extensive within the basin concerned. Corals and calcareous algae in limestones are conspicuous examples. Marker horizons may also have a distinctive lithology.

Ideally each system, or subdivision of a system, is defined in relation to a standard succession where the zonal sequence is best developed and with which less complete successions may be compared. The actual exposure or exposures form the type section. Sometimes the succession and section can be established within the region or country where the system was first named, but it often happens that better sections, more complete or more fossiliferous, are found elsewhere. All the Palaeozoic systems except the Permian were first named from some area in Britain (or, in the case of the Carboniferous, from a conspicuous character of the British rocks); the successions available for some systems, however, such as the Devonian or Cambrian, leave a good deal to be desired and better ones are known elsewhere.

All the Phanerozoic systems were first named and defined somewhere in Europe, but they have been successfully adopted in most parts of the world. There is one major exception, and that comprises the great non-marine successions of late Palaeozoic and early Mesozoic age of the Southern Hemisphere and peninsular India. Because correlation with the northern systems is exceptionally difficult, individual system names have been used in the different countries for these beds. They include the Gondwana System of India, and (probably the best known) the Karroo System of Africa. The latter includes formations from late Carboniferous to Jurassic age, and the fossil vertebrates allow an approximate correlation through continental beds and faunas in the Northern Hemisphere, such as those of Russia and North America.

Correlation of unfossiliferous strata

Where no reliable fossils are found, whether in a barren Phanerozoic formation or among Pre-Cambrian rocks, a variety of methods may be attempted, based on inorganic characters. The use of radiometric dating has already been described, but it operates on a relatively coarse scale and there are several rock-types, including most Palaeozoic and Pre-Cambrian sediments, in which it is as yet of little use.

Lithological marker horizons or thin beds of unusual lithology can be used for local correlation, provided that the rock is sufficiently exceptional to be identified with certainty, or forms part of a regular and widespread lithological sequence or cycle. Tuff bands, or groups of tuffs, of varied and characteristic composition, exemplify the first type and have proved effective for local correlation in Ordovician pyroclastic sequences of Ireland and the Lake District. Evaporite cycles, which form as a regular succession of salts throughout a saline sea, have also been used in comparing Permian strata in various outcrops of northern England. On a wider scale it is probable that thick groups of tillites found among late Pre-Cambrian formations in many countries are due to broadly contemporaneous glacial conditions.

Another aspect of correlation concerns the relative ages of two neighbouring, unfossiliferous rock groups which are not seen in contact, or which have a faulted junction. If there is a conspicuous difference in the amount of folding they have undergone, or metamorphism, or intrusion, then it is probable that the group that is the more affected is the older. Nevertheless, any or all of these differences may be misleading if there has been extensive movement on an intervening fault. It is then possible that the two groups were once much farther apart than they are now, and the differences of folding, metamorphism or intrusion are not related to age but are simply part of a broad regional or tectonic variation. If one rock group contains recognizable pebbles or minerals derived from another then clearly it is the later of the two; the proviso here is that the identification must be unequivocal, and thus the derived rock or minerals have some highly exceptional characters.

These lithological methods of deducing mutual age relations have been applied to British Pre-Cambrian outcrops, for instance in Shropshire and Scotland; but at their best they indicate a probability rather than a certainty.

Diachronism

In view of the lateral variation in modern sedimentary environments, especially those of coastal waters and shallow seas, and of the analogous variations in lithological facies in the past, it will be seen that there is no particular reason why a boundary between lithological groups, or units, should coincide exactly with a time-plane. It is quite as probable, and perhaps more so, that there will be a gradual shift of the facies boundary

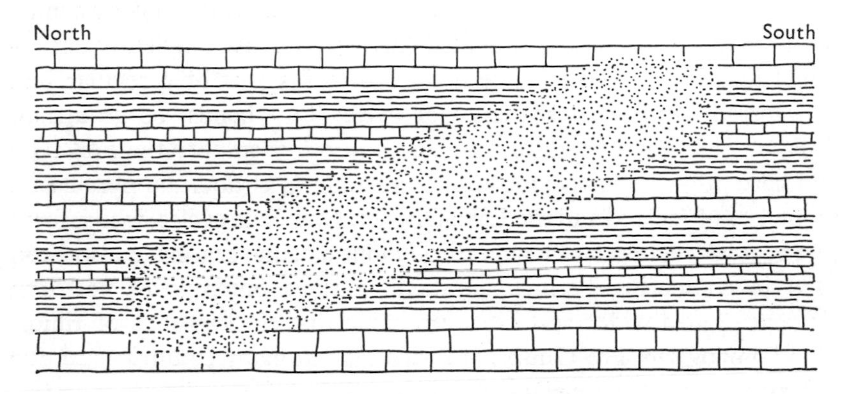

Fig. 3. Diachronism. A series of shales and limestones is interrupted by a diachronous sandy unit which is laid down first in the north of the area and later shifts gradually southwards.

as one environment encroaches on another: a sand bank gradually overwhelms the mud flats, or, with increased load, river muds spread farther and farther into the lagoon. A lithological unit, or boundary, that for similar causes transgresses the time-planes is called diachronous, and the phenomenon is diachronism (Fig. 3). Even though we may suspect that it is fairly common, demonstrable examples of diachronism are not very many. The difficulty lies in the establishment of a detailed time-sequence that is itself securely independent of the shifting conditions. This demands an exceptionally fine, reliable series of zones and sub-zones; a classic Jurassic example, based on an ammonite sequence, is described on p. 285, and other, lesser, ones mentioned elsewhere.

THE TECTONIC SETTING

On a world-wide scale tectonic foundations are divisible into two types—regions of relative mobility and of relative stability. These descriptions are not absolute, since there are intermediate zones of limited mobility and over long periods of time the boundaries shift or the characters change. In particular ancient mobile belts, such as those of Pre-Cambrian age, become stable areas and new mobile belts may develop on their margins. The great stable regions of the world have also been called shield areas, or Pre-Cambrian shields, on which any later rocks have escaped major deformation. 'Craton' also refers to a stable area within a continent and is commonly contrasted with neighbouring geosynclines or lesser subsiding basins.

The term geosyncline has had a long and valuable history in stratigraphical writings, despite the difficulty of defining it; more recently many variants have grown up around the term. It was first applied in the 1870s to the Appalachian region of the eastern United States—a long downwarped belt of heavy sedimentation which, persisting throughout much of Palaeozoic time, was uplifted and folded toward the end of that era; the facies belts and subsequent folds both lie parallel to the downwarp. The Appalachian geosyncline is still a good type example and the term has since been applied to linear belts of deep subsidence and thick sediments in many parts of the world, even though among them there are many dissimilar characters in sedimentary, volcanic and tectonic history. The more complex belts include a series of elongate, subparallel troughs which have somewhat different histories. It is then a matter of choice whether one refers to the component troughs of a geosyncline, or the component geosynclines of a mobile belt.

The theoretical geosyncline is bounded by more stable borderlands or forelands. Topographically these may take the form of a landmass, whence detritus is carried into the downwarp, or, lying below sea-level, become a shelf sea in which subsidence and sedimentation are less; an orthoquartzite and carbonate facies is characteristic of certain forelands and shelf seas. In some cases a 'hinge area' can be recognized between geosyncline and shelf where there is a zone of changing thickness and facies.

Orogenic movements, or orogenies, are those types of deformation that affect the mobile belts and which, ultimately obliterating the geosynclines as sedimentary troughs, are responsible for the great folded mountain chains of the world. In addition to being restricted in space such movements are restricted in time, although the greatest orogenies, such as the Alpine, have had a very long history, extending through several systems. They are also contrasted with epeirogenic movements—the broad, slow, up and down warping effects of the stable areas. This term has also been applied in regions, like Britain, that are dominated by mobile belts, to the smaller warping movements occurring outside those belts or during periods of relative quiescence between the major orogenies. The latter, however, are difficult to separate in practice from the growing or dying orogenic phases, and the distinction here into orogenic and epeirogenic is probably rather artificial.

The ultimate effects of tectonics on sedimentation are very far-reaching. The régime of a sedimentary basin depends primarily on the detritus that is carried in from the neighbouring land and the extent to which subsidence keeps pace with it. Without renewed uplift of the land the supply will gradually decrease, in quantity and grade, and in some circumstances chemical deposits replace clastic. If sedimentation overtakes subsidence the sediment surface is built up above water level and either a continental facies develops or erosion ensues. Thus many of the major facies-changes in stratigraphy can ultimately be traced to tectonic causes, and some types of cyclic sequence have been attributed to repeated tectonic warpings in areas of supply and deposition.

The effects of erosion are seen in unconformities and other types of breaks in a succession. An angular unconformity, in which there is a discordance of dip in the beds above and below, results from uplift accompanied by folding or tilting; disconformity refers to a similar break without discordance. Although the larger breaks are due to subaerial erosion and often represent considerable periods of time, minor ones (sometimes called diastems) can result from nothing more than prolonged dispersal of sediment, and we have already noted that these conditions may also produce a palaeontological non-sequence or the local absence of a zone.

Unconformity with overlap (Fig. 4) is caused by uplift, followed by the transgression of the sea, laying down a later series of beds farther and farther over a sloping, submergent shore or land surface. Overlap is valuable in palaeographic reconstruction, because when the succession is well exposed the position of the retreating shore-line can be traced. No such inference can be drawn from unconformity with overstep, where a single later formation transgresses, or oversteps, more than one underlying one, because there is no indication of how much farther the sea extended before the shore-line was reached. In neither case do the later, transgressing strata necessarily remain horizontal, since all may have been subjected to later earth movements. In complicated country the interpretation or distinction of these two types of unconformity may be debatable, and as a result the palaeogeographic implications may also be uncertain. Such an example, affecting Ordovician geography, is given on p. 101.

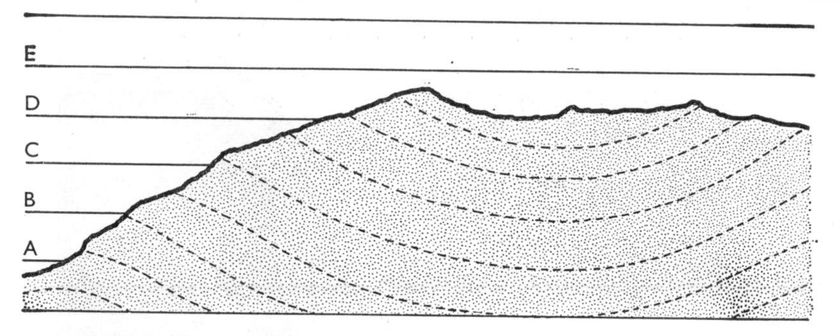

Fig. 4. Overlap and overstep. A land mass of older, gently folded strata is gradually submerged and a newer series, A–E, is laid down unconformably on its margins. Beds A–C show overlap against the submergent mass and the successive positions of the shore during their deposition can be determined. Bed D, however, oversteps several of the older strata, and the shore positions for the upper part of D and Bed E are not determinable.

Structural regions of the British Isles

The one, large, stable region of Europe is the Fennoscandian Shield, where Pre-Cambrian rocks are exposed for many hundreds of miles and on which later formations lie almost undeformed. The British Isles are part of Western Europe which is, broadly, a region dominated by mobile belts. Within Phanerozoic time three major geosynclines have in turn been established in the 2,500 miles between north Norway and the African shore; although at various times and places there have been regions or units of relative stability, the geology of Western Europe is largely linked with that of geosynclines, their borderlands and shelf seas, and the orogenies which transformed them into mountain ranges.

The major structural divisions of the British Isles can be summarized in relation to these episodes; in addition brief descriptions of structure serve to introduce the stratigraphy of certain areas as they appear in later chapters. Each of the three orogenies—Caledonian, Hercynian and Alpine—have given rise to folds and faults with dominant, though not ubiquitous, trends. 'Caledonian' refers to structures of that trend and of Caledonian age; 'Caledonoid' to structures of any age with the same trend. This system of suffixes is applied (though rather less often) to other orogenies.

Although they are not the oldest the Caledonian movements take pride of place. They affected nearly all of what is now upland Britain—the hilly or mountainous regions of Scotland, north-west England, Wales, north-west and south-east Ireland. The dominant strike is north-east–

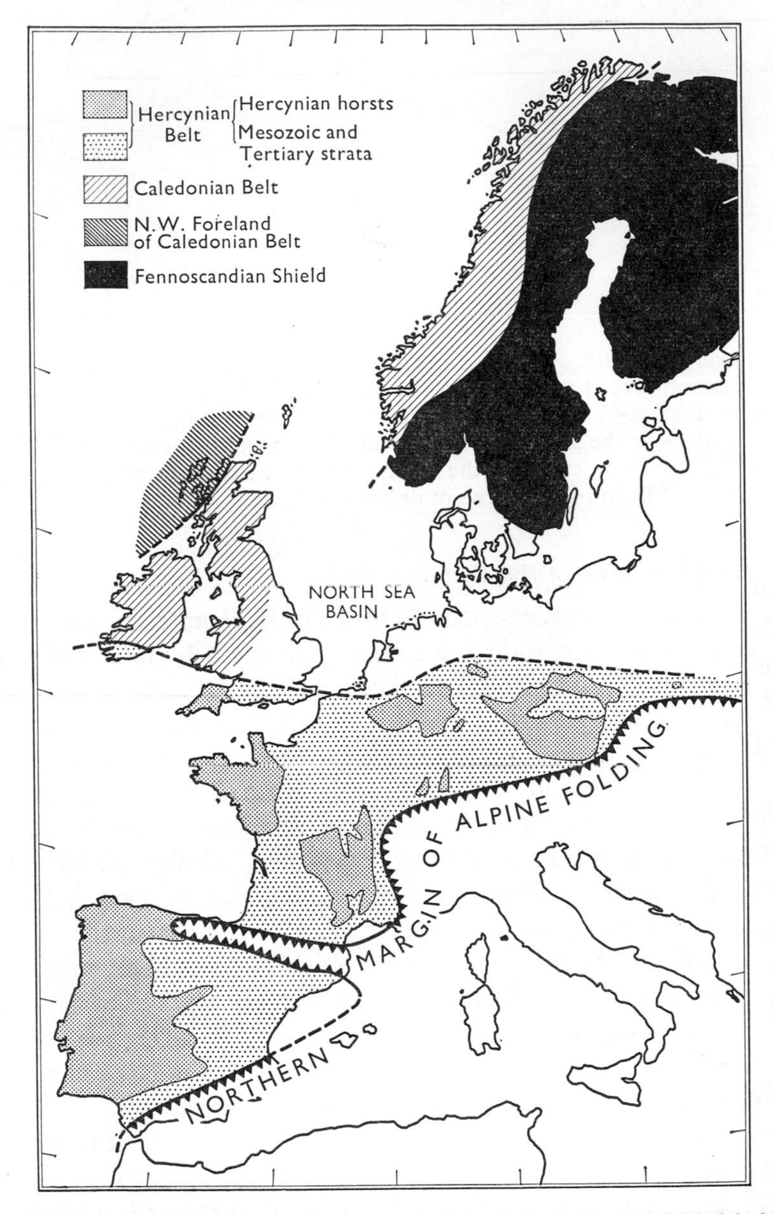

Fig. 5. A sketch map showing the principal tectonic units and folded belts of Western Europe, north of the main region of Alpine folding. Around the southern part of the North Sea Basin and eastwards to the Polish plain the deeper structures are largely hidden under a Mesozoic, Tertiary or Quaternary cover. In the Netherlands, North Sea and parts of eastern England these structures may be Caledonian and possibly have a north-westerly trend; this is by no means certain, however.

south-west, and in one area or another the orogenic phases extended from approximately Middle Ordovician to Middle Devonian times.[1] The rocks affected were chiefly those of the Lower Palaeozoic geosyncline, but the Pre-Cambrian Moine Schists of northern Scotland owe some, and perhaps most, of their structures to the Caledonian movements. The name is justifiably taken from Scotland, where structures of Caledonoid trend govern much of the country. The folded ranges extend to western Ireland in one direction and to northern Norway in the other; there are also ranges of similar age in Spitsbergen, Greenland and eastern North America.

Both folding and metamorphism were more vigorous in Scotland and north-western Ireland than farther south-east, and four great faults or fault belts divide the former country into five geological units whose distinction is evident even on a small-scale map. The Moine Thrust Belt, which extends from the north Sutherland coast to Skye, forms the western margin of Caledonian folding; beyond it lies the unmoved foreland composed of Pre-Cambrian rocks and structures. The thrust belt consists of a number of piled-up slices, carried westward on easterly dipping planes, the uppermost of which is the Moine Thrust itself. The whole is probably a late Caledonian structure. The Great Glen Fault, which is directly responsible for the through valley between the North-West Highlands and the Grampian or Central Highlands, is essentially a transcurrent, or wrench, fault; in its lateral displacement the western block has been moved southwards by 65 miles. The main movement was probably later than the Middle Devonian and earlier than Upper Carboniferous beds of the neighbourhood, but it is still a line of crustal weakness and a focus of earthquakes.

The Midland Valley of Scotland is a rifted structure, or graben, between the Highland Border Fault and the Southern Upland Fault. Some of the movement on these fractures is later than the Upper Carboniferous rocks of the valley, but it had an earlier Caledonian history. The Midland Valley continues south-westwards across into Ireland, but its margins are not seen so continuously there (the Southern Upland Fault is only deduced) because of a more extensive late Palaeozoic cover. The Highland Border Fault reappears on the south side of Clew Bay, in County Mayo. Although Caledonian structures dominate the remaining

[1] This is one interpretation of 'Caledonian'. Some authorities refer certain Ordovician effects to the Taconic orogeny, whose type area is in the eastern United States; others would extend Caledonian to include Pre-Cambrian episodes where, as in Scotland, these affected the Caledonian belt.

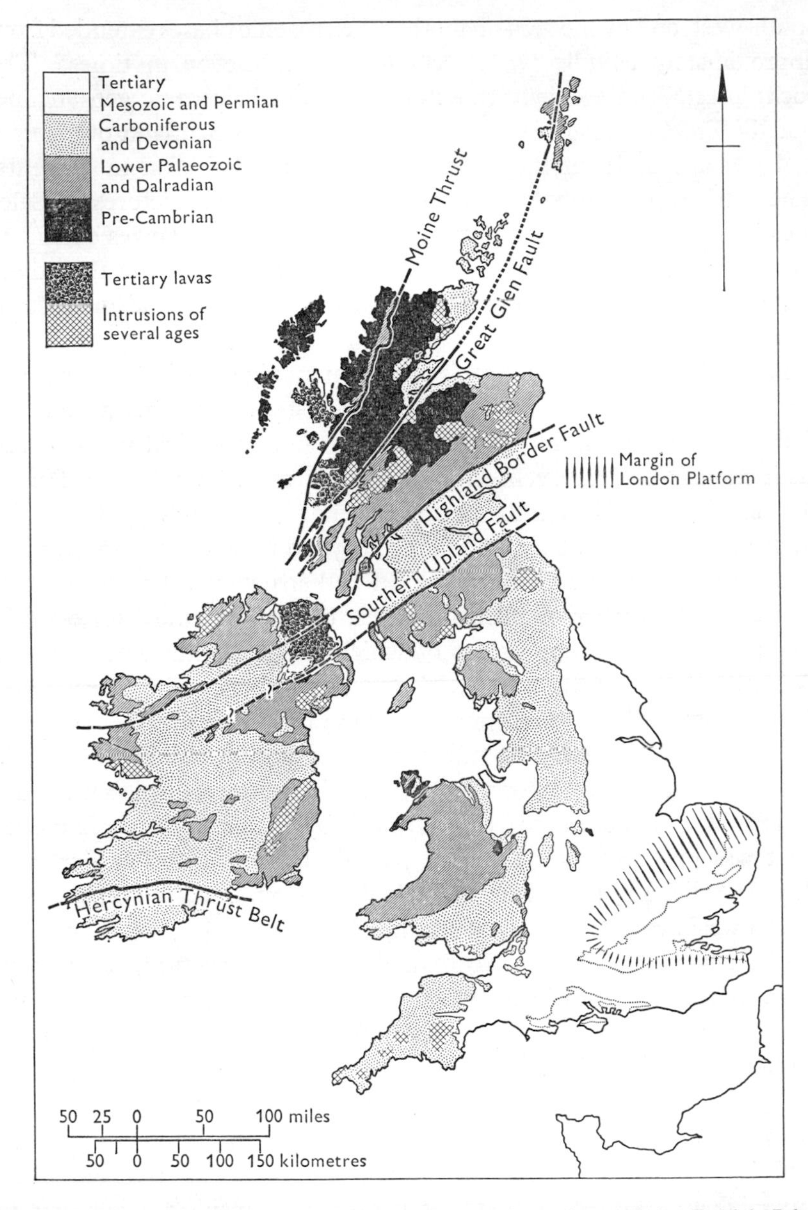

Fig. 6. The principal stratigraphical and tectonic regions of the British Isles, together with the more important thrusts and faults of Scotland, northern Ireland and southern Ireland. The Great Glen Fault is interpreted as continuing to the Shetland Islands and in south-west Scotland it displaces the Moine Thrust; its continuation in Ireland is problematical. The Shetlandian metamorphic rocks belong to the Caledonian belt but their precise correspondence is uncertain.

Lower Palaeozoic outcrops of Wales, the Welsh Borders, the Lake District and south-eastern Ireland, they are not so complex as those farther north and there is only low grade metamorphism.

The Hercynian geosyncline stretched across Middle Europe from southern Ireland through Cornwall and Brittany to northern France, Belgium and Czechoslovakia; the name is taken from the Harz Mountains. To the south there are further massifs with Hercynian structures in Spain, the Atlas mountains, Central France, the Vosges and the Black Forest. In Britain, which is on the northern margin of this belt, the strike varies a little on either side of east–west. Hercynian structures are dominant in the south-west, from the northern borders of the South Wales and Pembrokeshire coalfields to Cornwall; they become increasingly complex southwards and in south Cornwall there is strong deformation and metamorphism.

From Pembrokeshire the folded belt continues westwards into southern Ireland, where there is a partial northern boundary in a line of thrusts and aligned structures extending eastwards from the south of Dingle Bay to Dungarvan in Co. Waterford. More gentle folds affect the Devonian and Carboniferous rocks of the central Irish plain. Eastwards from Somerset the Hercynian structures disappear under the Mesozoic cover of southern England, where they govern the underground Kent Coalfield and its continuation in the Pas de Calais. It is probable that they are also effective on the London Platform, which is the major positive element or massif in south-east England although it does not appear at the surface. Its southern margin trends east–west through the southern London suburbs, and there is a much less well-defined northern and north-western margin, from East Anglia into the south Midlands. Beneath a Mesozoic cover Carboniferous, Devonian and Lower Palaeozoic rocks are known from several borings. The London Platform is probably continuous under the southern North Sea with a similar underground structure in Belgium, and the whole is sometimes called the London–Brabant Massif or London–Ardennes Massif.

Extending down the centre of England, from the Pennines, through the Midlands and as far south as Bristol, there are various Carboniferous outcrops whose structures are largely Hercynian in age but which follow older directions. They are well exemplified in the Pennine and Midland coalfields, in which the structures range from north-east (Caledonoid) in Shropshire and north Staffordshire to northerly in the west Midlands and north-west (Charnoid) in Leicestershire. The term Charnian refers to the north-westerly Pre-Cambrian structures of Charnwood Forest and this trend affects much of the east Midlands.

The northern Pennines comprise the Alston and Askrigg Blocks, fault-bounded on their northern, western and southern margins and on which a Carboniferous cover is gently tilted eastwards. The pre-Carboniferous basement is composed of Lower Palaeozoic rocks, a Caledonian granite and the Ingletonian Series (?Pre-Cambrian); their structures are chiefly Caledonian in age but not in direction.

In its influence on this country the Alpine orogeny is only partly comparable with the other two. The geosynclinal tract and orogenic focus lay far to the south, and the only phase of much importance in Britain took place in the Mid-Tertiary, possibly in the Miocene period. The effects are largely confined to southern England, and include the folds, minor faults and rare thrusts of the Weald, Isle of Wight and Dorset. It seems probable that most of the minor structures and regional tilting exhibited by Mesozoic rocks in central, eastern and north-eastern England are also due to these movements.

REFERENCES

(List of full citations on p. 417.)

General. Atlas of Britain; Donovan, 1966; Evans and Stubblefield, 1929; Grabau, 1913; Holmes, 1965; Read, 1958; Read and Watson, 1962; Weller, 1960; Wills, 1929.

Geochronology. Giletti *et al.* 1961; Hamilton, 1965; Harland *et al.* 1964; Holmes, 1965; Sabine and Watson, 1965.

Sedimentary rocks, etc. Grabau, 1913; Greensmith, Hatch and Rastall, 1965; Krumbein and Sloss, 1963; Pettijohn, 1957; Shrock, 1948; Weller, 1960.

Tectonic and regional geology. Anderson, 1947; Bailey, 1929; Bailey and Holtedahl, 1938; British Regional Geology (Geol. Surv.), see p. 418; Charlesworth, 1963; Craig, 1965*a*; Coe, 1962; Falcon and Kent, 1960; Flinn, 1961; George, 1960*a*, 1962*a*,*b*, 1963*a*,*b*; Holtedahl, 1952; Johnson and Stewart, 1963; Kennedy, 1946, 1958; Kent, 1949; Turner, 1949; Wills, 1948, 1951.

History and biographies. Adams, 1938; Fenton and Fenton, 1945; Geikie, 1905; Woodward, 1907; Zittel, 1901.

2

PRE-CAMBRIAN ROCKS

Most of this book is concerned with the stratigraphy of fossiliferous strata—or of strata which, if lacking fossils in one area, may ultimately be correlated with a fossiliferous sequence elsewhere. Many of them have relatively simple structures and have undergone little or no metamorphism. In these circumstances it is usually possible to apply successfully the principles set out in the preceding chapter, especially the correlation of rocks by fossils, and to interpret the original environment from the composition and structures of the sedimentary rocks and such fauna or flora as they may contain. It is true that, as will appear in later chapters, many questions remain unanswered, but for the most part they are on a relatively modest scale. In the British Isles, with its preponderance of fossiliferous facies, a rock formation can usually be attributed to its place in the geological systems—Upper Ordovician or Middle Coal Measures—and its environment broadly assessed. Of the Phanerozoic stratigraphy (of Cambrian and later systems) only parts of the Upper Palaeozoic regions in Devon and Cornwall and southern Ireland, and of the Lower Palaeozoic over a wider field, still present substantial gaps in our knowledge. In other parts of the world, more affected by, for instance, Tertiary orogenies, the unravelling of later stratigraphy may entail severe difficulties but by and large the same criteria are sought for and applied.

When we turn to the study of Pre-Cambrian rocks the picture is rather different. It is not only that fossils are lacking (the few accepted Pre-Cambrian records have all the interest of problematic rarities but little or no stratigraphical value), or even that most of the rocks are more highly metamorphosed, but that the whole scale of events is so much vaster. Phanerozoic and recent sediments cover more than three-quarters of the earth's land surface, but in time they represent only 600 million years, or less than one-fifth of the known Pre-Cambrian time-span. During this enormous stretch of time there were many periods of orogeny, intrusion and metamorphism so that, with a few exceptions, the rocks are much deformed and altered. They are best seen in the great Pre-Cambrian shield areas of the world, such as the Canadian and Fennoscandian shields or the several stable cratons and mobile belts that make up the Pre-Cambrian basement of the African continent.

In such regions, where very ancient rocks are exposed for thousands of miles, it is possible to recognize certain structural entities. In particular there are orogenic or mobile belts, linear or arcuate, along which there is a dominant structural trend, certain rock types are characteristic, and from which metamorphic minerals give roughly similar ages. The best known of these resemble the later orogenic belts, such as Caledonian or Hercynian, and presumably reflect similar mountain-making movements. Some appear to have been comparable also in that they were previously the sites of geosynclines; elsewhere associated sedimentary strata are not recognizable, and it is possible that the rocks deformed along the orogenic belt did not result from a preceding period of sedimentation but were rather those of an earlier orogeny, which were reconstituted in the later.

In some regions the relative ages of these mobile belts can be deduced by the truncation of the earlier ones by the later, comparable to the way in which the Hercynian folds cut across the Caledonian in South Wales, or the Caledonian structures of north-west Scotland cut across the north-westerly ones of the adjacent Pre-Cambrian outcrops. Some indication of time sequences has also been deduced from the relation of individual belts to major periods of intrusion. Such comparative methods, however, have been overshadowed by the great advances in isotopic dating in the last dozen years, and these techniques are now an essential tool in the unravelling of Pre-Cambrian history. Already it is possible to plot the broad sequence of events in some of the major shield areas and much more will be done in the future.

One difference between Pre-Cambrian and Phanerozoic studies emerges from this brief comparison, in the data on which the time sequence is based and way it it is described. In the Cambrian and later systems the history of the earth's surface is normally recorded in, and interpreted from, rock successions, because somewhere it has been possible to compile a complete succession of all the systems. Such a composite stratigraphical column can be compiled even within the small area of the British Isles (though they are exceptional), with only minor gaps such as that in the Mid-Tertiary. Certain events, metamorphic or tectonic, are overprinted on this ideal succession; their relation to the sedimentary processes is made more precise if isotopic determinations are available, but these are rarely crucial. Exceptions in Britain probably include the age of the Tertiary intrusions in Scotland and Ireland, and of the Caledonian metamorphism of the Scottish Highlands. But normally Phanerozoic history is envisaged in terms of sedimentary rocks, and as far as possible rocks with a reliable, marine fauna.

With certain, late Pre-Cambrian, exceptions the reverse holds for the previous five-sixths of known geological time. There the history is indeed interpreted by means of rocks—their composition, metamorphic grade and structural disposition—but that history largely depends on and is defined by *events* or *episodes*. We thus read of a metamorphic event 1,500 or 2,500 million years ago. Sometimes these events are far later than the original accumulation or intrusion of the rocks, whose early history is lost because they have been affected by more than one period of metamorphism. Several belts in Africa yield a date of about 500 million years; they are part of the Pre-Cambrian basement of the continent, but have been re-activated (and therefore reconstituted mineralogically) by a metamorphic event early in Lower Palaeozoic times. The distinction follows from the antithesis between history interpreted from events on the earth's surface—that is, the deposition of successive beds—and events that affect rocks below the surface, by metamorphism, deformation or intrusion. The depositional type of history can be carried back into Pre-Cambrian geology where the rocks are little modified, but it cannot be interpreted on world-wide basis without the use of fossils.

In the British Isles only the fringes of these rocks and their problems appear. In England and Wales Pre-Cambrian outcrops are all small and are commonly inliers adjacent to the larger Lower Palaeozoic tracts. In Scotland the various series are much more important and occupy much of the North-West Highlands and Outer Hebrides, where they form what remains of the foreland to the Caledonian geosyncline.

In northern Ireland there are a few small areas that come within the subject of this chapter, but will not be further considered. They include the Lewisian-like gneisses found on the island of Inishtrahull, 6 miles north-east of Malin Head, County Donegal, and rocks comparable with the Moine Schists of the Grampians between Loch Derg and Ballyshannon, east of Donegal Bay. The metamorphic rocks of south-east Wexford are more closely related to those of Wales. The Dalradian Series, of Scotland and Ireland, contains Lower Cambrian trilobites in Perthshire, and though its lower beds are probably late Pre-Cambrian in age it is essentially a part of Caledonian history and reserved for the next chapter.

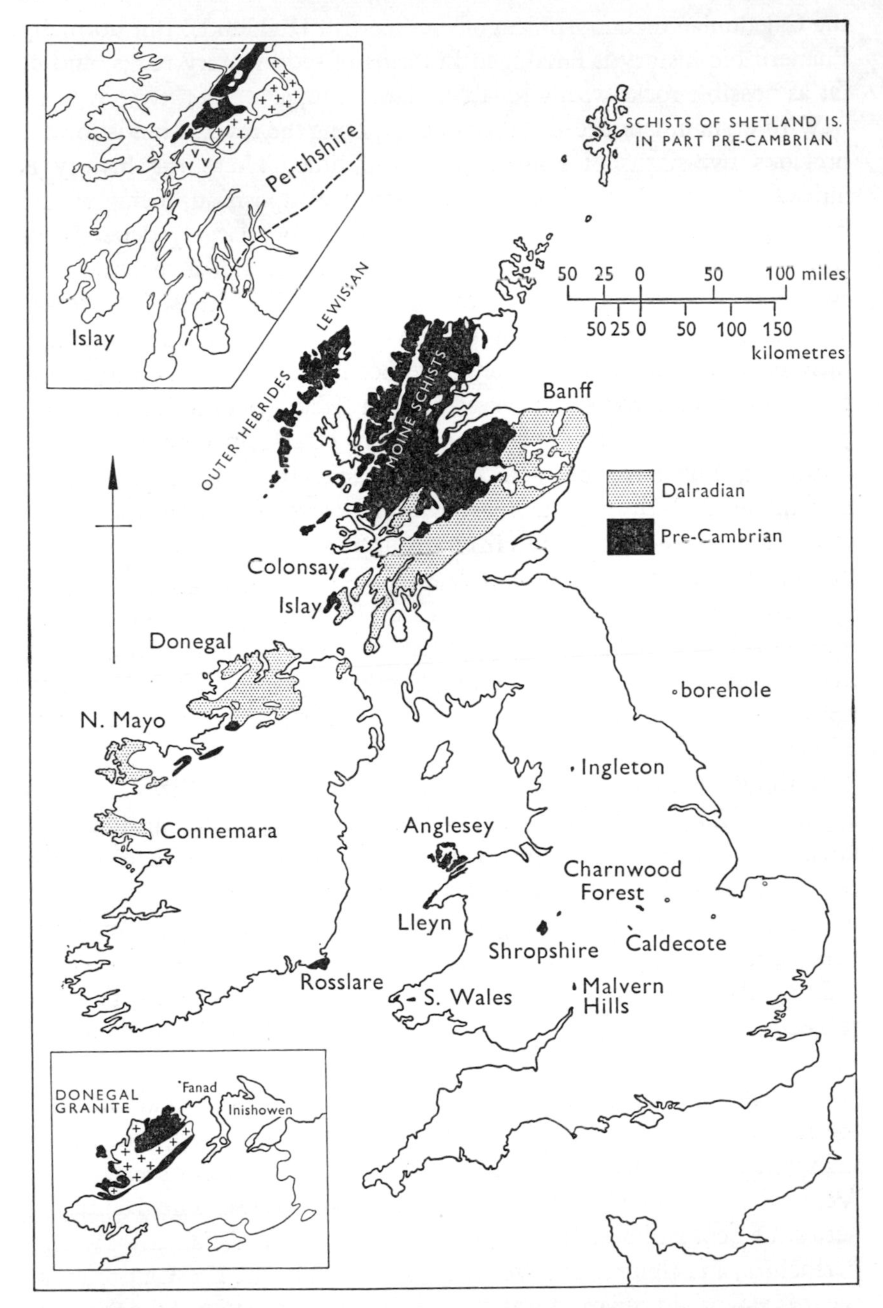

Fig. 7. Pre-Cambrian and Dalradian outcrops. Those of the North-West Highlands are shown in Fig. 8; Moine Schists also occur in part of the Grampian Highlands and probably in Ireland. A few boreholes in eastern England reach rocks assumed to be Pre-Cambrian. Insets: the Ballahulish succession (black) in Argyll and Donegal, with adjacent Caledonian igneous rocks.

NORTH-WEST SCOTLAND, BASEMENT AND COVER

West of the Moine Thrust[1] there are two strongly diverse rock groups—the Torridonian Series overlying the Lewisian complex. The former is largely unaltered and is late Pre-Cambrian; the latter includes a variety of much older rocks. Both are overlain with striking unconformity by Cambro-Ordovician beds and the three together exemplify a highly metamorphosed basement complex and the remnants of its sedimentary cover. East of the Moine Thrust, the Moine Schists are affected by Caledonian folding and metamorphism; thus while they are also part of the cover, through which small inliers of Lewisian rocks appear, they are not part of the unmoved foreland to the Caledonian geosyncline.

The Lewisian complex

These ancient rocks have often been called, rather loosely, the Lewisian Gneiss; they form all the Outer Hebrides, the islands of Coll and Tiree, and are exposed along the western seaboard of the mainland (Fig. 8). The commonest rock-type is a grey, rudely foliated orthogneiss dominantly composed of quartz, feldspar and one or more ferromagnesian minerals; some is derived from highly metamorphosed igneous rocks and from sedimentary rocks which have been intimately mixed with granitic material (migmatites). Rocks whose original sedimentary character is still recognizable are more restricted but are found particularly in the southern part of the mainland outcrop, from Loch Maree southwards, and in South Harris. These rocks, which include the oldest in Britain, were originally sandy and shaly sediments, with some carbonate and ferruginous types. The Lewisian complex also includes many intrusions, layered basic and ultrabasic masses, granites and pegmatites (especially in South Harris) and a widespread group of basic (tholeiitic) minor intrusions, which form dolerite dykes on the mainland and dykes and sills in the Outer Hebrides.

The great majority of these rocks have undergone metamorphism and migmatization and deductions on their ages stem from several sources. Most important are the isotopic dates which, as emphasized in the Introduction, can only give the time at which the mineral analysed became a closed chemical system, and hence only the latest metamorphism. Before this technique had been employed on Lewisian rocks, however, it had been shown that the foliation and other structures of the

[1] This and other structures which bound or characterize the various structural regions of Scotland are briefly described in ch. 1 (p. 23).

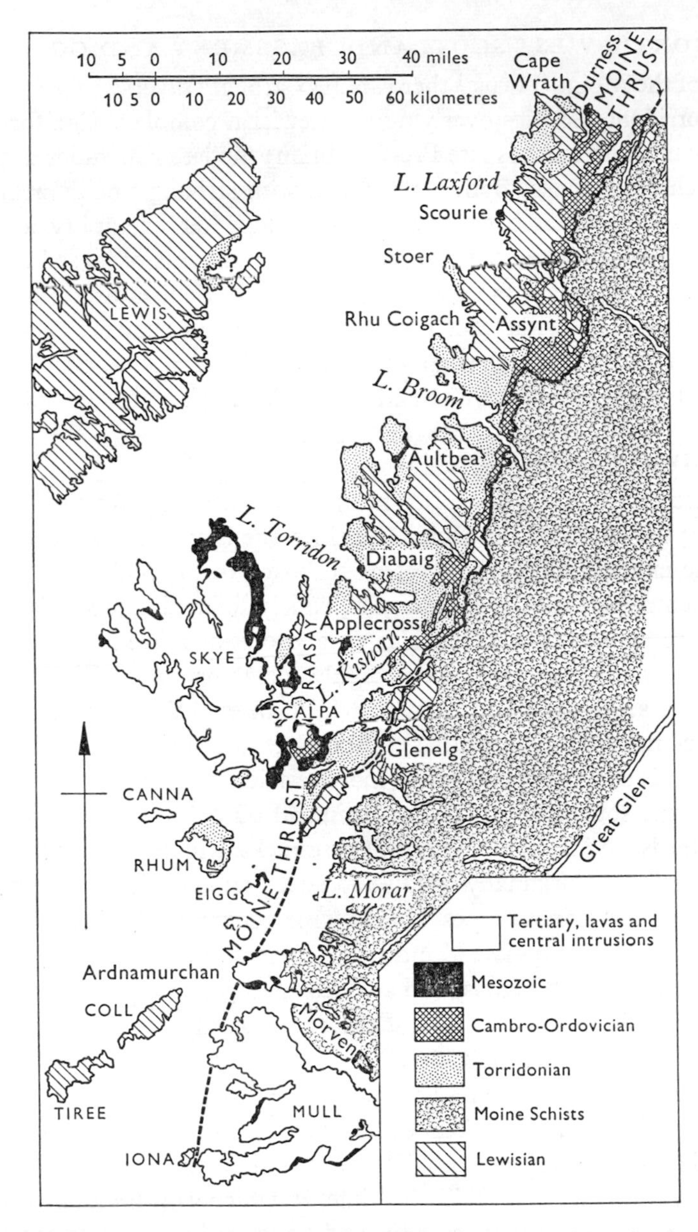

Fig. 8. Geological sketch map of the North-West Highlands of Scotland, with part of the Outer and Inner Hebrides. Intrusions and certain Lewisian inliers in the Moine Schists omitted. (Redrawn from Phemister, 1961, pl. II.)

gneisses in certain parts of Sutherland differed in direction, and further that the rock groups so distinguished were also separable by a period of dyke intrusion. This distinction has been substantiated by isotopic determinations and the two widely separate metamorphic phases are called Scourian and Laxfordian (about 2600 and 1600 million years respectively). They have also been recognized in the Outer Hebrides.

The Scourian metamorphism and folding affected the earliest known gneissose complex, the early sediments and the layered basic and ultrabasic intrusions. In several places, such as the type area on the west coast of Sutherland, the resulting structures and foliation have a north-easterly trend, but Scourian structures are only dominant in areas where they have escaped Laxfordian effects. The next recorded event was the intrusion of basic dykes and other minor bodies, with a dominantly north-westerly trend, some of which have yielded a date of about 2,200 million years. Then much later parts of the Lewisian complex were affected by Laxfordian metamorphism and folding, which also produced north-westerly structures.

The two widely separated events can also be considered as the Scourian and Laxfordian orogenies, for they have much in common with later orogenic effects, but, whereas the Scourian was preceded by a period of sedimentation, no widespread traces of Laxfordian sediments are recognizable, and these earth movements appear to have acted mainly on earlier Scourian rocks and the dykes which cut them. It might also be thought that the term 'Lewisian', embracing rocks so far apart in age (a period longer than all the Phanerozoic), should be replaced by the other two; nevertheless it remains a convenient group term for the gneissose rocks of the Pre-Cambrian basement and also can be applied to outcrops whose position is still uncertain. Moreover, while the isotopic dates are not disputed, nor the existence of Scourian and Laxfordian ages, there is not entire agreement on other aspects of their differences and definition, especially their separation by a single (or broadly unified) period of dyke injection.

The Torridonian Series

These rocks overlie the Lewisian Gneiss of the foreland in various regions from Cape Wrath to Skye and are themselves unconformably overlain on their eastern margin by Lower Cambrian beds; they also form part of the islands of Rhum, Iona, Colonsay and Islay. No direct isotopic date is available from them as yet, but a late Pre-Cambrian age, already suggested by their stratigraphical relations and lack of metamorphism, is reinforced by a probable equivalence with the Moine Schists (p. 38). No macrofossils have been found in the group but spore-like bodies and other poorly preserved microflora are known from several places and show some resemblance to those from the Riphean rocks of Russia (p. 52).

The considerable lateral facies change seen in the Torridonian rocks and their several, discrete outcrops make it convenient to divide them into four regions, from north to south:

(1) north-west Sutherland near Cape Wrath;

(2) from the Stoer peninsula of Sutherland, through the Coigach and the shores of Loch Broom and Loch Torridon to Applecross in western Ross-shire;

(3) from Loch Kishorn to Skye and Rhum (in the first two much of the outcrop is involved in the Moine thrust-belt);

(4) Colonsay and Islay.

The islands of the fourth region lie east of the presumed course of the Great Glen Fault and their original position is accordingly deduced to have been some 60 miles farther south of the other outcrops than at present. Torridonian rocks thus extended over an area that was some 270 miles from north-north-east to south-south-west; whether there was a comparable east–west facies change is unknown, because the outcrop is so narrow.

Tectonic structures in these beds are usually simple and dips low; indeed as seen in the outliers of north-western Ross-shire and Sutherland, where they stand up as mountainous relics of red sandstones and arkoses with almost horizontal bedding, the Torridonian rocks give an impressive but misleading appearance of having suffered little but erosion in all Phanerozoic time. This illusion is dispelled by the eastward dip of ten degrees or more of the Cambrian quartzites (Fig. 15); during their deposition the Torridonian beds must have had a comparable tilt to the west, apart from any later tectonic effects. In some places the Lewisian surface on which they were laid down was no more than gently irregular, but elsewhere (e.g. the Slioch, near Loch Maree) it retained a pronounced relief with hills and valleys up to 2,000 feet in amplitude.

Any stratigraphical classification in these rocks must be lithological with all the possibilities, perhaps probabilities, of diachronous boundaries. The recognition of three facies groups locally in superposition on the mainland in the Applecross and Loch Broom outcrops resulted many years since in a threefold division:

1. Suilven, Sutherlandshire: an upstanding outlier of almost horizontal Torridonian rocks rising from an irregular platform of Lewisian gneiss.

2. The North-West Highlands, near Loch More, Sutherlandshire, looking northwards. The easterly dipping white Cambrian quartzites overlie the Lewisian gneiss, which forms the lower ground on the west.

Aultbea Beds
Applecross Beds
Diabaig Beds

Of these the central, Applecross Beds, are the thickest, coarsest, and largely red or brown; it was that they were, understandably, confused with Old Red Sandstone in an early stage of their investigation. Since

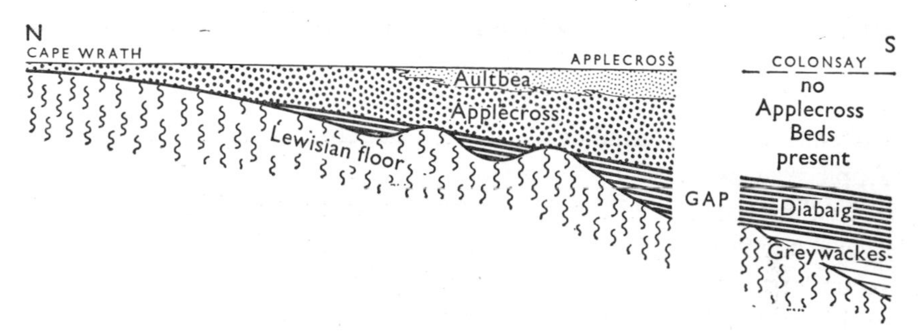

Fig. 9. Diagram to show the succession and inferred facies changes in the Torridonian rocks. Relative thicknesses are suggested but the section is not to scale.

then greywackes have been found at the base in the south and the major facies (with Aultbea as a laterial variant of the highest Applecross Beds) are broadly thus:

Applecross Beds	Rhum to Cape Wrath
Diabaig Beds	Islay to Stoer
Greywackes etc.	Islay and South Colonsay

In the Stoer peninsula, Sutherland, there is combined geophysical and field evidence for a substantially older 'Torridonian' group of coarse red sediments, otherwise similar to the Applecross above.

The graded greywackes of the southern two islands amount to a maximum of 6,000 feet and have been interpreted as turbidites, comparable to the more extensive and better known Lower Palaeozoic examples. They disappear northwards in the island of Colonsay, being overlapped by the Diabaig Beds which come to rest on the Lewisian Gneiss. Their unique position, as the lowest Torridonian division, depends on the correlation of the Diabaig Beds above with those of Skye and Rhum, but this is supported by a considerable degree of similarity in the individual stratal groups of the two regions, widely separate though they are.

The Diabaig Beds have a maximum thickness of about 8,000 feet in Skye, but become much thinner northwards, being less than a tenth of this at their type locality on Loch Torridon; in Sutherland they are finally overlapped by the Applecross Beds. Typical Diabaig sediments are siltstones and fine-grained sandstones, with rarer, thin beds of impure limestone; although red beds are known, more particularly in the north, the rocks are more usually grey and grey-green. The Epidotic Grits at the base in Skye contain pebbles of green epidotic gneiss and other Lewisian rocks; some of them are finely laminated or show graded-bedding on a small scale and there are many indications of shallow water—ripple marks and desiccation cracks. The sandstones are often current-bedded, with a dominant direction of foresets towards the north-east, or include distorted or convoluted structures similar to those in the Applecross Beds.

In the type area of Applecross the uppermost division amounts to over 8,000 feet, but becomes gradually thinner when traced northwards to under 2000 feet near Cape Wrath. Typical Applecross rocks are red and brown, feldspathic sandstones, arkoses and conglomerates, with a subordinate proportion of siltstones. There are many signs of shallow water, in ripple marks, current-bedding and channelling, which collectively point to fluvial and perhaps intertidal conditions. Impersistent black sands, dominantly of specular hematite, form beds up to 15 inches thick and must indicate winnowing in shallow water; more commonly, however, such concentrations are only thin laminae. The dominant directions of the foreset beds suggest a derivation from the west and north-west.

A cyclic sequence is beautifully displayed around Applecross and Loch Torridon, passing upward from the finer silts and sandstones, sometimes laminated, through current-bedded sandstone, and culminating in a massive convolute unit that may be as much as 30 feet thick. Such cycles are repeated time and time again. These and other types of disturbed or disrupted bedding appear to have been produced by abrupt shocks (probably earthquakes) while the sands were still uncemented and unstable. The Aultbea facies appears in restricted areas, chiefly near Loch Broom and Loch Ewe. It shows a return to finer sediments, dominantly red-brown, siliceous and fine-grained sandstones, usually convoluted and with only minor siltstones.

Over their long outcrop there is thus a conspicuous change in the Torridonian strata, from the red, very shallow-water sediments of the north to the grey, varied or turbidite facies of the south; in any one region there is also a tendency for a more limited upward change—Greywackes to Diabaig Beds, Diabaig to Applecross. If these lithological groups are also considered to be, broadly, stratigraphical units, which seems reasonable, then it follows that the Lewisian floor that was covered by this detrital complex had a gentle regional slope to the south, albeit one that retained considerable relief in places. Continued down-warping in that direction can also be deduced from the southerly increase in thickness of the various formations. Later there followed uplift, tilting, erosion and dissection, and still later the early Cambrian submergence. The Cambrian outcrop is so narrow that it is difficult to get a regional picture of the underlying surface, but it seems probable that the seas invaded a land of reduced relief, composed indifferently of Lewisian and Torridonian rocks.

The Moine Schists

This group of rocks occupies most of the country between the Moine Thrust and Great Glen Fault and is also found in the north-eastern part of the Grampian or Central Highlands. Apart from the last area, where they are in a somewhat debated relationship to the Dalradian Series (p. 66), their limits are either those of erosion or tectonic dislocation. Over most of their extensive outcrop the Moine rocks are a monotonous series of mica schists and quartz-feldspar granulites, and their metamorphic grade is high. Their structures are complex and multiple, incorporating several superimposed fold systems, but there appears to be a tendency for the larger folds to have a dominant Caledonian trend (around north-north-east) and, at least in the west, axial planes that dip south-east. This attitude agrees with the suggestion that the dominant movement of the rocks during folding was to the west, which is also the direction of that on the Moine Thrust itself. The ages of the successive phases of metamorphism and folding are not entirely clear; several samples from the schists have given an age of about 420 million years (i.e. Lower to Middle Silurian), but possibly some of the earliest effects may be late Pre-Cambrian. Pegmatites in Morar, on the west coast of Inverness, are at least 740 million years old and since they cut the Moinian rocks, the deposition of the latter must be older than this. This determination confirms the Pre-Cambrian age of the Moine Schists and it is also the nearest that we can get to an isotopic age for the Torridonian.

Most of the stratigraphical detail that has been worked out in the Moine Schists has been derived from the same south-westerly outcrops, particularly in Morar and around Glenelg, on the Sound of Sleat; here the regional metamorphism is unusually low and the underlying Lewisian rocks are exposed. Throughout the whole outcrop there are two principal areas where Lewisian-like rocks appear. In the north of Sutherland there are concordant sheets within the schists, not separable from them on structural or metamorphic characters, and the age of which is unknown. On the other hand in the second group, in the western parts of Ross-shire and Inverness, two types are recognizable: thrust wedges of Lewisian basement brought up among the schists, and those which, though also affected by thrusting, show an unconformable junction (locally with conglomerate) of Moine on Lewisian. In this type the Moine cover is locally autochthonous—that is, it is still attached to the underlying basement, and the two have been disrupted together by the earliest movements affecting the schists. The distinction between basement and cover is reasonably clear in these westerly exposures, but farther away to the east, where the Moine metamorphism is higher, the distinction has become blurred because, through metamorphic convergence, the two groups more resemble one another, structurally and mineralogically.

In the Morar region a good deal can be made out of the sedimentary rocks of which the series was originally formed. The psammitic and pelitic schists represent a variety of sandy, silty and shaly sediments, and among the coarser types there are still traces of depositional structures, such as lamination, ripple marks, current-bedding, slump units and small-scale erosion channels. The whole appears to have been a shallow water sedimentary complex, and the direction of the foreset beds suggests tentatively (because the observations are rather restricted, and have to be interpreted cautiously in rocks so deformed) currents from some southerly direction, between south-east and south-west. On a larger scale it is possible in the Morar and Glenelg regions to establish a crude stratigraphical succession of stratal groups that are either dominantly psammitic or pelitic, and which alternate with one another. Six such groups have been described, which (if there is no repetition by thrusting) bring the total thickness to something like 22,000 feet.

The relation of the Moine Schists to the other, neighbouring, formations in the Scottish Highlands has been discussed for many years. Some theories, such as an equivalence with part of the Lewisian Gneiss, are now obsolete. Another idea was that the schists intervened in age between Lewisian and Torridonian. The correlation that has gained increasing acceptance in recent years is that they are the metamorphosed equivalent of the Torridonian Series; on this view they have been brought from some region of prolonged Caledonian metamorphism and folding and carried westwards over the edge of the foreland into a region where the sedimentary cover was not so affected. This theory is supported by the date of 740 million years as a minimum for the deposition of the Moinian sediments, which at least places them firmly in the late Pre-Cambrian, and by the many sedimentary resemblances to Torridonian structures. Moreover, where the latter group is affected by local metamorphism in the Kishorn nappe of Skye (one of the under slices of the Moine thrust-belt) it takes on some Moine-like characters.

The lithological similarities between the two series are by no means identities. In bulk, the Moine Schists are generally finer, with originally more shale and silt than the bulk of the Torridonian, and pebbly rocks are rarer; to that extent they compare better with the lower facies than with the Applecross. There is some indication also that the current-bedding directions are opposed. These differences need not disturb the postulated equivalence too seriously, however. The difference in metamorphism alone presupposes a considerable distance between their original milieux; the Torridonian in itself shows some significant lateral changes in facies, and there is no reason why an increasing fineness to the east should not be an extension of these. Current directions, or foreset dips, in modern and the better known fossil sediments, frequently show wide swings of arc. It is probably enough to postulate that, in the rather broad type of correlation applicable to these rock groups, they show many signs of deposition in the same sort of sedimentary régime, they are both late Pre-Cambrian, they bear the same relations to the Lewisian basement, and are probably generally equivalent.

If we pause for a moment to review the north-west foreland of Scotland, it appears as a narrow remnant, taking the eye out over the waters of the North Atlantic; and in fact this is not a barren freak of the imagination. Each of its three component formations has some link with the North American continent. There are fold belts of the same age, approximately, as Scourian and Laxfordian; they cover much greater areas, and the long gap after the second is partly filled by the metamorphism and folding of the Grenville belt (800–1,000 million years), of which we have no counterpart. The clearest source of certain Torridonian pebbles, the varied and distinctive cherts in particular, is the Pre-Cambrian rocks of the Great Lakes and Labrador. The Cambro-Ordovician carbonate sequence (p. 74) has marked American affinities in facies and fauna.

ENGLAND AND WALES

South of the Scottish Highlands the Pre-Cambrian outcrops are few, mostly small, and give very little coherent view of any floor to the Caledonian geosyncline or a 'south-east foreland'. Some rocks are highly metamorphosed but in many areas they are not much altered. Isotopic determinations have so far not proved as useful as in the more extensive Scottish rock groups; many of the outcrops are in regions much affected by the Caledonian orogeny, and even though there is certain (or strongly presumptive) stratigraphical evidence of their Pre-Cambrian age, their minerals have been sufficiently reorganized to give Caledonian, or occasionally even Hercynian, dates. It follows that at present we are still chiefly dependent on other lines of evidence—field contacts and the relation to overlying Cambrian beds, structural and metamorphic history and lithological resemblances. Correlation between the various outcrops is often tentative, or scarcely attempted. The grouping adopted below is on the basis of convenience, and of some geological similarity, but not on very secure grounds.

Across the English Midlands there seems to be a broad region where Pre-Cambrian rocks, in places at least, are not so deeply buried as elsewhere or appear at the surface. It stretches from Shropshire and the Welsh Borders on the west to Norfolk on the east and extends northwards into the southern Pennines. The rocks are not highly metamorphosed and the few available dates suggest a late Pre-Cambrian age; nevertheless, the structures and directions they exhibit differ considerably from one outcrop to another. This region is probably the most that can be envisaged as any kind of south-eastern foreland to the Caledonian geosyncline in Britain. Nevertheless, it compares poorly with the north-western, as seen in Scotland or the south-eastern, as seen in Sweden, being fragmentary, ill-defined and, as will appear in the next chapter, not overlain by a typical Lower Palaeozoic foreland facies such as that of Scotland.

North-westwards from Shropshire, beyond the thick Lower Palaeozoic cover of North Wales, Pre-Cambrian rocks reappear in Anglesey and the Lleyn Peninsula and with them should probably be linked certain metamorphosed rocks of south-east Ireland. More dissociated geographically (though with some resemblances in their volcanic groups) are the outcrops of the Malvern Hills and Pembrokeshire. Lastly the Ingletonian rocks, commonly assumed to be Pre-Cambrian, fit in with none of the foregoing. If it is difficult to unite the Scottish Pre-Cambrian rock groups in any coherent synthesis, those of England inevitably degenerate into little more than a catalogue. From this probably one item should be removed. The highly altered rocks of the Lizard, adjacent to a belt of Hercynian deformation in south Cornwall, were once thought to be Pre-Cambrian. Their range of dates, however, from about 350 to 500 million years, does not support this ascription.

Anglesey, Lleyn and Wexford

Rather more than half Anglesey (Fig. 12, p. 58) is made up of the Mona Complex or Monian Series. Although the oldest fossiliferous beds found above are Ordovician, the Pre-Cambrian age of the Monian rocks is supported by several lines of evidence. Detritus from them has been recognized in the Cambrian beds of the mainland; in places they are much more highly metamorphosed and intricately folded than the overlying Ordovician sandstone and shales; a minimum isotopic age of 611 ± 16 million years has been obtained from schists in the Gwna Group.

Most of the rocks are comprised in a series of sediments with repeated volcanic intercalations which have been intruded by granites at a late stage in the folding. The whole is a small remaining fragment of a major Pre-Cambrian sequence of geosynclinal sedimentation, folding, intrusion and partial metamorphism. It was presumably completed and the complex deeply eroded before the Lower Palaeozoic seas were established in North Wales, but it is interesting to note that the dominant structural direction is the same. It seems likely that this part of the Caledonian mobile belt followed a pattern already impressed in the Monian orogeny.

The Monian sedimentary series (Table 4, p. 42) is well exposed in the cliffs of Holy Island, where the rocks are strikingly folded. The original thickness is very difficult to assess, but the total is of the order of 20,000 feet—comparable to the accumulation in the Caledonian geosyncline. The succession in Table 4 is deduced from sedimentary structures, especially graded-bedding. The mélange is a jumbled mass of angular blocks, very various in size, in a schistose matrix. The origin of this assemblage is somewhat obscure; it has been interpreted as a large scale thrust-breccia, but no accompanying thrust is known. A rather more likely view is that it is a true though unusual sedimentary breccia, formed from pre-existing strata and accumulating at the base of a steep, unstable slope; in places at least it resembles a vast underwater scree. The gneisses that crop out in various parts of Anglesey, and which were once thought to be a much older group, are probably the same sedimentary series, now highly metamorphosed and migmatized by the granitic intrusions so that they have become coarse-grained, banded rocks, acid and basic.

Table 4. *Monian succession, Anglesey*

FYDLYN GROUP: acid tuffs and possibly lavas.
GWNA GROUP: varied sediments including carbonaceous shales and subordinate limestones; spilitic lavas, including pillow structures, bedded jaspers; the mélange (see text).
SKERRIES GROUP: massive sandstones with pyroclastic debris.
NEW HARBOUR GROUP: alternating varied sediments (quartzites to shales); spilitic pillow lavas and jaspers.
RHOSCOLYN GROUP: graded greywackes and shales.
HOLYHEAD QUARTZITE: chiefly a white pure quartzose rock with little bedding.
SOUTH STACK GROUP: graded greywackes, sandstones and shales.

Thicknesses are difficult to estimate but the Gwna Group is very thick and contributes substantially to the geosynclinal proportions of the whole.

All the principal Pre-Cambrian formations of Anglesey can be found in the Lleyn Peninsula of the Welsh mainland. South-east of the Menai Straits another group—the Arvonian—appears in the Bangor and Padarn ridges. The rocks here comprise a variety of tuffs and lavas, mainly rhyolitic. They are much less metamorphosed than most of the Monian Series and on this account, and because they are overlain on the Padarn ridge by the Lower Cambrian of the Llanberis outcrops, they are generally taken to be a late Pre-Cambrian formation; nevertheless, the extent of the unconformity has been questioned, and the 'Arvonian' might conceivably be a basal Cambrian group or underlie the Lower Cambrian with little or no break.

The relationship of the Anglesey and Lleyn rocks with those of Wexford depends on lithological similarity and deductions made on their age. The south-east promontory of Ireland is composed of a rather variable suite, which is older than the overlying Ordovician and was once grouped with Bray Series farther north and provisionally considered to be Cambrian. They include, however, the highly metamorphosed Rosslare Migmatite Series—pelitic, semi-pelitic rocks and green hornblende schists—which resemble some of the Monian rocks. It is thus probable that we have, from Anglesey to Wexford, the remains of an old Pre-Cambrian ridge or narrow massif with a north-east–south-west trend; in Lower Palaeozoic times it comes into prominence again as the postulated 'Irish Sea landmass' and a possible source of the detritus that was carried eastwards into the Welsh geosyncline.

Shropshire

In the southern part of the county Palaeozoic and Pre-Cambrian rocks rise from under the New Red Sandstone of the Cheshire basin. They are largely governed by Caledonoid structures, and although the outcrops, especially of the pre-Silurian systems, are relatively small, this is one of the classic geological regions of the Welsh Borders. The Pre-Cambrian rocks form upstanding masses such as the dissected plateau of the Longmynd or the conspicuous hills of the Wrekin and the Caradoc range; not only do these uplands dominate the present scenery but, together with the Church Stretton Fault, they were important elements in Lower Palaeozoic sedimentation, as described in later chapters.

The Church Stretton Fault is really a Caledonian fault complex with a long history, certainly from Mesozoic back to Ordovician times and perhaps earlier. At its name place it follows a through valley which separates the two largest groups of Pre-Cambrian rocks, the Eastern Uriconian on the east from the Longmyndian on the west; the fault belt can also be traced far to the south-west into the Welsh Borders, where in places it brings up older rocks through a Silurian cover. There are other major faults in Shropshire that take a parallel course, important among them being the Pontesford–Linley Fault, west of the Longmynd. The Longmyndian beds for the most part dip steeply to the north-west; their succession is shown in Table 5. Small strips of Western Uriconian appear on their western margins, of which Pontesford Hill is the largest. The Eastern Uriconian rocks of Caer Caradoc and other uplands stretch north-eastwards to the Wrekin and the Wrockwardine hills (Fig. 10).

South of the Wrekin is the small, poorly exposed patch of the Rushton Schists; these are more altered than the other rocks in Shropshire, but whether they belong to a truly older series or are only a local metamorphosed variant is not known. Lower Cambrian quartzites overlie the Uriconian rocks unconformably on the Wrekin itself and among the Caradoc hills. Eastern Uriconian basalts from Wrockwardine have given an isotopic date of 638 ± 81 million years (or late Pre-Cambrian), but they have been folded and faulted and this should be taken as a minimum age.

Since the various outcrops in south Shropshire are commonly separated by tracts of later strata or by faults their mutual ages have been

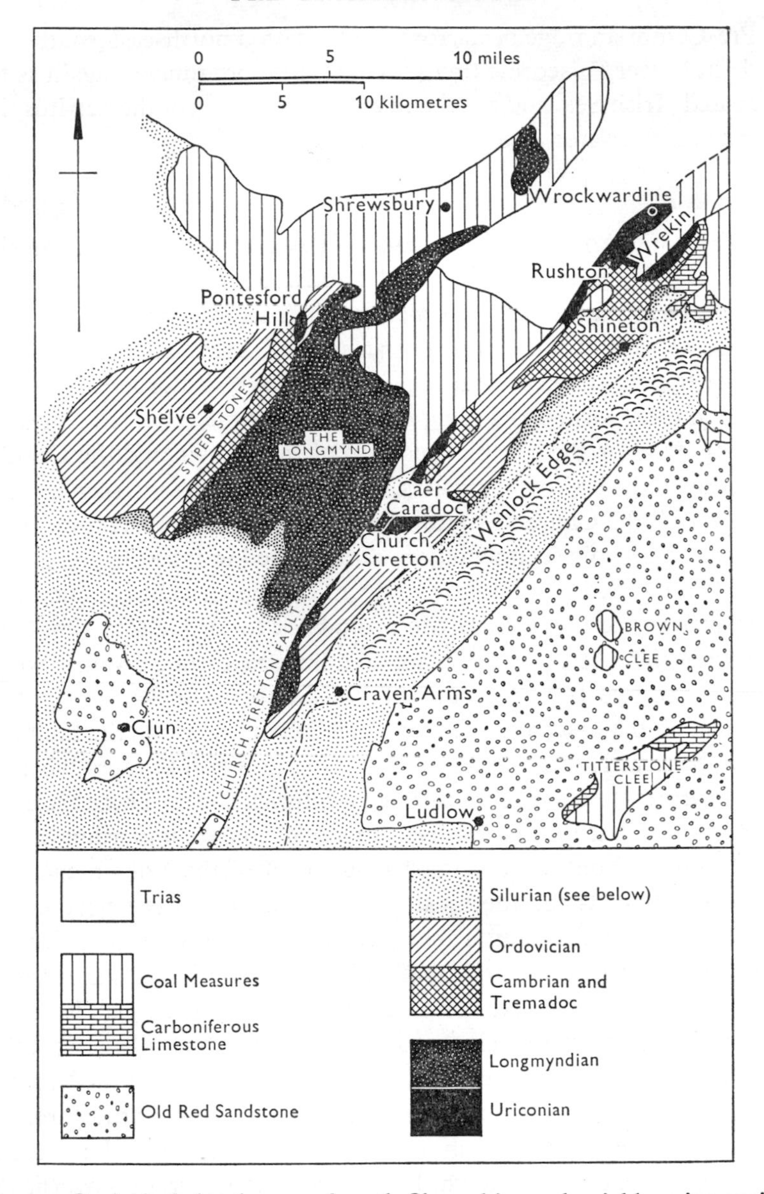

Fig. 10. Geological sketch map of south Shropshire and neighbouring regions. Silurian subdivisions are shown to the east of the Church Stretton Fault (together with the belt of 'ballstones' along Wenlock Edge) but not to the west; littoral Upper Llandovery facies in denser stippling.

Table 5. *Pre-Cambrian rocks of Shropshire*

		Succession and lithology	Approximate thickness (feet)
LONGMYNDIAN:			
'Western Longmyndian'	WENTNOR SERIES	purple sandstones and conglomerates, with some grey-green siltstones and shales	15,700
		major unconformity	
'Eastern Longmyndian'	MINTON SERIES	purple and green shales and sandstones with basal conglomerates	3,800
		unconformity	
	STRETTON SERIES	grey, green siltstones, sandstones and shales with subordinate purple beds and volcanic rocks	11,400
		unconformity, or disconformity	
WESTERN AND EASTERN URICONIAN:			
Rhyolitic lavas, tuffs, agglomerates and minor intrusions (base not seen)			

much discussed, but the version given on Table 5 has gained a good deal of acceptance. The time-relations set out there depend on various lines of evidence. The structures in the Longmyndian sediments (especially graded-bedding) suggest strongly that the western part of the Wentnor outcrops are inverted, and it is accordingly deduced that the major structure of the range is a slightly overturned syncline (Fig. 21), the Minton Series being below the Wentnor and separated from it by an unconformity; the Minton Series is probably also underlain by a break. The strong petrological similarity between the Eastern and Western Uriconian suggests that they are of similar origin and broadly contemporaneous; fragments are incorporated in the basal Stretton beds and also in the Wentnor conglomerate, so that the Uriconian as a whole is older than the sedimentary series. The latter was presumably laid down on an eroded surface of the tuffs and lavas and incorporated debris from it at various levels.

Longmyndian rocks are not widespread in the Welsh Borders but can be recognized at a few other places; small inliers appear along the line of the Church Stretton Fault at Pedwardine and Old Radnor, and they possibly also form a small part of the May Hill inlier in Gloucestershire. Uriconian rocks (or volcanics of similar type) have a wider distribution.

The relationship between the Pre-Cambrian rocks of Anglesey and Shropshire is problematical. The fold directions of the Monian Series and the Longmynd are similar, but their facies are quite unlike. Then whereas there is some resemblance in rock-type between the Uriconian and Arvonian groups, both being rich in rhyolitic tuffs and lavas, their stratigraphical positions are not comparable. Thousands of feet of Longmyndian sediments (on the classification given above) intervene between the Uriconian lavas and Lower Cambrian beds in Shropshire, whereas in North Wales the latter are directly underlain by the Arvonian. The isotopic determinations do not, as far as they go, support any direct equivalence and there for the present the matter rests.

Charnwood Forest and the Midlands

Charnwood Forest is an irregular, hilly region north-west of Leicester where Pre-Cambrian rocks protrude through a cover of Triassic marl (Fig. 43, p. 221) in the form of a much faulted anticline, trending north-westwards and plunging to the south-east. This, the Charnian trend, locally swinging nearer to north or west, affects much of the later rocks in the region, especially the Carboniferous of the east Midlands and concealed coalfield.

The rocks of Charnwood are chiefly pyroclastic, interbedded with some sediments, and are divided into:

Brand Series
Maplewell Series
Blackbrook Series

These are intruded by a number of porphyritic and syenite masses. The immediate contacts of both are with Trias, and in the north for a short stretch with Carboniferous beds, but their strong cleavage compared with uncleaved neighbouring Cambrian shales, the relation of very similar syenites to the Cambrian beds at Nuneaton (Warwickshire) and the age of 684 ± 29 million years for the intrusive rocks confirm their long-supposed Pre-Cambrian position.

In the southern and eastern outcrops the Charnian rocks include a great variety of bedded pyroclastic rocks, from agglomerates to tuffs, accompanied by lesser proportions of conglomerates and tuffaceous sandstones, siltstones and slates; the volcanic components are most marked in the lower part of the Maplewell Series and least in the Brand Series. In the north-west the stratigraphical succession is obscured and agglomerates, some very coarse, predominate; it seems likely that the volcanic centres were in this direction. Though the syenites and other intrusions cut all the bedded series they are petrologically similar and belong to the same period of vulcanicity.

Charnian rocks are unique in the British Isles for the impressions of the Pre-Cambrian fossil, *Charnia*, collected from fine-grained tuffaceous siltstones in the uppermost Maplewell beds. These organisms appear to have consisted of some kind of frond-like structure attached to a disc; although their biological position is uncertain (both primitive coelenterates and algae have been considered), there is no doubt that the genus represents a true fossil, as several specimens have been collected in Charnwood Forest and others are known in Australia.

Comparisons have been made between the Stretton Series (Eastern Longmyndian) and the Charnian, with the implication that they might be of the same age but that the pyroclastic components did not reach so far as Shropshire. There seems little positive support for such an equivalence, however, and the isotopic ages are opposed to it, in that they suggest an earlier age for the Charnian intrusions than for the Uriconian lavas and the latter underlie the Longmyndian sediments. Pre-Cambrian rocks are found in small outcrops associated with other Midland coalfields. Uriconian-like volcanic rocks crop out at Lilleshall on the west side of the Coalbrookdale Coalfield and at Barnt Green in the Lickey Hills of south Staffordshire; it is possible also that they underlie part of the southernmost Pennines, for similar types were met in a boring near Buxton. Moreover, a much greater expanse of such rocks than the present inliers is suggested during Permian times by the large proportion of Uriconian blocks and pebbles that are found in some of the Midland breccias.

The Caldecote Volcanic Series of Nuneaton is rather different and more like those of the Charnwood Forest. It forms a small, much faulted, strip of volcanic and intrusive rocks alongside, and locally underlying, the larger Cambrian outcrops. The intrusions include a variety of syenite (markficldite) very like certain Charnian types; they are intrusive into the volcanic series and also supply large boulders in the basal Cambrian quartzite above. These exposures thus not only prove the Pre-Cambrian age of the local series, but are a strong reinforcement for that of Charnwood. A further extension, possibly forming some kind of Pre-Cambrian ridge, is suggested by the Charnian-like rocks found in deep borings to the east—on the borders of Lincolnshire and Leicestershire, near Peterborough and in north Norfolk (Fig. 47, p. 230). At the last

(North Creake) the Pre-Cambrian lies at a depth of about 2,500 feet, below Trias; nearer the Pennines the cover is more usually Carboniferous.

Malvern Hills

The north–south ridge of the Malvern Hills, rising up on the west side of the Trias of the Lower Severn valley (Fig. 23, p. 106), is the most impressive of all the English Pre-Cambrian masses. Two rock groups are present. Most of the ridge is formed of the Malvern gneisses, a metamorphosed Pre-Cambrian basement locally overlain unconformably by Lower Cambrian quartzites; on the eastern side there are smaller outcrops of the Warren House Volcanic Series overlying the gneisses—rhyolitic tuffs and lavas very like the Uriconian of Shropshire and often correlated with them.

In its structures and regional setting the Malvern range is something of a puzzle. The eastern junction has been interpreted both as a great fault, or primarily as a Triassic overlapping margin (similar to that at Charnwood) with only minor adjustment faults. On the west the Cambrian and Silurian beds are strongly folded, in some places inverted and also overthrust by the Pre-Cambrian rocks. Farther north along the Malvern line Coal Measures exhibit similar vigorous structures, so that these effects are likely to be late Palaeozoic and due to some phase of the Hercynian orogeny. The Pre-Cambrian gneisses are largely altered igneous rocks; their foliation shows a wide variety of strike-directions, but in places a preference for north-west, or west-north-west. This combination of direction and degree of metamorphism does not conform with any other outcrop in England and Wales and enhances the possibility that a narrow zone of folding and faulting has brought up a deep-rooted ridge of Pre-Cambrian basement older than any seen south of the Highland border.

South Wales

All the Pre-Cambrian outcrops in this region (Fig. 19, p. 90) are confined to Pembrokeshire and the most important of them are found north of St Brides Bay, especially around St Davids, the best sections being along the coast to the south and west of that city. There are two rock groups—the volcanic Pebidian and intrusive Dimetian; the Lower Cambrian conglomerates contain pebbles of the former and can be seen overlying both. The Pebidian rocks are somewhat similar to the Uriconian, rhyolitic and trachytic lavas and tuffs with bands of flow-breccias and lava-conglomerates. These form the main part of several outcrops of which those at St Davids and Hayscastle are the largest. They are cut by the Dimetian intrusions, chiefly of granite, granophyre and quartz-diorite.

A smaller group of volcanic and intrusive rocks is found in south Pembrokeshire where it is brought up on one of the major Hercynian structures, the Benton Thrust on the south side of the Pembrokeshire Coalfield (Fig. 20, p. 91). These outcrops are not in contact with Cambrian strata but are locally overlain by Silurian, and since they resemble the northern types their Pre-Cambrian age is generally accepted. The Benton Volcanic Series comprises acid lavas, tuffs and breccias, which probably correspond to the Pebidian of St Davids; they are associated with a group of quartz diorites (the Johnston Series) but the field relations between the two are not clear. The Johnston Series is also foliated, but whether this was induced by the Hercynian movements responsible for the thrusting, or was much earlier, is again uncertain.

The Ingletonian rocks of West Yorkshire

These beds form part of the pre-Carboniferous foundation of the Askrigg Block, the southern of the two block-structures of the northern Pennines (Fig. 37, p. 179). They are exposed only in two neighbouring valleys in the south-west and, apart from the unconformable Carboniferous Limestone above, the adjacent strata are late Ordovician and probably all the junctions are faulted. The isoclinal folds of Ingletonian rocks strike west-north-west and the group is estimated to be about 2,500 feet thick; it consists of slates, siltstones which are sometimes current-bedded, and a lesser proportion of greywacke and arkose. There are some signs of shallow water deposition, with contemporaneous erosion and local conglomerates. The detrital fragments of the coarser rocks suggest derivation from an igneous and metamorphic terrain.

The ascription of a Pre-Cambrian age for the group depends largely on the somewhat greater degree of folding, metamorphism and dyke injection than that in the neighbouring Lower Palaeozoic formations. Nevertheless, in the nearest comparable outcrop, the Lake District, there were extensive Ordovician (pre-Bala) earth movements, and so it is just possible that these differences might reflect a similar event, and that the Ingletonian rocks belong to an earlier Ordovician or Cambrian formation. The lack of recognizable fossils is not crucial, for the exposures are restricted, and very few fossils have been found in, for instance, the Manx Slates, part of which are early Ordovician.

The Ingletonian is thus considered to be a possible, but not proven, Pre-Cambrian group. The low metamorphic grade suggests late Pre-Cambrian but the outcrops are so isolated, geographically and structurally, from the most similar rock types, such as the Longmyndian, that no close comparison can be made; nor is there any indication whence came the detritus. The group takes its place as part of the heterogeneous foundations of the Pennines.

REFERENCES

Scotland and Ireland. Anderson, 1953; Baker, 1955; Black and Welsh, 1961; Brown *et al.* 1965; Charlesworth, 1963, ch. 2 and 3; Dearnley, 1962; Johnson, 1965*a*; Kennedy, 1955; Ramsay, 1963; Ramsay and Spring, 1962; Selley *et al.* 1963; Stewart, 1962*a*, *b*; Sutton, 1963; Sutton and Watson, 1951, 1960, 1964; Tremlett, 1959; Watson, 1963, 1965.

England and Wales. Dunham *et al.* 1953; Ford, 1958; James, 1956; Meneisy and Miller, 1963; Pocock *et al.* 1938; Shackleton, 1954*a*; Watts, 1947; Whittard, 1952.

General. Anderson, 1959.

British Regional Geology. The Northern Highlands, North Wales, South Wales, the Welsh Borderland, Central England.

3

THE CAMBRIAN SYSTEM AND THE DALRADIAN SERIES

In the British Isles the three Lower Palaeozoic systems are linked together in several ways: in their history, which is largely that of the Caledonian geosyncline, in their allied areas of outcrop, and in their terminology. The origin of the names—Cambrian, Ordovician and Silurian—is worth a short explanation, because it illustrates the difficulties that arose in the early days of stratigraphical definition, in regions of some complexity and where fossils were often sparse. Briefly, in the 1830s Sedgwick and Murchison were working, respectively, in the Cambrian country of North Wales and the Silurian of the Welsh Borders; these names date from 1835. During the succeeding years it became clear that both authors were laying claim, in the intervening areas, to rocks of the same age—as 'Upper Cambrian' or 'Lower Silurian'—and a long controversy ensued. The Ordovician System was put forward as a solution by Lapworth in 1879 to include most of the disputed strata. It was much more than just a compromise, however, and brought together rocks and faunas of reasonable unity and convenient definition. Moreover, it is somewhat ironic that this system—the interloper—proves to represent a much longer period than the Silurian, and is comparable with the Cambrian.

The Cambrian System is the earliest to contain abundant fossils that can be safely attributed to the major invertebrate groups. Among these the trilobites are the most important, and their genera and families form the basis of the three subdivisions. Although the system was named from Wales, the faunas are more prolific in the thinner, finer and less disturbed outcrops of Scandinavia, particularly near Oslo and in south Sweden. The Cambrian rocks of Scandinavia, England and Wales, New England and parts of south-eastern Canada, form a single faunal province, characterized by the following trilobites:

Upper Cambrian: olenids and agnostids
Middle Cambrian: *Paradoxides* and agnostids
Lower Cambrian: *Callavia* and *Holmia*

In most of North America, Labrador, Greenland and Spitsbergen these groups are replaced by others to form another, American, faunal province. To this also belong the outcrops of the Scottish North-West Highlands, where in particular *Callavia* and *Holmia* are replaced by *Olenellus*.

To some extent the problems of Cambrian geology are different from, or additional to, those of the two succeeding systems. The outcrops are fewer, smaller and less fossiliferous. The upper limit is bound up with the position assigned to the Tremadoc Beds, which in this book are taken as Ordovician, and described in the next chapter. The base of the Cambrian System poses a perennial problem, essentially because diagnostic fossils are available on one side only, and are often inadequate there, as lower Cambrian faunas are notoriously sparse. The system as a whole records the appearance of nearly all the invertebrate phyla, but only by degrees. Where Lower Cambrian fossils are available, and are underlain at no great distance by a major unconformity the matter is not serious: the beds above the unconformity may be called Cambrian and those below Pre-Cambrian. This happy combination is present in places in the English Midlands, Shropshire and in the North-West Highlands of Scotland. But there are many other situations not so easily resolved. Chief among these is the existence of a thick, unfossiliferous sequence lying below fossil-bearing Cambrian rocks, but not separated from them by a marked stratigraphical or metamorphic break. Should such rocks be considered as Pre-Cambrian or unfossiliferous Cambrian? Over the world as a whole a variety of expedients has been employed and sometimes an extra name adopted, such as Infracambrian, Eocambrian (especially for such beds beneath the *Holmia* Series of Norway), the Greenlandian System, Sinian of China, or Riphean of Russia.

It may be that more isotopic age determinations will help to resolve this problem but some inconsistencies are likely to remain. In Britain the Dalradian Series is in this equivocal position and there are some thousands of feet of beds below either low Middle Cambrian or high Lower Cambrian horizons in North Wales. Other minor examples will emerge in the more detailed descriptions.

In their facies, structural and geographical relations the British Cambrian outcrops fall into four groups. There is the orthoquartzite-carbonate facies accumulating on the north-west foreland of the geosyncline and little affected either by its clastic influx or subsequent deformation. North and South Wales, the Welsh Borders and the English Midlands belong to the Welsh trough facies and eastern bordering shelf. The Dalradian Series represents the earliest geosynclinal sedi-

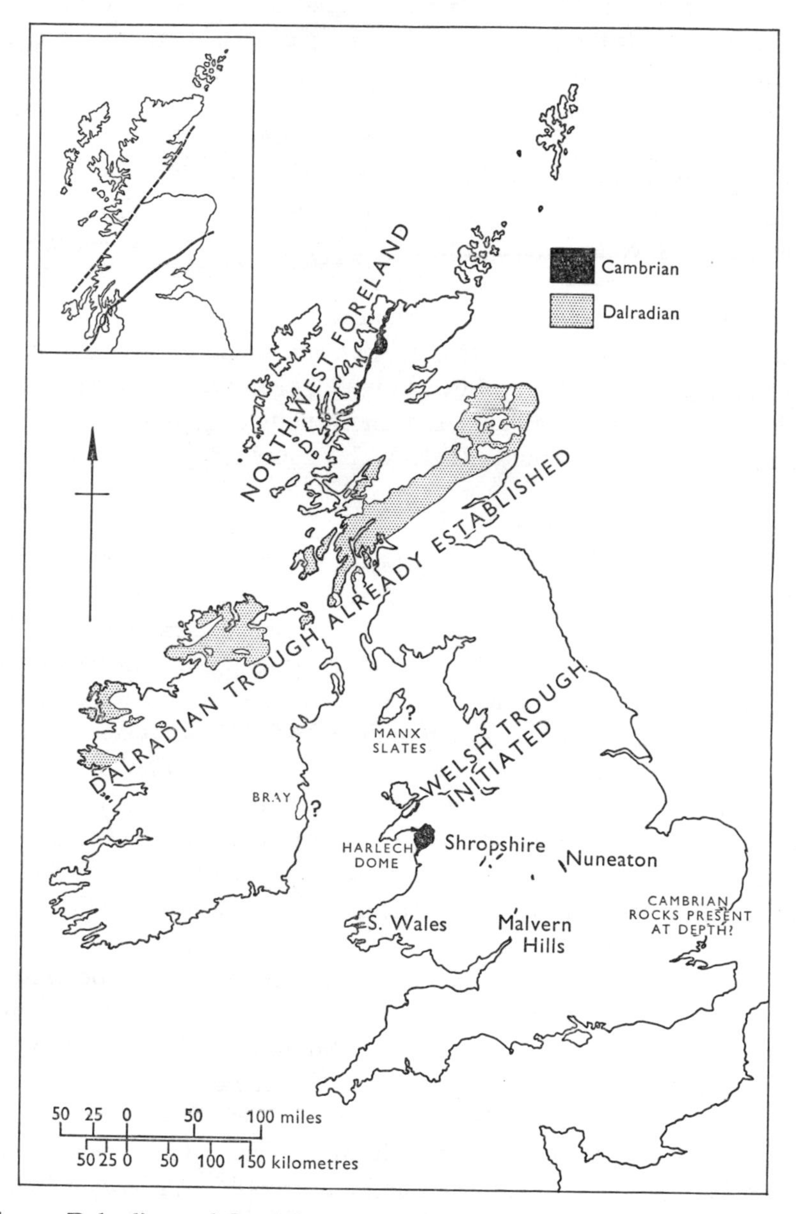

Fig. 11. Dalradian and Cambrian outcrops and their tectonic setting. The Bray Series of Ireland may include some Cambrian rocks; the Manx Slates probably do not. Inset: the deduced relative positions of the North-West and Grampian Highlands before movement on the Great Glen Fault.

CHRIST'S COLLEGE LIBRARY

ments, probably dating from late Pre-Cambrian times but containing Lower Cambrian fossils near the top. It is uncertain whether certain barren rocks in eastern Ireland are Cambrian or not.

THE EASTERN SHELF

Although small and separated at the present time there is some similarity in the Cambrian successions of Shropshire, Warwickshire and the Malvern Hills. None is really thick—that is, not over 5,000 feet. Where the base is seen the lowest beds lie unconformably on Pre-Cambrian rocks and are usually massive quartzites; glauconitic sands frequently occur in the succeeding Lower Cambrian and the later beds are dominantly argillaceous. Shropshire and Warwickshire combined have yielded a fairly complete faunal sequence, though not all the Scandinavian zones have been found. Anglo-Welsh zones are given on p. 415.

Shropshire and Malvern

The Cambrian rocks of south Shropshire appear in the same general region as the Pre-Cambrian hills and ridges described earlier (Fig. 10). It is uncertain whether any beds older than Tremadoc are present in the Habberley valley west of the Longmynd, where exposures are very poor, and so the best outcrops are those from the flanks of Caer Caradoc, north-eastwards across the Severn to the Wrekin. Drift is extensive here also and only the lower beds are well exposed, where they dip steeply off the Wrekin. In the south-west, at Comley, the details of the Lower and Middle Cambrian beds have been recorded only from excavations; nevertheless, the importance of the classic Shropshire sections is largely due to the faunas of these beds, particularly because they include Lower Cambrian trilobites.

The basal Wrekin Quartzite appears to have been deposited on a rather uneven submergent floor of Uriconian rocks and varies in thickness; for the most part it is a white, quartzose rock but near the base incorporates pebbles of the rhyolites and tuffs on which it lies. No trilobites are known from the quartzite, but the Lower Comley Sandstone, a glauconitic rock with calcareous and shaly bands, has yielded a few Lower Cambrian brachiopods, crustacea and one trilobite—*?Kjerulfia* (‘*Holmia*’). The calcareous character is much more pronounced at the top where the beds are separated off as the Comley Limestone and are the main source of the trilobites, including *Callavia* and *Protolenus*. The limestones are sandy with layers of pebbles and irregular phosphatic nodules. Their 6 feet include a number of successive, distinct faunas,

Table 6. *Cambrian rocks of Shropshire*

	Succession and lithology	Estimated thickness (feet)
	(Shineton Shales, Tremadoc)	
UPPER CAMBRIAN	Black bituminous and grey micaceous shales, incomplete and poorly exposed	?80+
MIDDLE CAMBRIAN	Shales with thin sandstones	?700+
LOWER CAMBRIAN	Comley Limestone	6
	Lower Comley Sandstone	500
	Wrekin Quartzite	up to 140
	(Uriconian)	

and the whole is probably a condensed deposit, which was laid down very slowly or intermittently in shallow water swept by gentle currents.

Middle Cambrian deposition followed after a phase of uplift, minor folding and erosion; the Comley limestones were removed locally and fossiliferous pebbles from them have been incorporated in the breccias at the base of the overlying beds. These beds themselves are not all of the same age, for they contain different *Paradoxides* species in different exposures. The Middle Cambrian shales are interspersed with sandy and calcareous bands and several *Paradoxides* zones are known (*groomi, bohemicus, hicksi, rugulosus, davidis* and *forchhammeri*) though some only from loose blocks. The isolated exposures make accurate estimates of thickness difficult, but there are about 700 feet of beds or more. Even less is visible of Upper Cambrian strata, and only two Scandinavian zones have been found—*Orusia lenticularis* in grey micaceous shales and *Ctenopyge* in black bituminous shales and limestones. The total of perhaps 80 feet represents only the uppermost, Dolgelly, beds of North Wales (p. 61). Thus there may be a break below the Upper Cambrian in Shropshire, and possibly in the Middle as well.

The Malvern outcrop lies south-west of the Pre-Cambrian ridge and is overstepped on the western side by Lower Silurian sandstones. The succession is not unlike that of Shropshire, but is less fossiliferous and possibly less complete. The Malvern Quartzite at the base corresponds to the Wrekin Quartzite, is arkosic at the base and unconformable on the Malvern gneisses. Both it and the succeeding Hollybush Sandstone are thicker than the Shropshire Lower Cambrian beds and though they contain a few fossils, lack the trilobites. The uppermost beds of sandstone may be Middle Cambrian but no *Paradoxides* species are known from the Malverns; the succeeding Whiteleaved Oak Shales probably correspond to the *Orusia* Shales of Shropshire.

Warwickshire and the Midlands

Among the heterogeneous, pre-Carboniferous rocks which form a basement to the Midland coalfields (Fig. 43) there is one important Cambrian outcrop and one small, probable, one. In the Lickey Hills, at the south end of the South Staffordshire Coalfield, the Lickey Quartzite forms a ridge, overlying Uriconian-like rocks; the quartzite is unfossiliferous and is overlain by Lower Silurian sandstones. The stratigraphical relations are thus like those at Malvern and the Lickey Quartzite is usually taken to be an isolated Lower Cambrian outcrop.

The quartzites and shales of Nuneaton form the most easterly of the Cambrian outcrops, which although small is one of the more important. It is a strip about 10 miles long on the north-east side of the Warwickshire Coalfield, being overlain by Coal Measures on the south-west and having a partly faulted, partly unconformable junction with the Trias on the north-east; for a short stretch the basal beds lie on the Caldecote Volcanic Series (Charnian). There is a simple lithological distinction into dominantly arenaceous beds below and argillaceous above—the Hartshill Quartzite (800 feet) and Stockingford Shales (2,800 feet). The former is largely unfossiliferous but a few Lower Cambrian fossils occur in a glauconitic sandy limestone near the top and *Callavia* has been found in nodules at the base of the shales above.

The Stockingford Shales contain abundant Middle Cambrian and rather sparser Upper Cambrian faunas; the fossil-bearing horizons are not regularly distributed, however, and many of them are crammed into the thin subdivision of the Abbey Shales, which lies 1,000 feet above the base and contains several *Paradoxides* zones. These are succeeded by an erosion level with pebbles and phosphatic nodules, and the lowest Upper Cambrian zone, of *Agnostus pisiformis*, is not present (nor anywhere else in Britain), possibly through a non-sequence. The remainder of the Upper Cambrian beds are shales with some coarser bands and a fauna including olenids and *Orusia*; the small thickness of Tremadoc beds follows conformably. While in general this is a less varied and less disturbed sequence than that in Shropshire, and possibly somewhat thicker, there is probably one non-sequence and may be more. Unfossiliferous beds resembling the Stockingford Shales, together with minor intrusions that are probably Ordovician, are known from several boreholes in the Midlands, but whether they are Cambrian or Tremadoc is uncertain.

No very extensive generalizations can be made about conditions in the Midlands and Welsh Borders from these scattered exposures. The basal quartzites, apart from locally derived pebbles, typically have well-sorted and well-rounded grains and a small assortment of stable accessory minerals. Such rocks are characteristic of secondary derivation from the well-worked waste of earlier sediments. A shelf area, based on an old landmass, was being established, in contrast to the subsiding trough of Wales. From time to time uplift or shallowing produced current-winnowing, condensed deposits, and in Shropshire was accompanied by minor folding and erosion. Throughout Lower Palaeozoic history south Shropshire has an especial place; it lay near the edge of the north Welsh trough and was particularly subject to uplift and tilting movements. Between the resulting periods of emergence there accumulated shelf sediments in shallow seas. No doubt it was not unique in such a position, but it is the region in which these processes are now best observed.

THE GEOSYNCLINE IN WALES

The distribution of the Lower Palaeozoic outcrops in North Wales can be related to the major anticlines and synclines, although there is much complication on a smaller scale. On the south-east flank of the Bangor and Padarn ridges, with their Arvonian and Cambrian outcrops, the Snowdon Syncline brings in an Ordovician tract that extends south-westwards into the Lleyn Peninsula. South-east of Snowdonia lie the Cambrian outcrops of the Harlech Dome, whence a traverse eastwards crosses the Central Wales Syncline and enters the Berwyn Dome. The Central Wales Syncline here is narrow, but it plunges southwards and then loses its singularity in a number of minor folds. In South Wales it reappears, separating a north-westerly, short Teifi Anticline from the longer Towy Anticline on the south-east; both have cores of Ordovician rocks. South Pembrokeshire is in the realm of Hercynian folding and thrusting, but the north of that county is largely composed of Ordovician rocks with Pre-Cambrian ridges locally flanked by Cambrian outcrops.

North of the Harlech and Berwyn structures the Silurian outcrops broaden into two synclines of Llangollen and the Denbighshire moors. The Vale of Clwyd, with Carboniferous and New Red Sandstone outcrops, bounds the latter on the east, but Silurian rocks are brought up again in the Clwyd range farther east by a major fault-belt.

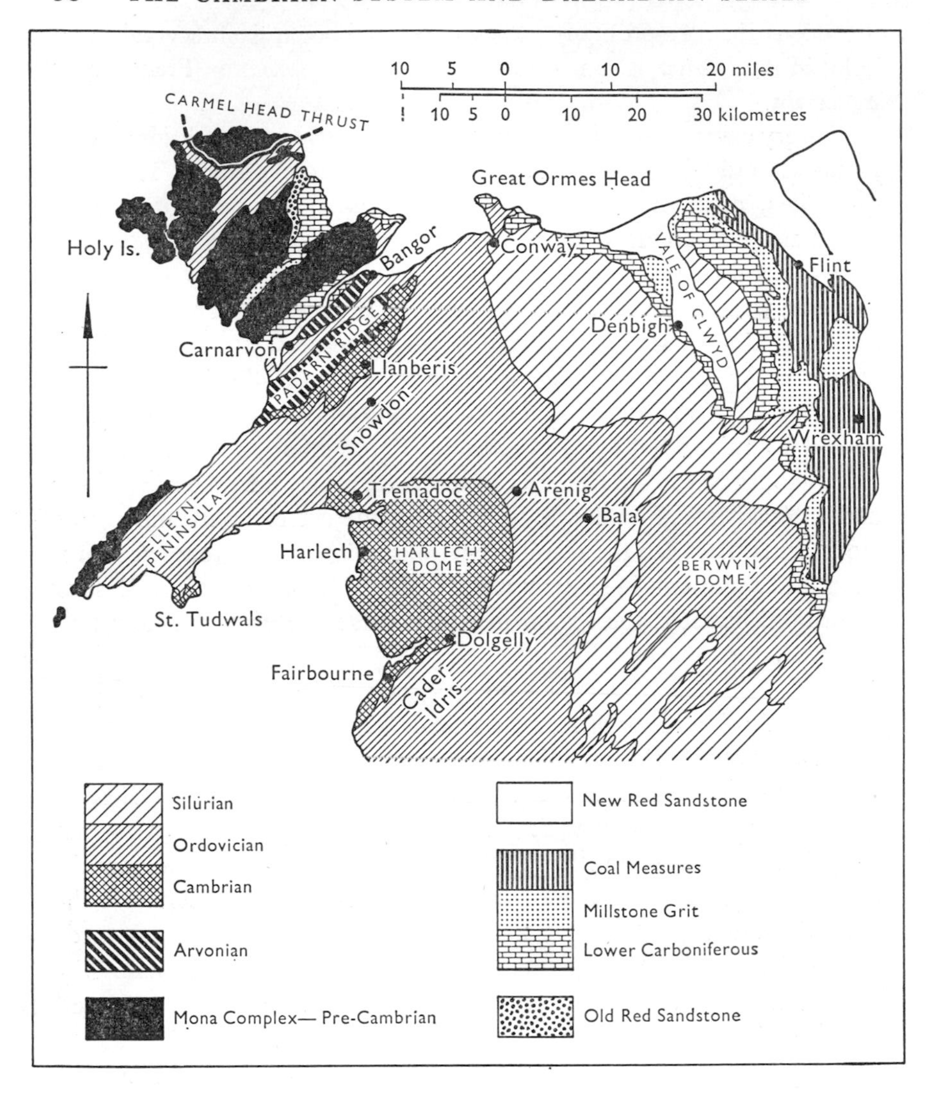

Fig. 12. Geological sketch map of North Wales; igneous rocks (largely Ordovician) not differentiated.

Most of these major structures date from some phase of the Caledonian movements, as does the Carmel Head Thrust of Anglesey, bringing Pre-Cambrian rocks over Ordovician (Fig. 13). Many of the subsidiary fold axes, however, are not Caledonoid in direction, and produce, for instance, north–south structures in the Harlech Dome and east–west in the Berwyns and north Pembrokeshire.

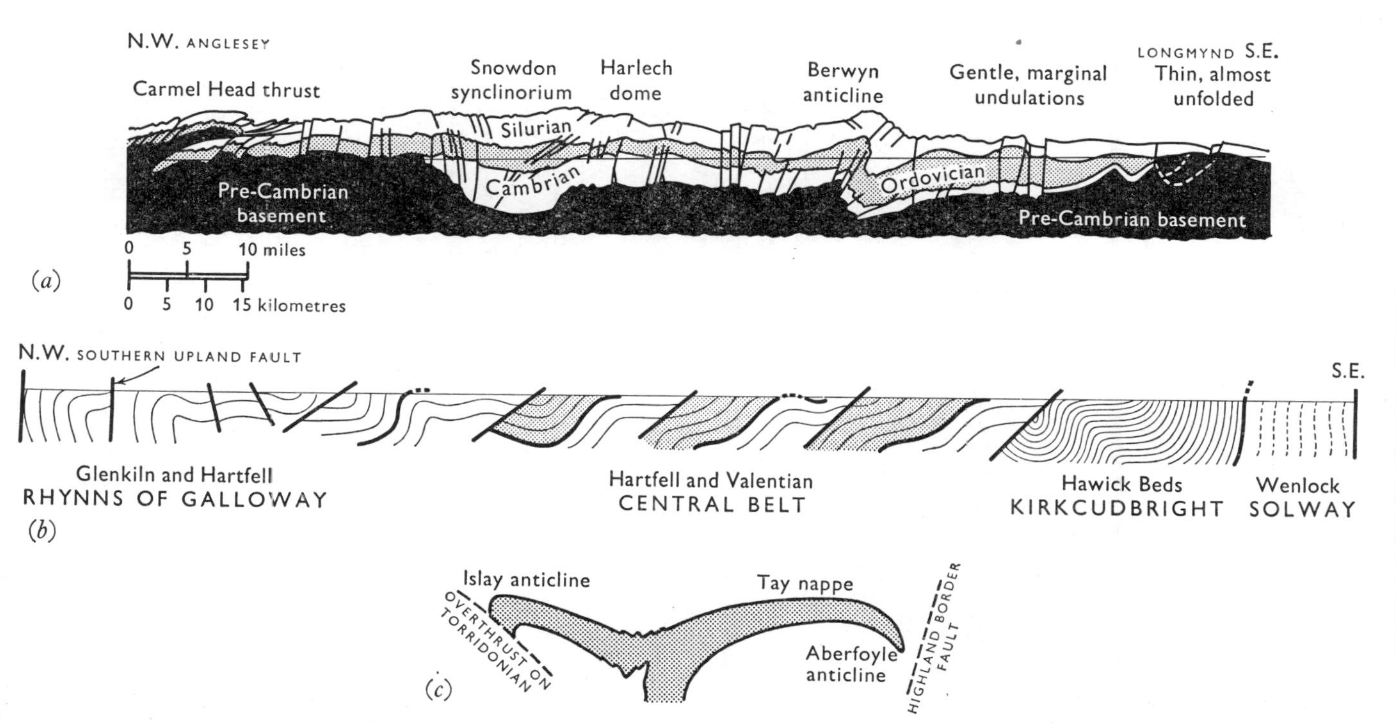

Fig. 13. Caledonian structures. (*a*) Section across North Wales from Anglesey to Shropshire, restored to the top of the Silurian (redrawn from Shackleton, 1954*b*, pl. 19).

(*b*) Diagrammatic section through the western part of the Southern Uplands, showing the combination of monoclines and faults (redrawn from Walton, 1963, p. 91).

(*c*) Diagrammatic cross-section of Dalradian folds in the south-west Highlands, to show the Tay Nappe and other structures (redrawn from Roberts and Treagus, 1964, p. 513).

North Wales

Cambrian rocks appear in the Harlech Dome of Merioneth, the Llanberis slate belt, and in a small inlier along the cliffs of St Tudwals on the Lleyn Peninsula. Early Ordovician erosion has removed much of the higher Cambrian beds from the last two, and conversely lower Cambrian beds are less well represented in Harlech and St Tudwals, where the base is not seen; even along the Llanberis strip the base is not unequivocal. On Anglesey no Cambrian rocks are known unless the small patch of Trefdraeth Conglomerate near the south coast belongs to this period. It follows that although Sedgwick's Cambrian System was named from North Wales, none of the outcrops forms a wholly satisfactory standard succession, and the extreme rarity of Lower Cambrian faunas is a further drawback.

The rocks are typical of geosynclinal sediments in the Welsh trough, but unless the lavas below the 'basal conglomerates' of the Llanberis succession are Lower Cambrian and not Pre-Cambrian, there was no vulcanicity. The whole is very much thicker than the deposits on the eastern shelf, perhaps three times as much or more, and the subsidiary limestones of that facies are absent. On the other hand there was at least one period of erosion, producing an unconformity in the north, and though the trough subsided many thousands of feet, terms like 'trough' and 'basin' do not necessarily betoken a very great depth of water. Here as elsewhere in stratigraphy depth of water is a very elusive factor.

There is much in common between the succession in the Harlech Dome, the largest outcrop, and at St Tudwals (Table 7), though faulting and lower Ordovician overstep have cut out most of the Upper Cambrian beds at the second. The age of the lower part of the Harlech Series (or Harlech Grits) is assumed from the lithological correspondence with the St Tudwals succession and particularly with the Hell's Mouth Grits, in which *Protolenus*, a genus characteristic of the upper part of the Lower Cambrian, has been found near the top. The junction between Lower and Middle Cambrian is thus some way down the Harlech Series, but its position cannot be determined at all certainly. The series is some 7,000 feet in its type area, out of a Cambrian total of perhaps twice that amount. The three grit divisions are dominantly graded greywackes, with sandstones and minor shale partings; the thickest of them, the Rhinog Grits, reaches 2,500 feet in the centre of the dome.

The coarser rocks are often pebbly and both rock detritus and mineral assemblages can be matched in the Monian rocks of Anglesey, though

Table 7. *Cambrian succession in North Wales*

	Harlech Dome		St Tudwals
	(Tremadoc slates)		
UPPER CAMBRIAN	Dolgelly Beds		
	Ffestiniog Beds		(little present)
	Maentwrog Beds		
MIDDLE CAMBRIAN	Clogau Shales		Nant Pig Mudstones
	Gamlan Shales	HARLECH GRITS	Caered Mudstones
	Barmouth Grits		Cilan Grits
	Manganese Shales		Mulfran Shales
LOWER CAMBRIAN	Rhinog Grits		Hell's Mouth Grits
	Llanbedr Slates		
	Dolwen Grits		(base not seen)
	(base not seen)		

with more emphasis on the gneisses, granites and other igneous components than in the present outcrops. This composition is probably the best clue to the source of the greywackes—a landmass to the north and north-west, part of which still survives in the present Pre-Cambrian rocks of Anglesey. Other deductions have been made, both at Harlech and St Tudwals, on current directions based on sedimentary structures such as current-bedding and sole (or bottom) structures on the under surfaces of the greywacke beds. The results, however, are somewhat varied, or even conflicting, and turbidity currents bringing detritus from both north-east and south-west have been deduced. It is possible that both directions operated, and the currents need not have been immediately related to margins of the trough or the ultimate terrestrial source of the sediment. The Manganese Shales, reaching a maximum of 800 feet, are exceptionally fine in grade among the Harlech rocks, being banded grey and greenish mudstones with only thin sandy bands. The manganese ores were in part a primary precipitate, together with clay and silica; they suggest a period of relatively quiet deposition perhaps in somewhat enclosed or lagoon-like conditions.

Above the Harlech Grits the rocks generally become finer grained. Middle Cambrian fossils are found in the Gamlan Shales and the black, pyritous Clogau Shales. The latter contain several species of *Paradoxides* (including *davidis* and *hicksi*) as well as agnostids, and correspond to the Menevian Beds of South Wales. Black mud sedimentation continued in part of Upper Cambrian times to produce much of the Upper Maentwrog and Dolgelly Slates, but the Vigra Flags (Lower

Maentwrog) and Ffestiniog Beds are coarser, with grey micaceous sandstones and silts, ripple marked and current-bedded; these harder, sandy beds are the 'ringers' of older descriptions. The black shales and slates contain olenids, agnostids and brachiopods—the last including *Orusia lenticularis* of the Dolgelly and *Lingulella* in the Ffestiniog Beds. This genus was responsible for the name 'Lingula Flags', often applied to the Welsh Upper Cambrian beds as a whole. But since the genus is not *Lingula*, and the rocks are varied and often fine-grained, the term is not really satisfactory on either count.

Although the Llanberis slate belt and Harlech outcrops are only a dozen miles apart they show considerable differences in lithology and uncertainties in correlation; nor is there an entirely straightforward comparison along the 25-mile slate belt itself, from Bethesda in the north-east to Nantle in the south-west. Particular problems concern the base, correlation of the slate facies with the Harlech Grits, and the absence of most of the Middle Cambrian.

Table 8. *Cambrian succession in North Wales: Llanberis*

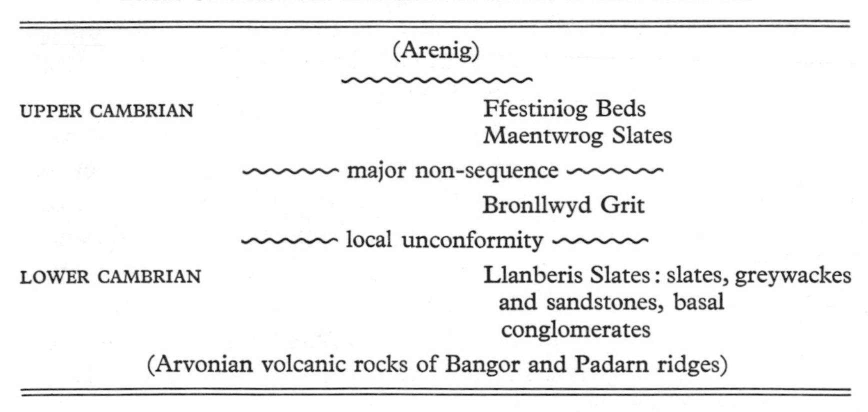

	(Arenig)
	~~
UPPER CAMBRIAN	Ffestiniog Beds Maentwrog Slates
	~~ major non-sequence ~~~~
	Bronllwyd Grit
	~~ local unconformity ~~~~
LOWER CAMBRIAN	Llanberis Slates: slates, greywackes and sandstones, basal conglomerates
	(Arvonian volcanic rocks of Bangor and Padarn ridges)

The Arvonian rocks of the Padarn and Bangor ridges underlie Cambrian conglomerates at the base of the slate succession. Doubt arises because there are several conglomerates above these volcanic groups, which are interbedded with slates like those above and also with tuffs and agglomerates like those below. It is possible that there is no major unconformity and the base of the Cambrian is consequently doubtful. If the volcanic groups are Lower Cambrian then they are unique in the British Isles where Cambrian vulcanicity is otherwise virtually unknown, but this is a somewhat negative point. On the other hand there is no reason why Lower Cambrian should not pass down into late Pre-

Cambrian rocks without a major unconformity, as they do in other parts of the world and most probably in the Dalradian Series.

The famous slate succession is composed of a great thickness of fine-grained rocks, rather variable in colour and containing grey, purple, reddish and green rocks. The only fossils come from the uppermost, green beds, where the trilobite *Pseudatops viola* is a late Lower Cambrian form. These beds are thus broadly of the same age as the *Protolenus* grits of St Tudwals and any representative of the Padarn volcanic facies would therefore be well below the surface outcrops in the centre of the Harlech Dome.

The major coarse unit at Llanberis is the unfossiliferous Bronllwyd Grit, with breaks above and below. The lower junction is conformable in places, but in the north-west there is transgression on to the lower beds of the Llanberis Slates. The gap above probably represents an even larger break for there are no fossiliferous Middle Cambrian Shales here, which are a conspicuous component in the other two outcrops. The fact that these breaks occur, and enlarge, to the north-west fits the general geographical setting pictured for north-west Wales, both in Cambrian and later times. Anglesey was part of a positive region, liable to uplift and erosion, which supplied much detritus to the coarser Cambrian and lower Ordovician sediments, and the uplifts are also recorded as unconformities. In the other direction the Welsh trough subsided, and its formations show, intermittently, both overstep and overlap to the north-west.

South Wales

The modest Cambrian outcrops in South Wales flank the Pre-Cambrian ridges of St Davids and Hayscastle in Pembrokeshire (Fig. 19). Much the best exposed is the coastal section south of St Davids along the north side of St Brides Bay. The base is unconformable, but Lower Cambrian fossils are very rare, and probably consist only of small crustacea (conchostracans); no trilobites are known in spite of an earlier record of '*Olenellus*'. At certain levels in the Middle Cambrian trilobite faunas, including the zonal *Paradoxides* species shown in Table 9, are fairly abundant and define the two subdivisions, Solva and Menevian. Both Middle and Upper Cambrian have some faunal resemblances to North Wales but the whole succession is thinner. Maentwrog and Ffestiniog beds can be recognized in the Upper Cambrian, the latter being again characterized by *Lingulella*, but it is doubtful whether the Dolgelly is present, the Ordovician rocks following above a major unconformity.

Table 9. *Cambrian rocks of South Wales*

		Estimated thickness (feet)
	(Arenig)	
UPPER CAMBRIAN	Grey mudstones and thin sandstones	2,000
MIDDLE CAMBRIAN	Menevian Beds: grey sandstones and mudstones, *Paradoxides davidis* and *P.* '*hicksi*'	750
	Solva Beds: grey, purple and green sandstones, *P. aurora* and *P. harknessi*	1,800
LOWER CAMBRIAN	Caerfai Beds: thin red shales, green sandstone, conglomerate	800
	(Pre-Cambrian volcanic rocks)	

In view of the isolation of the Pembrokeshire outcrops the amount that can be deduced about conditions in the south is limited. A Pre-Cambrian land surface, possibly slightly irregular, was submerged, and its debris incorporated in the basal conglomerate; a varied but dominantly sandy Lower Cambrian sequence followed. During the Middle and Upper Cambrian times there was free migration with the north Welsh seas and those of the Anglo-Scandinavian province generally; muds and fine sands were the dominant sediments and some of the arenaceous beds include greywackes. No evidence is available, however, on the source of the detritus or about neighbouring shelves or landmasses.

LEINSTER

No unquestioned Cambrian rocks are known in eastern Ireland. The traditional term, the 'Bray Series', includes certain beds that may belong to the system, on three counts. They underlie, in places, fossiliferous lower Ordovician slates and show a lesser degree of deformation than neighbouring rocks taken to be Pre-Cambrian. They also contain the problematical, ramifying marking called *Oldhamia*; this, though it cannot be ascribed to any known fossil group, is recorded from Cambrian beds elsewhere.

In its older use the Bray Series includes more than one group of rocks:

(1) A small outcrop forming the Hill of Howth north of Dublin Bay (Fig. 27, p. 124).

(2) The Bray Anticlinorium in the Wicklow hills south of Dublin.

(3) A long anticlinal region trending south-south-east from Cahore Point in County Wexford to the south coast.

(4) The Rosslare outcrops.

The last is the Rosslare Migmatitic Series, now considered to be Pre-Cambrian; the third outcrop contains slates and quartzites but is very poorly exposed and little known.

Oldhamia has been found at the Hill of Howth, Cahore Point and more than one place in the Bray Anticlinorium, together with other more dubious trace-fossils. The great thickness of rocks in the last outcrop includes slates, greywackes, current-bedded sandstones and massive quartzites. Near the western boundary they have been divided thus:

	Approximate thickness (feet)
Clara Group	8,500
Bray Group (in a restricted sense)	4,000
Knockrath Group	4,000
(base not seen)	

It is the slates of the Clara Group that chiefly concern us; they have been taken as Cambrian largely on account of an unconformity below, and also because they have yielded some very poorly preserved horny brachiopods. (The lower two divisions are thought to be Pre-Cambrian.) The field relation of these outcrops to those containing *Oldhamia* is, however, uncertain.

It will be seen that the Bray Series adds little to our understanding of the Cambrian period over the British Isles as a whole. There are some lithological resemblances to the rocks of North Wales, but the possibility of a Lower Palaeozoic landmass in the Irish Sea (p. 97) is a point against any close link between the histories of Wales and Leinster. A more cogent comparison has been made with the Manx Slates, which lie along the north-easterly Caledonian strike from the Bray outcrops. Fossils are extremely rare in the Manx rocks; such as there are suggest a late Tremadoc or early Arenig age and thus they belong to the next chapter. Nevertheless, it must be admitted that the main similarity between the Manx rocks and Clara Group (if we accept these as the most probable Irish continuation) is that both consist of thousands of feet of virtually unfossiliferous slates, and the link is thus somewhat negative.

THE DALRADIAN SERIES

This series occupies much of the highlands of Scotland between the Highland Border Fault and the Great Glen Fault, being found from the Banffshire coast on the north-east to the Kintyre Peninsula and island of Islay on the south-west. Farther in that direction Dalradian rocks make up the highlands of Donegal, north-west Mayo and the Connemara mountains; the last outcrop is unique in lying south of the Highland Border Fault, from which it is separated by a tract of Ordovician and Silurian rocks. Along its 450-mile belt the Dalradian Series maintains remarkably similar lithological characters, and to some extent structural ones. No fossils have ever been found in the great majority of Dalradian sediments, but in Perthshire, near Callander, assiduous collecting produced some specimens of the trilobite *Pagetides* from the Leny Limestone near the top of the series (Table 10); this genus is only known from beds of Lower Cambrian age (in North America) and the uppermost Dalradian strata are therefore Cambrian, but how much is uncertain and they are generally assumed to go down to include late Pre-Cambrian. The junction with the Moinian Central Highland Granulites is debatable; by some authorities the two are thought to be essentially conformable and that there may be a lateral transition from a Moine type of rock to a Dalradian type.

The Dalradian nappes and rocks of Scotland

It is a truism that in all complex mobile belts investigations into structure and stratigraphy are bound together, and this is excellently demonstrated in the Dalradian highlands. Only by determining the orientation or 'way up' of the individual beds can the recumbent folds be recognized and their upper and lower limbs identified; conversely the broader picture of succession and conditions of accumulation cannot be made out until the structure is deciphered.

Metamorphism has not seriously obscured the original types of the Dalradian sediments. They consist of a varied succession among which quartzites, greywackes, limestones, boulder beds and lavas still retain many of their original features; and in which slates, phyllites and schists represent finer grained (pelitic) clastic rocks, or occasionally tuffs. The boulder beds, limestones and quartzites have been particularly useful in correlating local sequences with one another and the graded and current-bedding of the greywackes and quartzites are invaluable.

The dominant structures are large-scale overfolds, often called by the alpine term 'nappes'. The most important of these is the Tay Nappe, a very large, flat-lying, recumbent fold found along the south-eastern margin of the highlands (Fig. 13). The nose of the anticline points in that direction (that is, the fold is said to 'face' south-eastwards) and near the Highland Border it turns downwards. Like other highland units the Tay Nappe has a number of later folds superimposed upon it, but it can be recognized from Kincardineshire south-westwards to Kintyre, and since generally dips are low, this zone has also been called the 'flat-belt'. Over much of this stretch the upper limb has been removed by erosion and the visible rock sequences belong to the under limb and are inverted. To the north-east in Aberdeen and Banffshire there is a similar flat belt which is probably a continuation of the Tay Nappe, but here parts of both upper and lower limbs have been preserved.

In a smaller area to the west around Ballachulish and Fort William there are the two smaller Ballachulish folds. These have generally been considered as north-westerly facing structures and their limbs are partly replaced by low-angle thrusts, or slides. They are separated from the Tay Nappe by some large intrusions (of Glencoe, Glen Etive, etc.) and by another major slide, but also by an intermediate structural zone that is not easy to interpret.

From this brief and generalized description, which takes no account of secondary folds and many complex or uncertain problems, we may turn to the stratigraphical succession. This was first securely established in Perthshire, but satisfactory lithological correlations can be made in Banffshire, the south-west Highlands and in Islay (Table 10). The amount of deformation the rocks have undergone does not allow thicknesses to be estimated as they may be in later systems; even the quartzites may vary substantially in thickness on the two limbs of a fold. But in individual sections it appears that the limestones are normally thinner than the other beds listed in the table. The whole formed a geosynclinal pile of sediments many thousands of feet thick, perhaps over 30,000.

There are some differences of facies in the upper and lower beds; in Perthshire the latter are a varied succession of quartzites, impure limestones and black or calcareous pelites. The quartzites in particular show signs of shallow water accumulation, being well-sorted rocks with locally abundant current-bedding; alga-like structures have been found in the Islay Limestone, but their origin is not known for certain. In the upper beds a greywacke facies predominates, which may perhaps have set in earlier in the south-west, in beds below the Loch Tay Limestone. Here too, in the Loch Awe region, is the principal area of Dalradian volcanic

Table 10. *Dalradian successions in the Scottish Highlands*

	South-western Highlands (Islay, Jura, Loch Awe)	Central Highlands (Perthshire)
UPPER DALRADIAN	Loch Avich Grits	Ben Ledi Grits (including Leny Limestone)
	Loch Avich Lava Group	
	Tayvallich Slates and Lavas	Green Beds (volcanic) Aberfoyle Slates
	Tayvallich Limestone	Loch Tay Limestone
MIDDLE DALRADIAN	Crinan Grits and Slates (greywackes)	Ben Lui Mica-schists (pelites)
	Shira Limestone	Ben Lawers Calcareous Schists
	Ardrishaig Phyllites	
	Easedale Black Slates	Ben Eagach Black Schists
	Scarba Transition Group (phyllites and quartzites)	Carn Mairg Quartzite
		Killiecrankie Schists
	Islay Quartzite, with pebbly dolomitic and pelitic intercalations	Schiehallion Quartzite
	Portaskaig Boulder Bed (tillite)	Schiehallion Boulder Bed (tillite)
LOWER DALRADIAN	Islay Limestone	Blair Atholl Series (limestones and black schists)
	Mull of Oa Phyllites with dolomitic beds	

rocks; they include pillow lavas and the 'Green Beds' of many areas represent tuffs.

The boulder beds in the lower Dalradian succession (Schiehallion of Perthshire or Portaskaig of Islay) have many features of a tillite, containing an unsorted medley of boulders in a sandy matrix. Two of the crucial land-ice features, striated and faceted boulders and striated pavements are, however, lacking, and an alternative solution is that the boulders were dropped by melting icebergs. It has been suggested that this boulder bed represents the same glacial period deduced from 'tillites' in many countries around the North Atlantic, for instance, Scandinavia, Greenland and Spitsbergen. If substantiated this correlation is a further indication that much of the lower Dalradian Series is late Pre-Cambrian, as are comparable sequences in Spitsbergen and Greenland.

The rocks of the Ballachulish folds contain the same lithological types —schists, quartzites, phyllites, slates and limestones—but as yet no detailed correlation has been demonstrated with the Tay Nappe. It is conspicuous in the latter that the lateral continuity of beds observable along the strike (north-east to south-west) is not found across the strike, and the lack of correlation with the Ballachulish succession may be an example of this. The lateral continuity might be interpreted as resulting from facies-belts lying approximately parallel to the shore-lines of the period, in which case these would have a Caledonian direction also, and we might begin to think in terms of a Dalradian trough or geosyncline—the earliest of those that made up the Caledonian mobile belt. Any such margins are, however, dubious. It is true that somewhere to the north-west there accumulated from early Cambrian to early Ordovician times the orthoquartzite and carbonate facies, possibly with gaps, of the

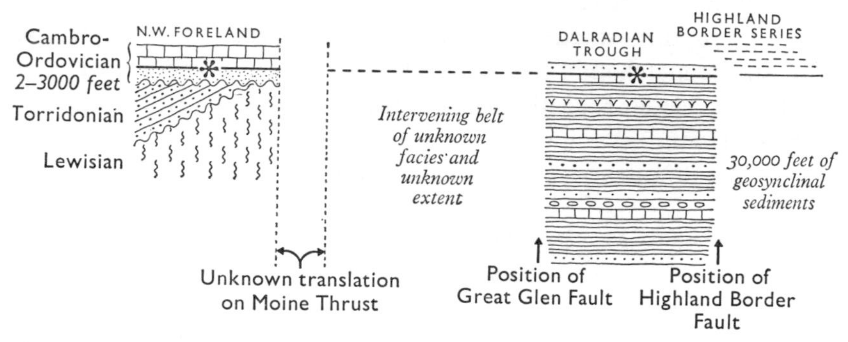

Fig. 14. Diagram to show the possible relationships between the orthoquartzite-carbonate facies of the foreland and the Dalradian geosynclinal succession. The (?) Arenig Highland Border Series may well have extended north-westwards over the Dalradian region. Asterisks denote the positions of the Lower Cambrian genera that allow the correlation to be made—*Olenellus* in the foreland facies and *Pagetides* in the geosyncline (Leny Limestone).

North-West Foreland (p. 72). These beds have a marked unconformity at their base, so that immediately preceding their deposition there was a long period of emergence and erosion. To that extent during part (perhaps all) of Dalradian time some limit to the geosynclinal trough lay to the north-west. But its distance and position are problematical. Between the present Dalradian outcrops and the foreland there intervene the Great Glen Fault and the Moine Thrust and movements on these have brought the two closer together. During Cambrian and late Pre-Cambrian times, therefore, the foreland was probably farther away, by some unknown amount, and how the Dalradian facies passed into the orthoquartzites and carbonates is quite unknown.

In the opposite direction there is no clue to the margin at all, short of the Welsh Borders and English Midlands. The way in which the Tay Nappe turns down just north of the Highland Border Fault is complementary to the upturning of the newer rocks (chiefly Old Red Sandstone) on the other side. In the latter this monocline dies out away from the fault, so that the Dalradian rocks may possibly flatten out also, as part of an unseen basement to the Midland Valley of Scotland. At all events the area of deposition cannot be envisaged as stopping abruptly at the present fault-line, which was largely effected later than the Lower Old Red Sandstone. Indeed Dalradian rocks *do* appear to the south in Connemara, but beyond that all is conjecture.

The Dalradian Series in Ireland

The best known successions in Ireland, and those most like the Scottish, are in northern Donegal. In particular the rock structures of Islay can be matched in the peninsulas of Inishowen and Fanad. The Kintyre succession is continued into north-east Antrim where the Tay Nappe is continued as a very similar recumbent fold, facing south-east and slightly down-turned in that direction.

Table 11. *Comparison of Dalradian successions in Ireland and Scotland*

Fanad–Inishowen	Islay	North-east Antrim
Green beds, slates and grits	Loch Avich grits, green beds and slates	Schists and green beds
Limestone, slates and grits	Tayvallich Limestone	Torr Head Limestone
Crana Quartzite	Crinan Grits and Slates	Schists
Schists	Schists and phyllites	
Malin Head and Fanad Head quartzites and dolomites	Islay Quartzite, and dolomites	
Fanad Boulder Bed	Portaskaig Boulder Bed	
Fanad Limestone	Islay Limestone	
Fanad dark phyllites	Mull of Oa phyllites	

These two successions are compared in Table 11 with that of Islay and Loch Awe. As in Scotland the ascending trio—limestone, boulder bed and quartzite—is a particularly widespread and recognizable combination; similarly there is continuity of rock type along the strike and changes across it. The Fanad–Inishowen succession can be traced along the south-east side of the Donegal Granite (a complex, elongate Caledonian intrusion) to the Slieve League peninsula of west Donegal. In most of this region the strata exposed are right way up, but in Antrim erosion has removed much of the 'Tay Nappe' upper limb and the exposures are chiefly of the lower, inverted, limb.

The sequence on the north-west side of the Donegal Granite has several analogies with that of the Ballachulish nappes; the folds similarly face north-westwards and the formations cannot be correlated with those of Fanad. Moreover, although for most of their outcrops these two successions are separated by the granite, where they meet at either end the junction is a tectonic dislocation.

There is thus no doubt of the similarity between the Dalradian Series in Scotland and in this part of Donegal, nor that they were originally laid down in similar conditions. The remaining Irish outcrops are not known in such detail. These include those in Londonderry, Tyrone and south-east Donegal, around Lough Derg; in the last-named, there is the same apparent continuity between Moine Schists and Dalradian rocks that has been observed in the northern Grampians. In northern Mayo the Dalradian sequence is apparently a continuation of that from Fanad to Slieve League, emerging from beneath a Lower Carboniferous cover and the waters of Donegal and Sligo bays. The formations include a trio of limestone, boulder bed and quartzite like those of Fanad and Islay.

The ancient rocks of Connemara have long been known to resemble the Dalradian of Scotland. Though mica schists are abundant (they are often called the Connemara Schists) there are also massive quartzites, limestones and ophicalcites ('Connemara Marble') and boulder beds, though these are not very conspicuous; the whole compares well with Donegal and Islay, both in general and in individual successions. The metamorphic grade is high and structure complex; quartzites form the mountains of the centre, and are flanked to north and south by more pelitic rocks.

Much remains to be discovered about the history of the Dalradian orogenic belt. The data include a large number of isotopic dates—with additions every year—and the stratigraphical evidence from the Leny Limestone of Perthshire and the Lower Palaeozoic rocks of Connemara. The latter show that a major metamorphic phase was completed by Arenig times (early Ordovician), because metamorphic minerals from the Connemara schists are incorporated in the nearby basal Arenig sediments. Some of the earlier isotopic dates fit in with this evidence and may be either Cambrian or late Pre-Cambrian. Others, however, are later (e.g. 440–460 million years) and give mid-to-upper Ordovician ages or Silurian. This group probably reflects relatively minor events—perhaps a late retrogressive metamorphism or the slow cooling of the Connemara massif so that there was a significant delay in the final closure of isotopic diffusion.

In Scotland a comparable range of late ages (430–470 million years) is also conspicuous and similar causes have been put forward. There seems little reason to doubt that the Leny Limestone is an integral part of the Upper Dalradian sequence, and so a major phase of folding and moderate metamorphism in the group must be later than Lower Cambrian.

Whether any of the deformation of the Scottish Dalradian rocks is Pre-Cambrian is uncertain, but what is fairly clear is that in such a highly complex orogenic belt—not necessarily uniform throughout its length—the isotopic dates commonly do not give the age of maximum metamorphism; and that there are several, complex, interacting factors producing a range of dates, some of which may be much later than the main deformation.

THE NORTH-WEST FORELAND

The quartzite and carbonate facies of the foreland includes Ordovician as well as Cambrian rocks but, since they form a unified sequence, on the edge of a stable mass, it is convenient to describe the whole in this chapter. The principal outcrop is a long strip in north-west Scotland from Loch Durness on the north coast of Sutherland to the mainland opposite Skye, and there are further areas in the south of that island.

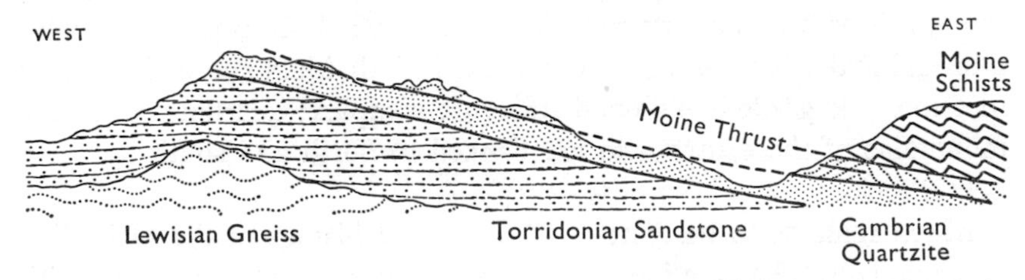

Fig. 15. Section to show the relationships of Cambrian rocks (here largely the Arenaceous Series) to the underlying Torridonian Sandstone and Lewisian Gneiss, and also to the overthrust Moine Schists. A subsidiary, lower, thrust has here carried fragments of Lewisian rocks over the Cambrian quartzites; elsewhere along the Moine Thrust Belt wedges of Cambrian and Torridonian rocks may also be caught up between the thrust planes. (Redrawn after Peach and Horne, 1907, p. 541.)

The beds dip eastwards, overlying at different places Torridonian and Lewisian with marked overstep. On the east the Palaeozoic beds are overthrust by the Moine Schists, and wedges of them are locally caught up among the thrust slices. Probably all the outcrops in Skye are of this type, because they are overlain as well as underlain by thrust masses of Torridonian Sandstone.

The Pre-Cambrian basement had been subjected to a long period of erosion, and reduced to a fairly level surface before being invaded by Lower Cambrian seas; these seas, as deduced in the last section, had already been established in the Dalradian geosyncline that lay some uncertain distance to the east and south-east. The Palaeozoic rocks of the foreland are divided lithologically into an arenaceous series below and a carbonate series above, the latter known as the Durness Limestone. There is a tendency for both to thicken southwards, the former being about 300 feet in the north and 500 feet in Skye. The Durness Limestone is some 1,600 feet in the north, and most of the groups are probably represented in Skye. The succesion is given on p. 41.

The most important fossils are the trilobites, especially the species of *Olenellus*. *Salterella* is a small mollusc of doubtful affinity, but which is known from Lower Cambrian beds in several countries; its range in north-west Scotland shows that nearly half the succession is Lower Cambrian. The only other reliable fauna is the Ordovician assemblage in the upper part of the Durness Limestone; this is not easily compared with the European series (p. 77) but may perhaps be correlated with Arenig or even Llanvirn—that is in the lower part of the system but not right at the base. In between there are some unfossiliferous limestones and dolomites, and the few, poorly preserved specimens of the Sailmhor Group. Opinions have differed on the age of these, and consequently on how much of the carbonate sequence is Cambrian and how much Ordovician. What does appear, however, is that there is very little, if any, of the Middle Cambrian beds, and possibly not much Upper Cambrian.

The thin beach pebble-beds of the invading Cambrian seas were succeeded by current-bedded, well-rounded and sorted quartz sands, now in the form of quartzites. The vertical tubes of the 'Pipe Rock' above are probably analogous to the structures formed by marine worms of shallow water and intertidal environments at the present day. There was then a gradual reduction in the clastic components and an increase in the carbonate, perhaps as the seas deepened slightly and extended; both components go to form the mixed dolomitic and silty beds at the top Arenaceous Series, and the grey Grudhaigh dolomites. Even in the white Eilean Dubh above there are rounded sand grains, which, like much of the other detritus in this early Cambrian sea, had had a long history of earlier abrasion.

In addition to the shells, mostly molluscan, which were embedded in the fine calcareous muds, there are several algal horizons in the Durness

Limestone. These are mounds or incrustations with concentric layers typical of algal growth, but in which the microscopic filamentous or cellular structures have not been preserved. Other rock-types include oolites and calcarenites (detrital limestones of sand grade). At present these carbonate rocks consist of limestones and dolomites; some at least of the dolomite is secondary, as is some of the chert, the latter perhaps being derived from sponge spicules, traces of which remain. The original fine calcareous muds must have been deposited very slowly and are typical of a carbonate shelf or foreland facies, which was very much thinner than that in the contemporary geosyncline. There may well be an unconformity within the Durness Limestone; breccias occur at several levels, but as in other carbonate sequences it is not easy to determine whether any one represents a major break in sedimentation or only a minor phase of erosion and re-cementation.

Table 12. *Cambro-Ordovician succession of North-West Highlands*

	Lithological groups	Faunas and age	
DURNESS LIMESTONE	Durine Group	Fossils rare, age uncertain	
	Croisaphuill Group Balnakeil Group	Gastropods, cephalopods; a few brachiopods and fragmentary trilobites. Low in ORDOVICIAN	
	Sangamore Group	No fossils found	
	Sailmohr Group	Poorly preserved cephalopods, gastropods, one trilobite ?LOWER ORDOVICIAN or ?UPPER CAMBRIAN	
	Eilean Dubh Group	*Salterella*	LOWER CAMBRIAN
	Grudhaigh Group	*Salterella*	
ARENACEOUS SERIES	Alternating dolomitic shales, sandstones and quartzites	*Salterella*, *Olenellus* and other trilobites	
	Massive quartzites with vertical tubes ('Pipe-rock')	*Salterella*	
	Current-bedded quartzites	No certain fossils recorded	
	Conglomerate		

The Cambro-Ordovician succession in the North-West Highlands is remarkably dissimilar to those elsewhere in the British Isles. It is true that there is a basal, Lower Cambrian quartzite in parts of the Welsh Borders and Midlands as well as in Scotland, but this is because both reflect the submergence of an old landscape by a sea that at first re-distributed much abraded and well-sorted sediments. Thereafter there is nothing in the Anglo-Welsh geosynclinal or shelf facies at all like the

Scottish sequence and, more important, the faunas are substantially different. The Anglo-Scandinavian province is characterized by such genera as *Holmia*, *Callavia*, *Paradoxides*, *Agnostus* and *Olenus*, and these genera are also found in a relatively small portion of the Cambrian rocks of North America—New England, New Brunswick, Nova Scotia and south-eastern Newfoundland. *Olenellus* and the lower Ordovician fossils of north-west Scotland belong to a much more widespread American province, which also includes Labrador, Spitsbergen and Bear Island. Moreover, in these countries they are commonly found in a carbonate facies very like that of Scotland.

These outcrops are now separated from one another by the North Atlantic; the positions they may have held in Lower Palaeozoic times are directly related to the problems of continental drift and as such considered in chapter 13. Meanwhile we may summarize late Pre-Cambrian and Cambrian history in the British Isles as the first stages in the development of the Caledonian geosyncline. Parts of the margins of this structure are seen, but not very precisely defined, in north-west Scotland and the English Midlands. Directly along the north-east strike the Caledonides —or mountain ranges resulting from the geosyncline—reappear in Norway and north-western Sweden. In Scandinavia the eastern margin of the geosyncline is represented in the extensive Pre-Cambrian block of the Fennoscandian Shield.

REFERENCES

Cambrian of England, Wales and Scotland. Basset, 1963; Basset and Walton, 1960; Cox *et al.* 1930; Jones, 1938, 1956; Pocock *et al.* 1938; Shackleton, 1954*b*; Simpson, 1963; Walton, 1965*a*; Whittard, 1952.

South-east Ireland. Charlesworth, 1963, ch. 3; Tremlett, 1959.

Dalradian Series of Scotland and Ireland. Anderson, 1947, 1953; Brown *et al.* 1965; Charlesworth, 1963, ch. 2; Harris, 1962; Johnson, 1965*b*; Kilburn *et al.* 1965; Knill, 1963; Rast, 1963; Shackleton, 1958; Stone, 1957; Watson, 1963.

General. George, 1963*b*; Stubblefield, 1956, 1959; Sutton, 1963.

British Regional Geology. The Grampian Highlands, the Northern Highlands, North Wales, South Wales, the Welsh Borderland, Central England.

4

ORDOVICIAN AND SILURIAN SYSTEMS

The 100 million years of the Ordovician and Silurian systems saw the closing stages of the Caledonian geosyncline, from the great piles of lavas, tuffs and sediments of the lower and middle Ordovician to the calcareous siltstones or greywackes of the late Silurian. In certain regions a major phase of the Caledonian orogeny took place in late Silurian times, and the folded slates and greywackes are overlain, as in parts of southern Scotland, by gently tilted Lower Old Red Sandstone. But in many other places the dominant structures were imposed far earlier, in middle or late Ordovician times or early Silurian. Like other major orogenies the Caledonian earth movements were not a single upheaval, but their effects began far back in Lower Palaeozoic times and continued into the Upper Palaeozoic, to be separated by no very long interval (or quiescent period) from the early phases of the Hercynian orogeny.

In spite of this long time-span and complexity it is convenient to treat the Ordovician and Silurian systems in a single chapter. Their outcrops lie for the most part within the same major regions and often share the same intricate structures; their rock-types are similar, and the same major faunal groups predominate. Although in certain areas an unconformity separates the two systems, in others there are equally important breaks within the Ordovician. At the end of this geosynclinal history the seas retreated from most of the British Isles and a new pattern of sedimentary basins was established.

In one point there is a great difference between the systems. Silurian vulcanicity was essentially local and on a very small scale; Ordovician vulcanicity was vast, and great thicknesses of lavas and pyroclastic rocks —sometimes subaerial and sometimes submarine—are found from Scotland to South Wales, and Cumberland to western Ireland. The seas and sea floor were studded with local, and geologically speaking, impersistent volcanoes and volcanic islands. The sediments that accumulated in the fluctuating basins between them were sometimes affected by the widespread volcanic detritus. It is this admixture of facies, major earth movements, local emergence and local submergence, that make a general statement on Ordovician history so hazardous. In its shorter duration, much decreased vulcanicity, generally thinner sequences and, through the period, increased uniformity of facies, the Silurian System is in some sense the concluding act of the Ordovician drama.

The sedimentary facies of the two systems have a certain amount in common with those of the Cambrian; there are conglomerates and pebbly sandstones, greywackes (often with turbidite characters), black shales, grey mudstones, siltstones and fine-grained sandstones. Limestones are rare, and relatively pure limestones are largely restricted to the south-eastern shelf facies; even here they are not often more than 100 feet thick, though rich in fossils.

Table 13. *Principal divisions of Silurian and Ordovician systems*

SILURIAN	Ludlow Series
	Wenlock Series
	Llandovery Series
ORDOVICIAN	Ashgill Series
	Caradoc Series
	Llandeilo Series
	Llanvirn Series
	Arenig Series
	Tremadoc Series

The Tremadoc Series is sometimes included in the Cambrian System. Ashgill and Caradoc together are also called Bala. Llandovery is also called Valentian. Wenlock and Ludlow together are called Salopian. Zones are given in the Appendix.

The zones and larger units of the two systems are given on pp. 413 and 414. In stratigraphical palaeontology the outstanding event is the appearance of the graptolites (or strictly the Graptoloidea) during Tremadoc times; dendroid forms such as *Dictyonema* are also prominent and there is a series of rapid evolutionary changes among the Graptoloidea in the Arenig. Thereafter successive faunas characterize the major divisions of the Ordovician; monograptids cover almost all the Silurian System, appearing just above the base and in Britain (though not universally) disappearing somewhat before the end of the Ludlow times. The graptolites are commonest in the black, sometimes pyritous, shales but are known from all but the coarsest types of sediments. The Ordovician is characterized faunally by the abundance and variety of its trilobites, including asaphids and trinucleids which are confined to the system. In addition to these a few families continue from the Cambrian; such forms as olenids are found in the Tremadoc Series and agnostids in the Ashgill. Many others appear for the first time and

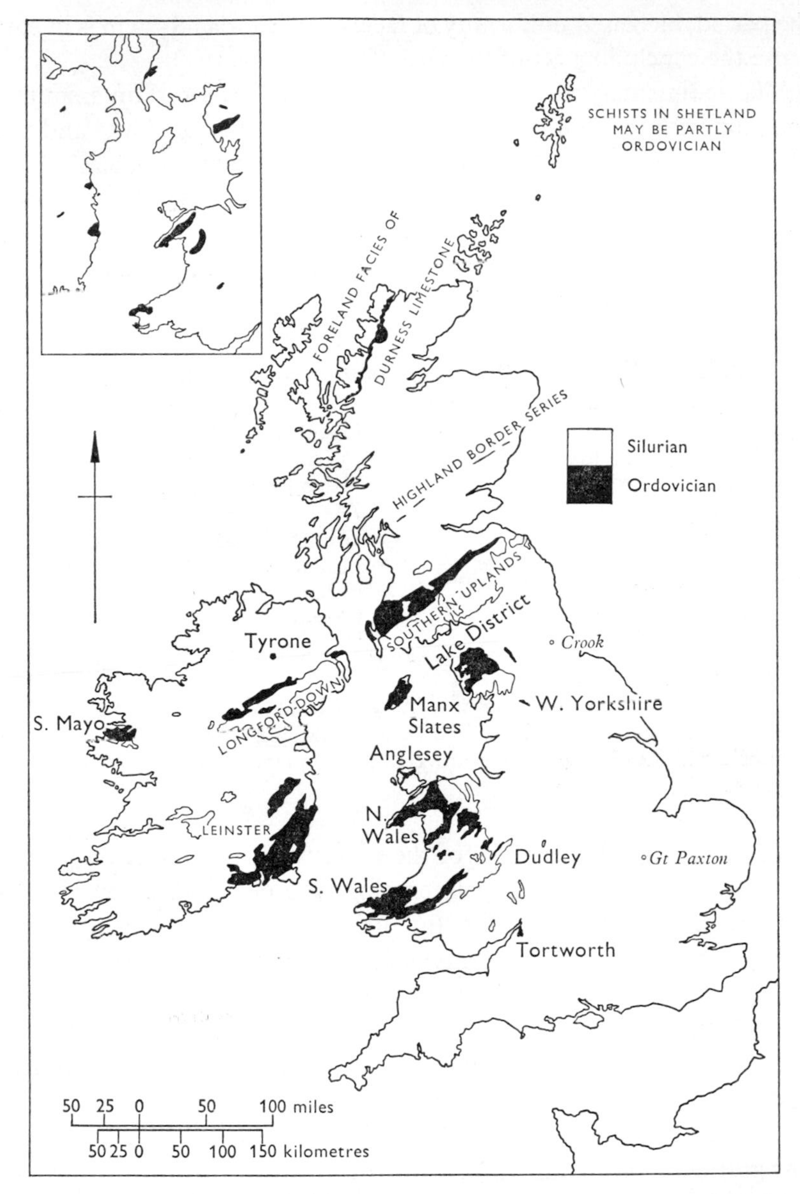

Fig. 16. Outcrops of Ordovician rocks (black) and Silurian (outlined). Inset: main areas of Ordovician vulcanicity around the Irish Sea.

Only selected boreholes appear in Fig. 17 and succeeding maps (see also note on p. 107).

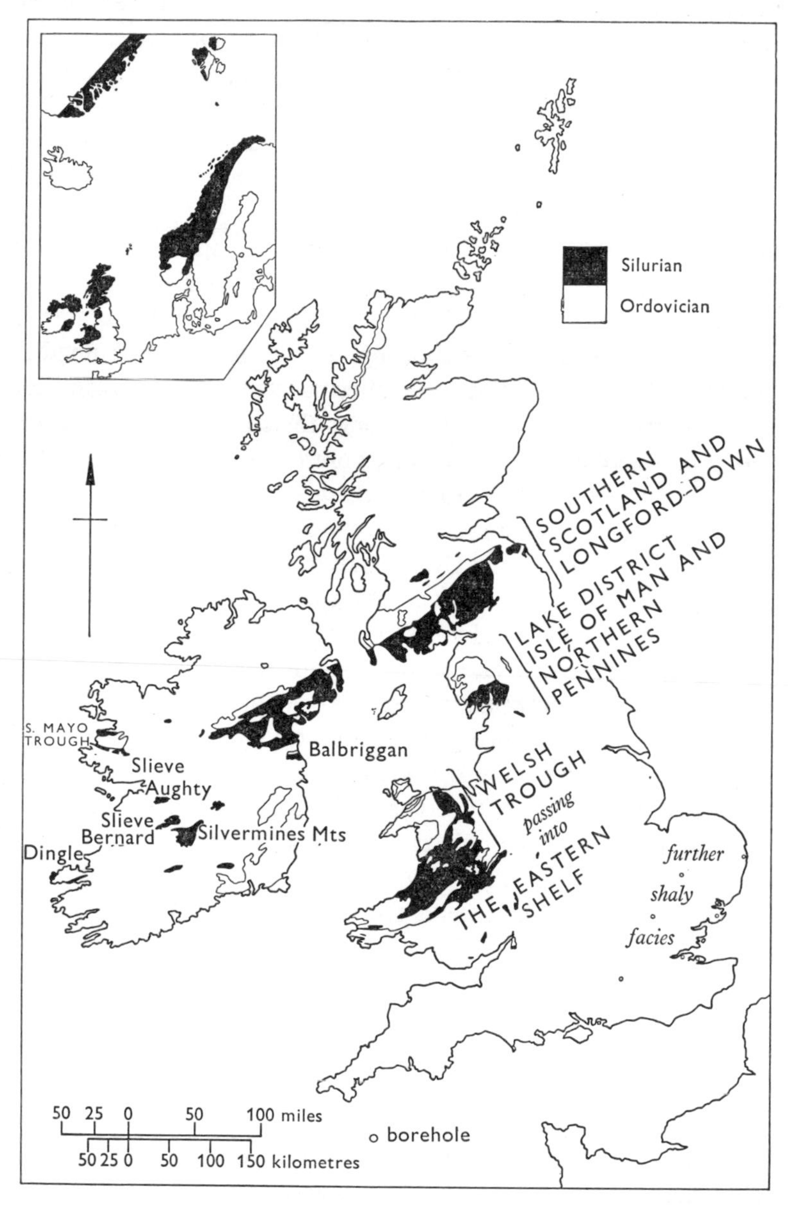

Fig. 17. Outcrops of Silurian rocks (black) and Ordovician (outlined) and the main Caledonian troughs south of the Highland Border. Silurian, or probable Silurian, rocks are encountered in several boreholes in south-east England. Inset: regions of Caledonian folding around the North Atlantic (after Holtedahl, 1952, p. 66).

continue on into the Silurian, such as the calymenids and encrinurids. Brachiopods rise to great importance and the calcareous types are varied —orthoids, clitambonitids, strophomenoids being characteristic. Mollusca are represented by lamellibranchs, gastropods and nautiloids, although they are not of outstanding importance. Among the corals tabulate forms are more important than rugose, and cystids and crinoids represent the echinoderms.

The shelly faunas of the sandstones, siltstones and grey mudstones include especially brachiopods and trilobites. In certain series, notably the Caradoc and Ludlow, they do more than supplement the graptolites, for when their assemblages are thoroughly collected and examined they are found to provide a finer, more detailed succession of stages and zones than do the graptolites. In the limestones the shelly faunas are accompanied by tabulate and rugose corals, polyzoa and occasionally echinoderms and calcareous algae. In the British Isles the first fragmental remains of vertebrates are known from the Silurian beds, although bony plates have been found in the Ordovician of the United States.

In outcrop the rocks of the two systems are much more extensive than the Cambrian and form by far the larger part of the Caledonian uplands south of the Scottish and Donegal highlands. Most of them can be grouped into regions of, broadly, common geological history and structure—including the principal geosynclinal troughs; a few do not align themselves so closely or are not as yet adequately known:

North and South Wales, together with Anglesey: the Welsh section of the geosyncline and the only one where Cambrian sedimentation is known to precede Ordovician.

The eastern shelf to the foregoing, comprising the Welsh Borders and inliers to the south and east.

The Lake District, Pennine inliers and Isle of Man.

The Southern Uplands of Scotland and its western continuation, the Longford–Down massif of northern Ireland; inliers in the Midland Valley of Scotland and of Ireland; the trough of south Mayo and north Galway.

A somewhat heterogeneous or little known group of inliers in southern Ireland; the Leinster Massif.

The transition in sediments and structures from the geosynclinal margin into the trough can best be traced from the Welsh Borders into Wales; elsewhere either the margins are not clear or else they no longer bear a Lower Palaeozoic cover. A traverse north-westward from Shropshire demonstrates a change from a facies dominated by mudstones, fine-grained sandstones and occasional limestones to one of slates, greywackes and volcanic rocks—a change accompanied by an increasing intensity in folding and the gradual appearance and intensification of cleavage. The thickness also increases, and though some of this is due to tectonic compression, there was also an expansion in the thickness of the original sediments from the shelf succession to that of the trough. The total thickness of the three Lower Palaeozoic systems in North Wales is of the order of 30,000 feet, or 5 miles.

The small amount of Durness Limestone that is of Ordovician age has already been mentioned in the preceding chapter, as it is essentially a continuation of the foreland carbonate facies established after the Lower Cambrian invasion. The sparse and poorly preserved uppermost faunas are closest to those of similar carbonate rocks in America; these belong to the lower part of the Ordovician System (p. 73) but the correlation is not very exact. How much longer this foreland facies accumulated is unknown because, except where overthrust by the Moine Schists, the top of the Durness Limestone is an erosion surface. None of the Dalradian formations is younger than Cambrian; consequently only the folding and metamorphism of this sector of the geosynclinal belt is part of later Caledonian history.

The Shetland Islands lie about equidistant from the northernmost Dalradian outcrops in Banffshire and the Norwegian coast, and they are usually assumed to be part of the Caledonian mountain belt. Old Red Sandstone is present on some islands but most of them are formed of metamorphosed rocks, originally both sedimentary and igneous, piled up into overthrust nappes. The Shetlands as a whole lack a modern, general description but certain resemblances have been noted to highly altered Ordovician rocks in Norway.

NORTH WALES

Within the regions summarized, the north Welsh outcrops (Fig. 12) are outstanding, though in virtue of their Ordovician successions rather than their Silurian. This is not because they form a standard succession, easy to study and describe, but rather because they exemplify typical Lower Palaeozoic problems—thick groups of clastic sediments, facies and faunal change, vulcanicity, unconformity, overlap and overstep, and the mutual unravelling of stratigraphy and structure that is necessary for their understanding. Only the bioclastic, fossiliferous limestones of the shelf seas are lacking.

Table 14. *Summary of representative Ordovician and Silurian successions in North Wales*

System	Series	Lithology	Approximate thickness or range (feet)
SILURIAN		(Upper Ludlow probably absent)	
	LUDLOW	Ludlow Grits: greywackes, siltstones and mudstones	5,000
		Nantglyn Flags: banded siltstones and mudstones	
	WENLOCK	Denbighshire Grits: greywackes, sandstones, conglomerates of Denbighshire Moors, becoming finer upwards	1,000–4,000
	LLANDOVERY	Chiefly mudstones and shales, thinner and finer in north, coarser and thicker in the south	300–5,000
ORDOVICIAN	ASHGILL	Absent in the west, mudstones and shales in the south-east	*c.* 5,000
	CARADOC	Volcanic series at Snowdon and Conway, underlain by sediments	*c.* 6,000
		Mudstones and thin impure limestones east of Harlech Dome	
		~~ unconformity in many places ~~	
	LLANDEILO	?Absent except for varied sediments and minor volcanic rocks in Berwyn Dome	*c.* 2,000
	LLANVIRN	Mudstones and shales with lavas and tuffs of Cader Idris, Arenig, etc.	*c.* 2,500
	ARENIG	Chiefly slates or shales, with lavas and tuffs on Cader Idris	1,000
		The Garth Grit (sandstones and conglomerates); Rhobell Fawr lavas and tuffs	*c.* 3,000
		~~ major unconformity ~~~	
	TREMADOC	Mudstones and shales or slates, absent in the west	1,000
		(Upper Cambrian)	

Here and in Table 15 thicknesses can only be given in round figures.

Tremadoc Slates

The position of the Tremadoc Series is somewhat controversial. On balance, British practice has been to take it as the uppermost Cambrian division; certainly over much of Wales there is a marked unconformity at the base of the overlying Arenig and elsewhere Tremadoc beds are only well known in Shropshire and the eastern shelf. But their faunas link them with the Ordovician and this is the more common grouping on the continent; it is increasingly used in this country and is accordingly adopted here.

In North Wales the Tremadoc outcrop is more restricted than those of later Ordovician beds because of the major pre-Arenig upheaval and consequent erosion. At their name place, north of Tremadoc Bay, the beds amount to about 1,000 feet and overlie the Upper Cambrian with little or no break. The rocks are chiefly slates, with minor, fine, sandy bands. The fauna is characterized by early graptolites, of which *Dictyonema flabelliforme*, near the base, is the best known; there are also trilobites such as asaphids, *Angelina* and *Shumardia*.

Tremadoc rocks are also present, in varying thickness and completeness, round the margins of the Harlech Dome, the pre-Arenig uplift having acted irregularly in this region. Thus at Dolgelly in the south the sequence is comparable in thickness and type to that at Tremadoc, but elsewhere there is a larger unconformity and the base of the Arenig steps down over a thinner, eroded sequence. In the opposite direction, to the north-west, the situation is very different. Within a few miles of Tremadoc, beds of this series are cut out completely. This is only one aspect of this major early Ordovician upheaval and erosion, but it was widespread, for there is also a complete absence of Tremadoc beds in South Wales.

Vulcanicity and sedimentation, the first phase: Arenig, Llanvirn and Llandeilo Series

In North Wales these three divisions are linked together in their sedimentary and volcanic history. After a basal arenaceous phase muddy sediments predominated in Arenig and Llanvirn periods; these are now seen as slates, shales and mudstones, with relatively rare coarser or calcareous bands. The Llandeilo beds are poorly represented or absent and the second phase in Ordovician history begins with the Caradocian transgression. There were also great outpourings of lavas and ashes in lower Ordovician times, especially in the belt of country flanking the Harlech

Dome, from Arenig on the north-east round to Dolgelly and Cader Idris on the south-east and the country south of the Barmouth estuary.

The extremely varied distribution of these rocks, and particularly of the volcanic products, makes a general estimate of thickness impossible, but in the Dolgelly country the Arenig series amounts to several thousand feet, perhaps four or five thousand. In most regions the early marine transgression of this period is marked by an arenaceous facies, the Garth Grit, a conglomerate or coarse sandstone. On the southern and eastern flanks of the Harlech Dome it may be as little as 200 feet, but much thicker basal sandstones (up to 1,500 feet) appear farther south, at Fairbourne. An even more spectacular increase in thickness takes place north-westwards, or in the same direction that saw the greatest previous uplift and erosion. From the Tremadoc country across Caernarvonshire the basal Arenig sandstones cut down successively on to Upper, Middle and Lower Cambrian beds, until on the north-west side of the Padarn ridge and on Anglesey they lie on Pre-Cambrian rocks. On Anglesey, this facies is much more than just a basal sandy phase to a marine invasion and amounts to over 3,000 feet. No zonal fossils have been found but on the mainland the facies contains colonies of the polyzoan *Bolopora*.

One important volcanic episode took place even beforc the Arenig invasion, for the Rhobell Fawr group, north of Dolgelly, is intercalated between Tremadoc Slates and the Garth Grit. This cycle includes andesitic lavas and coarse pyroclastic rocks, which formed a subaerial volcanic pile, spreading out on an eroded surface of Tremadoc and Cambrian beds. The extrusive rocks are cut by vent intrusions, and the whole was later submerged and the Garth Grit laid down over the volcanic rocks.

There follow the marine muddy sediments of the *Tetragraptus* shales (Arenig) and shales with *Didymograptus bifidus* and *D. murchisoni* (Llanvirn). These are varied locally by sandy bands or thin limestones, the latter with trilobites and orthids. There are also tuffaceous sediments in which sprinklings of ashes mingled with, but did not supplant, the normal detritus. Some of these are fossiliferous also, like the so-called '*Calymene* Ashes' of the Arenig country.

The principal activity of the Arenig period was concentrated around Cader Idris (Fig. 18(*a*)) and extended southwards to Fairbourne. Rhyolitic lavas and tuffs are characteristic and from their maximum of about 1,000 feet on Mynydd-y-Gader they become thinner both to north and south and until there are only thin bands of rhyolitic tuff among the contemporary sediments. In this region, the '*bifidus* slates' (also with sandy and tuffaceous beds) are followed by two further great volcanic groups, the Cefn Hir Ashes, fine-grained andesitic tuffs, and the Llyn-y-Gafr group of soda-rich tuffs, agglomerates and spilites, the last often showing pillow form. The former group extended as far north as Arenig, but the very thick upper group there—4,000 feet of andesitic and rhyolitic tuffs —is of different type and may be different in horizon; their only indications of age are the Llanvirn shales below and Caradoc above.

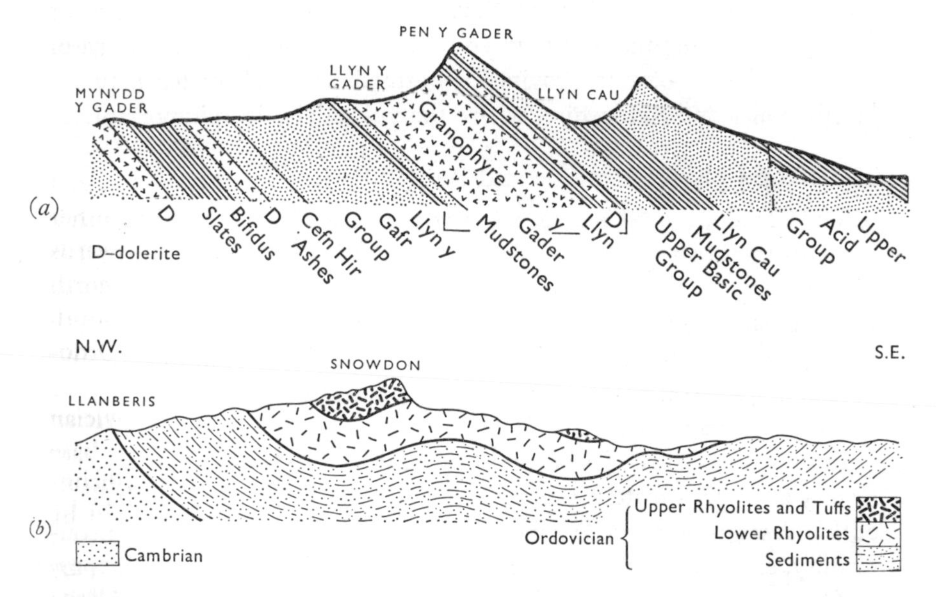

Fig. 18 (*a*). Section illustrating the geology of the Cader Idris range, to true scale (redrawn from Cox, 1925, p. 544). (*b*) Generalized section through the synclinorium of Snowdonia (from several sources).

In the Lleyn Peninsula there are also volcanic rocks among the Arenig and Llanvirn slates but they are not so thick as those just described. Similarly in the outcrops north of Snowdonia and in northern Caernarvonshire the rocks are chiefly slates. In the shaly Ordovician sequences of Anglesey all sign of vulcanicity has disappeared entirely.

The Llandeilo period in much of North Wales appears to have been one of emergence, or at least relatively little sedimentation. In some areas perhaps this resulted from positive uplift; in others the great piles of volcanic debris brought the level of accumulation, never very deeply submerged, above sea-level. The principal exception is in the Berwyn Hills. This is a structure east of, and in some ways complementary to, the Harlech Dome but is not so deeply eroded. Llandeilo beds are exposed towards the centre and amount to 2,000 feet, with considerable variation in facies. The dark-grey graptolite bearing shales are rare (though the zonal species, *Glyptograptus teretiusculus*, is found) and shelly faunas are common, in siltstones and sandstones, calcareous mudstones and impure limestones. In lithology these beds are more like the type Llandeilo of the Towy Anticline than the Ordovician slates of North Wales. They do, however, contain several bands of acid tuffs and some lava flows. None is so thick as those developed earlier at Dolgelly or Arenig, suggesting that the Berwyn outcrops lie toward the margins of the Ordovician volcanic province, intermediate in character between North Wales and Shropshire, where the vulcanicity is unimportant.

The second phase: Caradoc and Ashgill Series

The Caradoc period is especially important in North Wales because it includes the great volcanic series of Snowdonia. There were also other outpourings, on only a slightly lesser scale, in various places northwards to the coastal outcrops at Conway. The thin Ashgill beds of the north form little more than an appendage to Caradoc history, but to the south greater thicknesses come in, as a transition to the massive, late Ordovician greywackes of central Wales.

In very many areas, far beyond Wales, there is a break in Ordovician history at, approximately, the beginning of Caradoc times, or at the base of the *Nemagraptus gracilis* zone. The change may be marked by unconformity, erosion and submergence, by changes in facies or faunas, or by shifts in the centres of volcanic activity. This invasion of new territories is sometimes referred to as the '*gracilis* transgression'. Within our present region the stratigraphical break is most clearly seen in Anglesey where the *gracilis* shales overstep the lower Ordovician beds on to the Pre-Cambrian Mona Complex. They also contain large blocks of an earlier Ordovician limestone which is not now found elsewhere—blocks which are presumed to have slid from some submarine scarp, possibly in response to earthquake shocks. On the mainland Caradoc rocks lie unconformably on Arenig at the extremity of the Lleyn Peninsula and on Tremadoc near the village of that name. Elsewhere a lesser gap is represented by the partial or complete absence of Llandeilo beds.

The early Caradoc sediments show rather more variation than the muds of the Arenig and Llanvirn, from the greywacke incursions in the west, on the flanks of Snowdonia, to the fine black muds (*Dicranograptus* Shales) at Conway. The Cader Idris vulcanicity came to an abrupt end early in the period. The last phases produced a series of spilitic pillow lavas, followed by rhyolites and andesites with thin bands of sediments and pyroclastics. The succeeding 2,000 feet of Caradoc mudstones contain a fauna of the *gracilis* zone near the base. It is thought that the latest extrusion (the Upper Acid Group) was derived from the same magma that, under a relatively shallow cover, formed the thick granophyre sill of Cader Idris mountain (Fig. 18 (*a*)). The intrusion was accompanied by doming, and thus contributed to the emergence of a volcanic island, so that part at least of the final extrusive products was subaerial.

In Snowdonia the main period of vulcanicity was a little later. The lavas and tuffs overlie slates and greywackes with *Nemagraptus gracilis*, and shelly faunas of the Harnagian and Soudleyan stages (i.e. low Caradocian, see p. 414). On the eastern side fossils of the *clingani* zone have been found in mudstones both below and above the volcanic group; comparable volcanic rocks at Conway are restricted to a portion of lower Caradoc time, and all vulcanicity seems to have ceased by the upper Longvillian. Volcanic rocks of Caradoc age are also known from the Lleyn Peninsula but on Anglesey the shale sequence is uninterrupted; the uppermost zone found there is that of *Dicranograptus clingani*, but it may be that the Ashgill is represented in barren beds above.

The variety of volcanic rocks, in thickness and in composition, the way they pass laterally into tuffaceous sediments (which locally may even include fossils), their reaction to Caradocian earth movements, are all characteristic of the Snowdon Volcanic Series. On Snowdon mountain itself the group is over 3,500 feet thick, composed dominantly of acid, rhyolitic, lavas and tuffs and subsidiary intercalations of sediments. They form the beautifully peaked summit and are exposed on the flanking corries and arêtes. To the south and north the thickness of volcanic rocks falls off, and though only a few vents are visible it is probable that the volcanic centres were largely within Snowdonia itself.

One cause of complexity is the relation of the North Wales trough to sea-level at this time. It seems unlikely that any of the sediments were laid down in deep water; rather that in different regions and perhaps locally for considerable periods the volcanic piles maintained themselves above sea-level. On such a view, their products accumulated on the island slopes and at the shore passed laterally into bedded deposits, sometimes largely volcanic but elsewhere ashy muds and silts on which a bottom fauna could survive. It is also possible that upward doming, caused by the pressure of acid magmas below, contributed to the maintenance of these temporary islands or archipelago.

This view is supported by the widespread recognition of welded tuffs among the pyroclastic rocks. Characterized by welded shards, usually of volcanic glass, these are considered to be products of high temperature incandescent clouds (or nuées ardentes) from a Peléan type of eruption. The name ignimbrites has been given to such rocks; other Snowdonian types include volcanic mudflows. Ignimbrites cannot form below water and their abundance in Snowdonia is one of the strongest reasons for thinking the region to have been above sea-level, at least from time to time. Such a picture is far from that once envisaged of a central geosynclinal trough of deeper water flanked by shallower margins—but this is an over-simple concept which has been found increasingly inappropriate in several sections of the British Caledonian geosyncline; in Wales the diverse vulcanicity adds force to its abandonment.

There is nothing higher than Caradoc in Snowdonia, but Ashgill comes in to the east. The joint total of Bala beds (Caradoc plus Ashgill) at Bala itself, on the east flank of the Harlech Dome, is about 5,000 feet; they consist largely of mudstones and impure limestones, with a shelly fauna, but the Caradoc part includes several tuff bands, tuffaceous sediments and minor lava flows. Farther south the Bala beds expand into 9,000 feet of poorly fossiliferous mudstones south-east of Cader Idris, on the borders of north and central Wales.

The last, sedimentary, phase: Llandovery, Wenlock, Ludlow Series

Long continued erosion has removed Silurian, and particularly late Silurian, rocks from much of North Wales. Between the Harlech and Berwyn domes the Central Wales Syncline brings in only a narrow strip of Llandovery and Wenlock beds. Farther north, from Conway south-east to Denbigh, in the Clwydian range and in the Llangollen Syncline, there is a more complete succession.

Over the more westerly outcrops there is conformity between Ashgill and Llandovery Series and at Conway the latter comprises 300 feet of black shales, which nevertheless contain all the numerous graptolite zones and must result from exceedingly slow accumulation of mud in quiet conditions. Such conditions probably extended westwards, for though there are no Silurian outcrops on Anglesey comparable with the Ordovician, shales in a synclinal overfold on Parys mountain have yielded graptolites of several Llandovery zones.

Slightly coarser Lower and Middle Llandovery rocks persist southwards through the Bala country and the Central Wales Syncline; the upper division swells out with arenaceous additions and passes into the greywackes of the Aberystwyth region (p. 96). In the easterly outcrops there are signs of the marginal uplift that is much more apparent in the shelf regions of Shropshire; between the latter and the Berwyn Dome, in Montgomeryshire, the Lower Llandovery rests unconformably on Ashgill beds, and here and there on Caradoc. The 1,000 feet or so of silty mudstones are also intermediate in type, containing a shelly fauna and relatively few graptolites.

In the largest outcrop of the Denbighshire moors there is a maximum of nearly 10,000 feet of Wenlock and Ludlow beds, in the following groups:

Ludlow Grits Nantglyn Flags	Lower Ludlow
Denbighshire Grits	Wenlock

At their maximum (about 4,000 feet) in Denbighshire, the lowest group are greywackes interbedded with siltstones and mudstones. The whole is interpreted as the products of turbidity currents, flowing down into the deeper parts of the north Welsh trough. Fossils are rare, as they usually are in greywackes, but the finer beds contain pockets of shells and occasional graptolites. This thick greywacke facies gives place both in the north and south-east to the much thinner shales and mudstones of Conway, Llangollen and Welshpool, the latter being part of the grey muddy succession that lies off the edge of the Shropshire shelf.

The succeeding Nantglyn Flags of Denbighshire comprise calcareous siltstones, and laminated and homogeneous mudstones. The first have a mixed shelly and graptolitic fauna and in their graded-bedding and current-aligned structures these sediments also show some turbidite features. Certain mudstones have traces of a sparse indigenous and burrowing fauna. The Ludlow Grits recall aspects of the earlier grit group, though generally they are not so coarse, being greywackes and siltstones with interbedded mudstones on the Denbighshire moors and Clwydian range, and calcareous siltstones in Montgomeryshire. The northern outcrops are especially noted for their extensive spreads of slumped sediments, which were folded and corrugated as they slid down a submarine slope. The general direction of slumping, and therefore presumably of the slope, was to the south and south-east, and this appears to be more or less at right angles to current directions deduced from the uncontorted greywackes. The two types of sediment suggest that the turbid currents, in this case at least, moved dominantly along the axis of the trough, and the unstable muds and silts slipped in contorted sheets down its sides.

A 'Denbighshire trough' thus seems to be a reasonable interpretation through much of the later part of Silurian times, but the topography need not have been very spectacular, as both slumping and turbidity flow can take place on small angles of slope. Any late Ludlow history in North Wales is almost unknown, as erosion has removed any deposits there may have been; that some at least were laid down, however, is indicated by blocks with the Upper Ludlow *Protochonetes ludloviensis* in the basement conglomerates below the Carboniferous Limestone to the east.

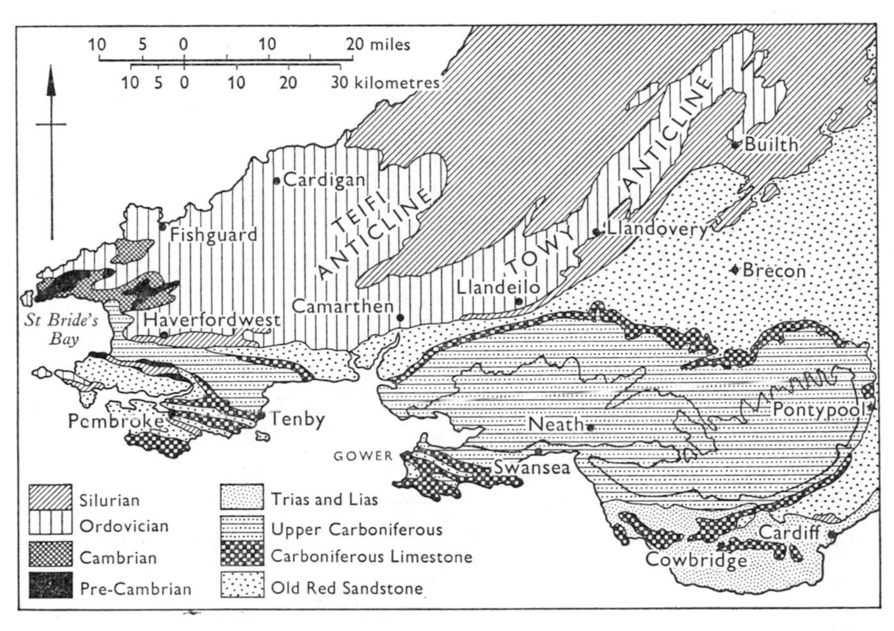

Fig. 19. Geological sketch map of South Wales. Igneous rocks (chiefly Ordovician) are not differentiated; the Upper Coal Measures of the central South Wales Coalfield are outlined.

SOUTH WALES

During Ordovician and Silurian times North and South Wales were part of the same geosynclinal area of sedimentation, and their histories have much in common; the main difference is the much lesser importance of Caradocian vulcanicity in the south, and the fuller sequence of Llandeilo beds. The geographical division can only be drawn artificially, where most convenient, and is here taken approximately along a line east-south-east from the Dovey estuary, in the centre of Cardigan Bay, to Clun in south-west Shropshire. This cuts through the very large Llandovery outcrops of central Wales, where, somewhat paradoxically, the

Central Wales Syncline is not clearly defined. To the south, however, a comparable structure is picked up again in Carmarthenshire where a central syncline separates the short Teifi Anticline on the west from the much longer Towy Anticline on the south-east—the latter being a belt of complex history, with Ordovician as well as Silurian uplift. Along much of the south-east flank of this structure, narrow outcrops of Llandovery, Wenlock and Ludlow rocks are succeeded by Old Red Sandstone. But westward from Llandeilo, on the middle Towy reaches, a major unconformity with overstep develops between Lower and Upper Palaeozoic rocks. Part of this is due to late Silurian or early Devonian

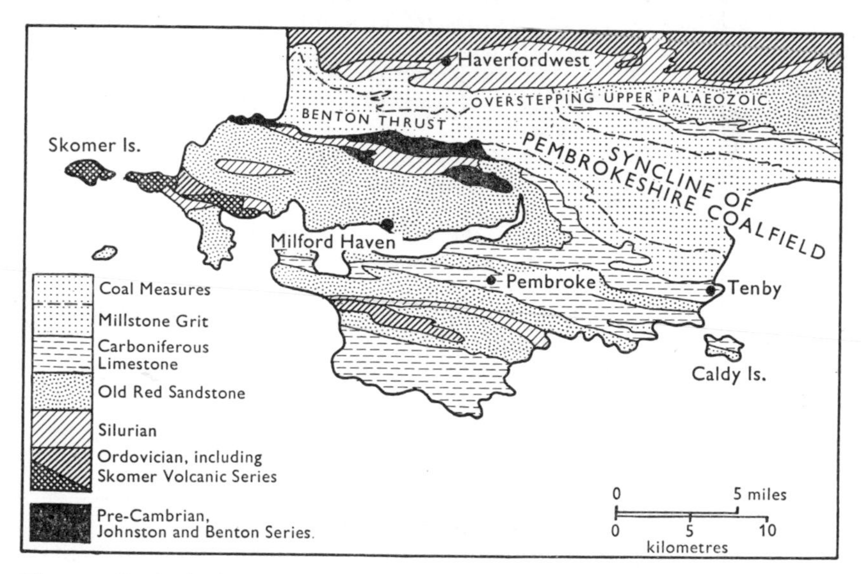

Fig. 20. Geological sketch map of South Pembrokeshire, illustrating the over-stepping relation of Upper and Lower Palaeozoic rocks in the north, and the Benton Thrust, and Lower Palaeozoic inliers farther south.

uplift and part to Lower Carboniferous, and the results are complex; near Carmarthen, for instance, Old Red Sandstone rests on Arenig beds, but Silurian reappears farther west, and near Haverfordwest Carboniferous Limestone rests on Llandovery mudstones.

South of this unconformity, in south-west Pembrokeshire (Fig. 20), there is a region of dominant, Hercynian, east-west folds and low-angle thrusts. Most of the outcrops are Upper Palaeozoic, but some Pre-Cambrian, Ordovician and Silurian rocks are brought up in anticlines or in conjunction with the thrusts; the Silurian rocks in particular are rather different from those of the main outcrops.

Table 15. *Summary of representative Ordovician and Silurian successions in South Wales*

	Series	Lithology	Approximate thickness or range (feet)
SILURIAN	LUDLOW	Grey siltstones and mudstones in the north-east; shelly mudstones in south Pembrokeshire; absent in the west	6,000–2,000
	WENLOCK	Siltstones, flaggy sandstones and shales and mudstones in the east; mudstones and limestones in south Pembrokeshire; absent in the west. Local unconformities within series	1,600
	LLANDOVERY	Shelly mudstones at Llandovery, greywackes, mudstones and conglomerates in Cardiganshire and central Wales. Local unconformities within series	*c.* 5,000
ORDOVICIAN	ASHGILL	Dominantly mudstones, with sandstones and conglomerates in central Wales	500
	CARADOC	Chiefly shales with thin impure limestones	800
	LLANDEILO	Shales in the west, impure limestones and sandy beds at Llandeilo in the east	2,500
	LLANVIRN	Shales with many groups of lavas and tuffs	*c.* 3,000
	ARENIG	Shales and extensive volcanic groups; Basal sandstones and conglomerates	*c.* 3,000
		~~major unconformity~~~	
		(Upper Cambrian)	

Ordovician sedimentation and vulcanicity

The early Ordovician earth movements were strong in South Wales and Tremadoc rocks, if they were laid down, have not survived. The lowest Arenig beds rest transgressively on some part of the Upper Cambrian, and, as in North Wales, there is a basal sandy facies; this is seen as conglomerates with *Bolopora* on Ramsay Island (off the north side of St Brides Bay), and as massive white quartzites, with red and green sandstone and detritus from local Cambrian and Pre-Cambrian rocks, on the north Pembroke coast. This basal sandy phase is succeeded by sandy mudstones with orthids, early ogygiids and calymenids, and then the *Tetragraptus* Shales, the latter including faunas of both the *extensus* and *hirundo* zones.

There was also extensive Arenig vulcanicity in Pembrokeshire. The Trefgarn Series (750 feet of andesitic lavas and tuffs) were piled up on a basement of Cambrian beds and later overlapped by *Tetragraptus* Shales. At Abercastle, on the north coast, keratophyres flowed out over the sea floor and incorporated the muds lying there. But by far the most impressive example is the Skomer Volcanic Series; it is 3,000 feet thick on Skomer Island, which stands out westward from the south side of St Brides Bay. In composition the lavas range from acid to basic, and although few pyroclastic rocks are known the flows are thought to be subaerial. Their Arenig age is based chiefly on their similarity to the Trefgarn lavas; on the mainland they become much thinner and disappear beneath unconformable Silurian rocks.

In sedimentary facies Llanvirn beds resemble the upper Arenig, and are black shales with didymograptids; the type area, from which great numbers of *D. murchisoni* have been collected, is a small farmstead in north Pembrokeshire. The Llanrian Series comprises 500 feet of rhyolitic tuffs and some lavas on the north coast, but expands to 1,400 feet on Ramsay Island. At about the same level is the Fishguard Series, including rhyolites, basic tuffs, agglomerates and spilitic lavas; the last amount locally to 3,000 feet, including pillow lavas and sediments rich in volcanic debris. Welded tuffs have been recognized near Fishguard, similar to those of North Wales and the Lake District, and it may be that they are more common than is apparent from some of the older descriptions.

A remarkably complete history of the growth and submergence of a volcanic island has been made out from Llanvirn rocks near Builth Wells, east of the upper Towy Anticline. At its most complete the sequence here is fourfold, with an unconformity between each unit:

Cwm Amliew Series	Rhyolitic tuffs and mudstones with *Didymograptus murchisoni*
Newmead Series	Beach sands and boulder beds
Builth Volcanic Series	Keratophyres, breccias, tuffs, spilites and agglomerates
Llandrindod Volcanic Series	Rhyolitic tuffs

(Shales with *Didymograptus bifidus*)

The main component (nearly 2000 feet) is the Builth Volcanic Series, which was built up on the sea floor and later emerged above sea-level. Erosion set in and the lavas developed a terraced landscape and the keratophyres rounded knolls. The short-lived island was then gradually submerged and the sands and boulders of the Newmead Series were banked up against the cliffed shores, surrounded stacks and filled in hollows and fissures. This remarkable Llanvirn topography can still be traced, laid bare by the chance effects of modern erosion. Moreover, all this local episode took place within the time of the *murchisoni* zone, and 7 miles away to the north it is represented only by an unrevealing unconformity in the shales of that zone.

Twenty-five miles to the south-west further Llanvirn rocks appear near Llandeilo in the core of the Towy Anticline and they also show signs of vulcanicity. There are tuffs, volcanic clays, pebbly sands and calcareous sandstones with a shelly fauna. The volcanic elements increase to the north-east, and it is probable they are the erosion products from the Builth volcanic centre, spread farther afield over the sea floor. The volcanic activity seems to have migrated southwards at the end of Llanvirn times, for rhyolites and tuffs appear at the top of this group at Llandeilo. Vulcanicity also continued and the shales are locally interbedded with tuffs.

After the Llanvirn activity, vulcanicity is slight in South Wales, and the principal feature of later Ordovician sedimentation is fairly simple lateral variation in facies. In the north-west Llanvirn shales are succeeded by a similar graptolitic facies, the 500 feet of Hendre Shales of Pembrokeshire; at Builth also there are Llandeilo graptolite shales. The Llandeilo flags and limestones are characteristic of the type area, in the southern part of the Towy Anticline. Here there is a varied series of sandy, flaggy beds and impure limestones; shales are subordinate and not graptolitic, and certain beds are rich in brachiopods, trilobites and polyzoa. There are many signs of shallow water and *Lingula* is present, but the outcrops are so narrow that it is not possible to say how far this facies extended to the south and east. There are, however, about 1,000 feet of similar rocks in the extreme south-west of Pembrokeshire.

Caradoc history in South Wales is one of quiet sedimentation in striking contrast to the spectacular vulcanicity in the north. The *Dicranograptus* Shales are graptolitic, dark-grey, and contain thin bands of impure limestone, the total being about 800 feet. The principal variation is the incoming of greywackes in a small area of the north coast between Newport and Cardigan.

A rather different régime set in with the Ashgill Series, for towards the end of Caradoc times there was local uplift and erosion in several areas. Near Haverfordwest, Ashgill mudstones and impure limestones contain a shelly fauna (including cystids) and there are minor unconformities at or near the base of the series. A more important break occurs over the crest of the Towy Anticline at Llandeilo, where locally Llandeilo, Caradoc and Ashgill beds have all been removed; this uplift may have been initiated in Ashgill times, or perhaps later; it was completed at any rate before the deposition of the Upper Llandovery.

To the west of the Towy Anticline there is a great thickness (5,000 feet) of Ashgill beds, mainly poorly fossiliferous mudstones; they become coarser towards the top, even to the appearance of conglomerates in the north-east. This massive development is also extensive in south Cardiganshire and northwards in central Wales. Over 3,400 feet of beds are present in the inliers of the Plynlimmon district, where the base of the group is not exposed; again there is an upward coarsening from mudstones and shales with *Dicellograptus anceps* to flags, sandstones and conglomerates. Late Ordovician times thus saw the influx of substantial amounts of detritus, much of which was presumably derived from the west, and accumulated in the downwarped region of central Wales. The evidence from the Towy Anticline suggests another, possibly minor, source in the east as well.

Silurian sedimentation and unconformities

Llandovery beds crop out in three principal areas: around the type locality on the Towy Anticline, near Haverfordwest, and in a very broad region, chiefly of Lower Llandovery, in central Wales. In the first two the rocks are similar—mudstones with small calcareous nodules, thin sandstones and an abundant shelly fauna. The junction with the Ordovician below is usually sharp but without any strong evidence of unconformity. Within the Silurian rocks themselves, however, there is ample evidence of breaks and instability. At the town of Llandovery the maximum thickness is 5,200 feet, but the three divisions (Lower, Middle and Upper) are separated from one another by local unconformities. The Middle Llandovery is absent from portions of the outcrop and the base of the Upper Llandovery is strongly transgressive. That division itself varies from 2,000 of mudstones to 40 feet of *Pentamerus* Sandstone and is overstepped by the Wenlock mudstones. Early Silurian movements are thus characteristic of the Towy Anticline, and at Llandeilo the Ordovician beds were thrown in small, sharp folds before the Upper Llandovery was laid down.

There is a comparable history in the outcrops near Haverfordwest, where the Middle Llandovery is again missing; the Upper Llandovery transgression is further emphasized by the overlap of these beds on to the Pre-Cambrian Benton volcanic rocks, 5 miles to the south. On the mainland near Skomer Island there is a similar overlapping relationship, but the Upper Llandovery strata (which include a basic lava flow and tuffs) lie on beds of different ages, including the Skomer Volcanic Series. In these various outcrops there is some rather fragmentary evidence of a south Pembrokeshire landmass or group of islands and shoals, at least in late Ordovician and early Silurian times. Its extent, however, is unknown; in particular between the Haverfordwest outcrops and those of the Benton Volcanic Series there is the Hercynian-trending Benton Thrust, on which the country to the south has been shifted northwards for an unknown distance.

In central Wales and along the coast of Cardigan Bay there are very wide areas of a more graptolitic Llandovery facies—not so much the black muds of Conway, as a variety of grey, green and purple mudstones and shales. Interrupting this muddy sedimentation there were periodic influxes of coarser detritus, from siltstones to sandstones and greywackes, and occasionally conglomerates. These are more prevalent in the west, but at Rhayader they include the 1,000-foot (and apparently local) mass of the Caban Conglomerate Group, which was laid down in a hollow eroded in the Middle and Lower Llandovery rocks below.

The greywacke facies is exemplified by the Aberystwyth Grits. As seen in the fine coastal sections of Cardigan Bay, these amount to some 5,000 feet of thin-bedded greywackes and mudstones. Fossils are rare, but tracks of bottom living organisms have been left in the finer beds. Some of the greywackes show laminations and ripple marks, convolute bedding and local small erosion features; others are the more usual graded rocks. From both the early appearance of greywackes in the south, and directional structures within them, a southerly or south-westerly source is postulated. Like other greywacke sequences the Aberystwyth Grits are attributed to turbidity currents, which spread their load out on the sea floor. Here and elsewhere, however, there is little evidence for the depth of the trough, though it was part of the major downwarp of central Wales that persisted from late Ordovician to Upper Llandovery times.

The outcrops of Wenlock and Ludlow rocks can be divided into those on the east side of the Towy Anticline, which expand north of Builth into the much broader area adjoining the Welsh Borders, and the inliers of south Pembrokeshire. Along the Towy Anticline the early Wenlock was a further time of uplift and overstep. Near Llandovery the grey Wenlock mudstones and shales pass from Upper Llandovery on to Lower, and to the north and south lie on Ashgill beds. At Llandeilo they lie in turn on Llandeilo and Llanvirn, and there is also a minor break within the Wenlock mudstones themselves.

Wenlock beds also surround the Builth inlier; they are thicker than on the Towy Anticline, and consist of 1,600 feet of grey shales, mudstones and flaggy sandstones. The Ludlow rocks in particular exhibit a thick 'basin' succession, from Builth northwards through Knighton in south Radnorshire to the Clun Forest. The siltstones, grey mudstones and shales of this facies are 5,000 feet or more and reach over 6,000 feet at Knighton—a remarkable contrast to the shelf facies to the east, at Ludlow itself (p. 104).

The lowest shales still contain graptolites but these become scarce in the later beds and disappear before the end of the period. The shelly faunas are locally sparse, particularly in the siltstones, where conditions appear to have been unfavourable for bottom faunas. Contorted siltstones, or slumped beds, are found at more than one level; they affect great thicknesses of strata (2,000 feet) at Builth, the unconsolidated sediments having suffered repeated slipping on the slopes of the submarine depression.

West of Llandeilo, along the Towy Anticline, the upper Silurian beds disappear under the Devonian and Carboniferous overstep. Where Wenlock beds reappear in the isolated outcrops of south Pembrokeshire (Fig. 21) they have a more calcareous facies and a shelly fauna. A coral-bearing limestone occurs at Marloes Bay, at a level comparable with the Woolhope Limestone of the Herefordshire inliers (p. 107). A maximum of 2,000 feet of Ludlow has been recorded above, but neither upper nor lower boundary is very well defined. The rocks are yellow to greenish mudstones and sandstones, sparsely fossiliferous. In some of the anticlines south of Milford Haven smaller amounts of similar rocks with bands of brachiopods are exposed.

The Silurian rocks of south Pembrokeshire are not very like those of Welsh trough facies as a whole; they have more in common with those of Herefordshire and south Shropshire, especially in the Llandovery overlap and Wenlock limestones. It is possible that they should be thought of as a westward continuation from the eastern shelf, described in a later section.

An Irish Sea Landmass

Before leaving the geographical setting of the Welsh geosynclinal facies we should consider how far anything can be deduced of the sources of sediment, or actual landmasses, to the west. Two lines of evidence point in this direction. The pre-Arenig uplift and unconformity was a major event over all the Welsh trough, but the most marked overstep is that from Harlech north-westwards through Caernarvonshire to Anglesey, and the latter was a temporary landmass, more effectively uplifted and eroded than the remainder of North Wales. Moreover, there is overlap within the Ordovician shales themselves on Anglesey—notably of the Caradoc over the earlier beds. A periodic uplift of a westerly or north-westerly mass, of which Anglesey remains as part, fits Ordovician history, and as in the Cambrian period it supplied recognizable detritus into the neighbouring seas of North Wales.

The second line of evidence is more indirect. At several periods, from mid-Ordovician to early Silurian (late Silurian history is a blank on the west side of Wales), tongues of coarse deposits, usually greywackes, appear in the more westerly sediments. For their source a landmass in that direction is also a possibility. The form of such a land is a much more difficult question. There is the similarity of the Pre-Cambrian rocks of Rosslare (County Wexford) to those of Anglesey to be considered, and a ridge running north-east to south-west linking the two would be in about the right position (Fig. 7). Unfortunately the Ordovician history of east Leinster on the far side, in so far as it is known, is not very helpful and Silurian rocks are lacking. In order to supply the great thickness of greywackes such as those of central Wales, however, any such ridge would have to have been frequently rejuvenated and persistently eroded.

THE EASTERN SHELF: SOUTH SHROPSHIRE

The Ordovician and Silurian outcrops of south Shropshire (Fig. 10), like those of earlier rocks, have a significance beyond their size. They exemplify the shelf facies that bordered the geosyncline on the east and south-east—a facies composed chiefly of mudstone and fine sandstones, often calcareous and rich in a shelly fauna. There were also local littoral sands and pebble beds and impressive, but not very thick, limestones. That these variable and shallow marine rocks extended much farther afield, at least during Silurian times, is shown by the several inliers that are found in the western Midlands and southwards as far as Gloucestershire. A few boreholes in the Midlands and south-east England carry our data a little farther.

The Tremadoc Series or Shineton Shales

As in North Wales the Tremadoc beds agree with the underlying Cambrian in their distribution and not with the Arenig. The best sequence comes from outcrops on either side of the River Severn. Here blue-grey shales, becoming sandy towards the top, rest on similar Upper Cambrian beds. The thickness is substantial, but owing to faulting and poor exposure estimates have ranged from 1,500 to over 3,000 feet. There are several fossiliferous horizons, which include, in upward succession, *Dictyonema flabelliforme*, *Clonograptus tenellus* and *Shumardia pusilla*; comparison with Scandinavia suggests that most of the Tremadoc Series is present. The Shineton Shales of the Habberley valley, west of the Longmynd are even less well exposed, but the junction with the overlying Arenig quartzites is probably unconformable. It is thus likely that the uplift at this time that affected all Wales had some repercussions in the Welsh Borders, though on a more modest scale.

Table 16. *Ordovician succession in the Shelve District, Shropshire*

Series	Formations and groups	Approximate thickness (feet)
	(Upper Llandovery)	
	～～～～～～～～～～	
CARADOC	Whittery Shales with Whittery Volcanic Group; Hagley Shales with Hagley Volcanic Group	2,000 +
	Aldress Shales	1,000
	Spy Wood Grit	300
	Rorrington Beds	1,000
LLANDEILO	Meadowtown Beds	1,100
LLANVIRN	Betton Beds	600
	Weston, Stapeley and Hope Shales, with Stapeley Volcanic Group	*c.* 3,500
ARENIG	Mytton Flags	3,000
	Stiperstones Quartzite	*c.* 400
	～～～～ unconformity ～～～～	
TREMADOC	Habberley Shales	
	(base not seen)	

Shelve and Caradoc Districts, Arenig to Caradoc Series

There is a remarkable difference in the overlying Ordovician successions in the two main outcrops, that of the Shelve and that of the Caradoc region itself (Fig. 21). At present they are separated by the Longmynd, the Church Stretton Fault and the Cambrian and Pre-Cambrian outcrops of the Caradoc range: their original relationship is somewhat uncertain.

The Shelve outcrop, which extends westward from the Habberley valley, includes a full complement of beds from Arenig to middle Caradoc (Table 16). Above the basal, unfossiliferous Stiperstones Quartzite there is a mixed facics. The sediments range from the fine-grained sandstones of the Spy Wood Grit, through siltstones and mudstones with shelly faunas to the dark-grey graptolitic shales of the Betton

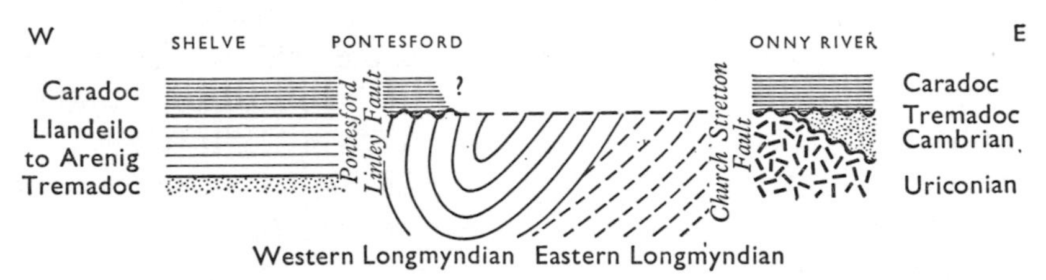

Fig. 21. Diagrammatic section to show the distribution of the Ordovician formations in south Shropshire; only the Caradoc Series is present at Pontesford and east of the Church Stretton Fault. The two rock groups of the Longmynd and their structure are also represented.

Beds and Rorrington Shales. Only minor volcanic influence was felt in Shropshire. Thin lavas, volcanic breccias and tuffs are found in the Whittery and Hagley volcanic groups; the Stapeley volcanic rocks (Llanvirn) are water-laid andesitic tuffs and the Meadowtown Beds contain tuffaceous additions among the flaggy sandstones. All these are a far cry from the thousands of feet of volcanic rocks on Cader Idris or Snowdon.

The type Caradoc or eastern succession is best exposed in the little Onny river that cuts across the southern part of the outcrop. The stages Costonian to Onnian (p. 414) were first described from here. The outstanding difference between this succession and that of the Shelve district is the total absence of Arenig, Llanvirn and Llandeilo beds. The basal Caradoc sands and pebbly beds (the Hoar Edge Grit) are strongly transgressive and lie on various formations from Uriconian to Shineton Shales. That this is another of the *gracilis* transgressions is shown by the few calcareous and shale bands that have yielded that zone fossil. After the initial submergence there followed some 2,000 feet of shallow water sediments, varying from muds to silts and fine sands; all are calcareous and at many levels there is a profusion of brachiopods and trilobites.

Graptolites are rare and probably do not rise above the *clingani* zone, nor is there any evidence of Ashgill beds.

A third small outcrop of Caradoc strata (*multidens* zone only), raises awkward questions by its geographical position. The Pontesford outlier lies west of the Longmynd and unconformably on Longmyndian rocks (Fig. 10), and is only a mile from the Shelve outcrops, though separated from them by the major Pontesford-Linley fault. Small though the outlier is, its existence affects the interpretation of the two larger ones and of Ordovician history. For the question arises—were earlier Ordovician beds laid down east of the Longmynd and later removed in pre-Caradoc erosion; or was the easterly area, and by deduction much of the eastern shelf, land from Arenig to Llandeilo times, to be submerged in the *gracilis* transgression? The second view is more usually held, with the implication that through much of the earlier Ordovician history a shoreline lay in the neighbourhood of the Longmynd. It does meet a difficulty, however, because it demands the overlap of all the lower strata (some 8,000 feet) between the Pontesford and the Shelve outcrops, which are at present only a mile apart. If however there had been much *lateral* movement on the intervening fault, then the distance in middle Ordovician times might have been greater.

The Caradoc beds of the type area are themselves not found farther east and this absence can again be interpreted as either an original absence of deposition—i.e. a land area behind a narrow Caradocian shelf—or the result of subsequent (late Ordovician or early Silurian) erosion. The distribution of all these beds in south Shropshire thus illustrates in a particularly cogent form the question that accompanies many unconformities in regions of repeated movement. Were the beds of the time-gap never laid down, or were they (or some part of them) laid down and later removed? An original absence of all Ordovician beds, except Tremadoc, from the Midlands of England is usually assumed, but it is difficult to prove positively. (See, however, note on page 107.)

The Upper Llandovery submergence

Whatever the answer to the question posed in the last paragraph, there was undoubtedly an important period of earth movement in Shropshire between late Caradoc and Upper Llandovery times. Faulting and folding affected the Shelve district in particular, and the Ordovician rocks here have a much more pronounced structure than have the Silurian.

When the seas returned it was to a region of considerable topographic relief. The main upstanding masses were the hardest rocks—the

Stiperstones ridge, the Longmynd, and the Uriconian ridge from the Caradoc hills to the Wrekin. The littoral sands and gravels of this sea are particularly well seen round the present margins of the Shelve and Longmynd, and from which they pass away into muds which were farther offshore. Modern erosion has allowed several coastal features to be discerned, such as the pebble beaches, bars, stacks and cliffed headlands at the south end of the Longmynd. It is the same kind of glimpse of an ancient landscape that is visible at Builth, where the beach beds are seen around the Llanvirn volcanic island.

Table 17. *Silurian succession around Wenlock Edge and Ludlow*

Series	Formations	Average thickness (feet)
	(Ludlow Bone Bed)	
LUDLOW	Whitcliffe Beds Leintwardine Beds (includes Amestry Limestone) Bringewood Beds Elton Beds	1,200
WENLOCK	Wenlock Limestone Wenlock Shales	1,000
UPPER LLANDOVERY	Purple Shales Pentamerus Beds Local arenaceous facies	600
	～～～～～	
	(Caradoc Beds)	

In these littoral and offshore conditions it is not possible to give a single thickness estimate for the Upper Llandovery beds, which vary from one outcrop to another. They were, however, thin compared with those of Wales, and perhaps amounted to 500 feet at most. Away from the marginal sands the sequence comprises the Pentamerus Beds below and the Purple Shales above. The former consist of calcareous mudstones and sandstones, which include layers close packed with the valves of *Pentamerus*. All the Llandovery beds were laid down in shallow water, and the abrupt change (though no visible unconformity) at the base of the Purple Shales may be due to a period of non-deposition though not of actual emergence. A local beach bed, thick with Uriconian pebbles, is intercalated with the Pentamerus Beds on the south side of the Wrekin.

Wenlock and Ludlow Series, the shelf facies

These beds succeed the Purple Shales with a more marked break, and the sharp lithological change is accompanied by the absence of certain zones above and below the lower Wenlock boundary. If there was actual emergence, however, it was not accompanied by tilting. In the western outcrops all the Wenlock is composed of grey, graptolite-bearing mudstones, which merge into similar rocks in central Wales and the Berwyn Hills. In the east there is the double succession of Wenlock Shales overlain by Wenlock Limestone. The latter, with its impressive scarp on Wenlock Edge, is so well known that it is a little surprising to realize that out of the 1,000 feet of the series as a whole, only a tenth at most is formed of the limestone.

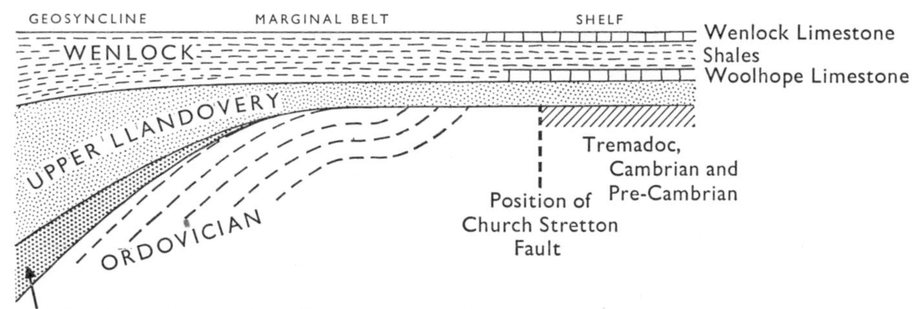

Fig. 22. Diagrammatic section to show the transition of Llandovery and Wenlock beds from the shelf to the geosyncline. The latter is not exemplified from any one area. Not to scale.

The lowest division of the Wenlock Shales is the grey-green calcareous Buildwas Beds in which the shelly fossils are mostly small and the larger valves often broken. It is possible that the sea at this time was so shallow that the bottom muds were easily stirred up, and conditions of life were somewhat unfavourable. The succeeding, more sandy shales, bear occasional rows of calcareous nodules, and these become increasingly abundant towards the top so that there is a gradual transition to the nodular, bedded type of Wenlock Limestone. Graptolites are absent from the limestone and rare in the shales, but enough are known to show that the former lies about at the junction of the Wenlock and Ludlow series, though the traditional name is retained.

Two facies in particular are characteristic of the Wenlock Limestone. There are the irregular, thin-bedded, somewhat nodular limestones with sporadically a rich fauna of tabulate corals (especially *Favosites*), polyzoa, brachiopods and simple rugose corals. Locally these are replaced by the 'ballstones'—large masses of unstratified limestone, chiefly of calcite mudstone but also with algae, polyzoa, stromatoporoids and corals; essentially they are reef-knolls, a term more commonly used for similar Carboniferous features. The bedded limestones pass laterally into, and arch over, the knolls, which stood up as low mounds on the sea floor. At the top a few feet of crinoidal rocks spread over both bedded and unbedded rocks, and the limestone phase was brought to an end by a further influx of muds, the lowest Ludlow shales. South-west along Wenlock Edge, towards Craven Arms and the Ludlow region the limestone becomes thinner and the ballstones disappear (Fig. 10).

South-east of Wenlock Edge, where the Ludlow beds rise from the dip-slope of the limestone, they are only about 600 feet thick; they increase to 1,200 feet at the town of Ludlow, and thence westwards to 2,000 feet (where complete) before their more marked expansion into the basin facies at Knighton, another dozen miles to the west. The lowest beds in the succession (the 'Lower Ludlow Shales' of earlier descriptions) are chiefly grey to yellow shales and mudstones with both a shelly fauna and graptolites; the middle and upper beds are dominantly calcareous siltstones. Graptolites become more and more scarce upwards and disappear well below the base of the Devonian; even the locally rich brachiopod, trilobite and lamellibranch faunas become more restricted, particularly the first two groups. It is partly this decrease in the marine faunas, and the rarity of species short-ranged enough to form reliable index fossils, that makes correlation of the Ludlow beds difficult, even where exposures are good and tectonic complications few. The divisions are based chiefly on shelly assemblages especially of brachiopods.

A second contributory factor is the diachronism or local absence of certain conspicuous facies. Chief among these are impure limestones with thick banks of *Conchidium* valves, tabulate corals and stromatoporoids, commonly called the Aymestry Limestone. South-west of Ludlow, at the village of Aymestrey, they reach 250 feet, but elsewhere beds of this type are much thinner and occur at more than one horizon, though always somewhere in the middle part of the succession. This type of limestone is scarcely developed at all on Wenlock Edge, being replaced by calcareous mudstones with a different fauna. West of Ludlow, near Leintwardine, another form of local facies-change is caused by channelling. The middle Ludlow beds are here cut into by several channels, approximately parallel and falling towards the west. Their infilling is marine, and differs only in detail from the surrounding shelf sediments, so that they are interpreted as submarine gullies, cut by currents flowing across the shelf into the deeper water of the basin to the west.

Towards the top of the Ludlow Series the calcareous siltstones become more sandy and pass upwards conformably into Downtonian beds, the lowest of the Old Red Sandstone. Conformity at this level is not common in Britain and is found principally in the Welsh Borders, where the late Silurian phase of the Caledonian movements had little or no effect, in contrast to the post-Caradocian folding and tilting. The base of the Devonian is taken at the Ludlow Bone Bed (p. 139). This is the best known of the remanié accumulations of vertebrate remains, but some earlier occurrences are known, even as low as Upper Llandovery.

INLIERS TO THE SOUTH AND EAST

The eastern shelf extended well beyond the major outcrops of the Welsh Borders and its deposits are seen again in several inliers that appear from under a varied, Upper Palaeozoic, cover. They also form part of the subsurface London Platform.

Eastwards from Wenlock Edge and Ludlow there rises the Old Red Sandstone mass of the Clee Hills and beyond that again the complex outcrops and structures of the various Midland coalfields (Fig. 43), separated from one another by New Red Sandstone. Southwards from Ludlow towards the Severn Estuary the Old Red Sandstone is interrupted by the Silurian inliers of Woolhope, Mayhill and Usk, while Tremadoc and Silurian beds are found on the western side of the Malvern range. Among the medley of outcrops and structure south of the Severn, inliers appear at Tortworth, 15 miles north of Bristol, and in the Mendips.

Within the Silurian area of the Welsh Borders older rocks appear along the line of the Church Stretton Fault at Pedwardine and Old Radnor where Longmyndian outcrops have already been noted. At Pedwardine these rocks are faulted against Tremadoc shales with *Dictyonema*, which are themselves overlain by sandstones and conglomerates that may be Caradoc. The inlier thus repeats on a very small scale some of the beds seen in the Longmynd region 10 miles to the north and can be considered as part of the same strip of country reappearing from beneath the Silurian overstep. At Old Radnor the Pre-Cambrian basement is overlain by sandstones (? Llandovery) and the Woolhope Limestone.

Elsewhere the shelf sediments most closely resemble those of the easterly Shropshire succession. Apart from Tremadoc shales, Ordovician strata are missing; the facies is often calcareous and the thicknesses moderate—the Silurian beds being normally 2,000–3,000 feet.

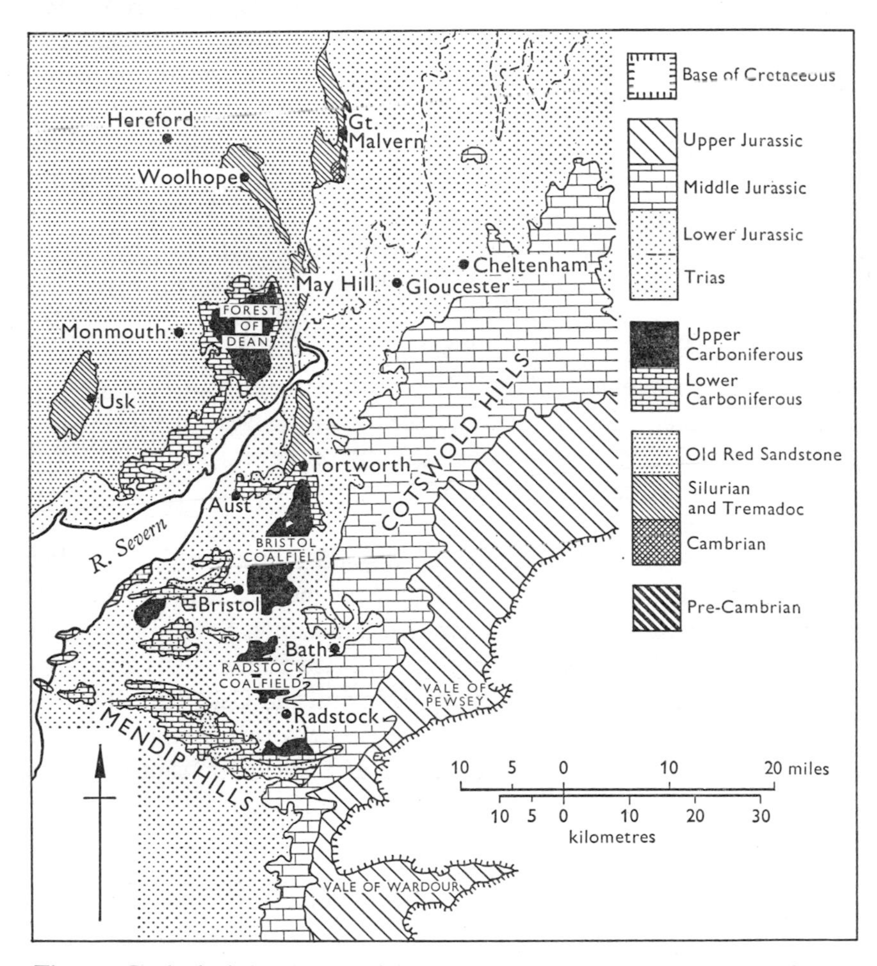

Fig. 23. Geological sketch map of the region around the Severn Estuary, Somerset and Gloucestershire. Triassic and Lower Jurassic (Liassic) outcrops differentiated in the north of the map but not in the south.

The base of the Upper Llandovery is only seen at Malvern, where it overlies all the earlier rock groups; in general the Llandovery beds are sandy and often termed the May Hill Sandstone. The Wenlock Limestone is thinner than in its type area, but some ballstone facies is known at Woolhope and Usk. The Woolhope Limestone is a widespread calcareous development in these south-eastern inliers, at the base of the Wenlock Series; in its type locality the rock is nodular and partly argillaceous, but at Old Radnor it becomes a fairly pure, grey crystalline limestone with abundant polyzoa and algae, especially *Solenopora.* Calcareous beds also appear among the Ludlow siltstones, especially in the middle, or about the Aymestry level, but massive limestones and *Conchidium* banks are missing. This facies is also characterized by very thin or condensed sequences, which at their extreme result in only 11 feet of Upper Ludlow beds between May Hill and Woolhope.

The shelf succession at Tortworth includes many similar characters: a thick sequence of Tremadoc shales, limestones (some rich in brachiopods and corals) in all three Silurian divisions and a conformable passage from Ludlow to Downtonian. But the exceptional feature is the two basalts intercalated with the Llandovery beds, the lower a sill nearly 200 feet thick and the upper a 100-foot lava flow. More igneous activity is known from the Mendips, where Silurian rocks occupy the core of the most easterly anticline and the 400 feet of pyroxene andesites are accompanied by lavas and tuffs. They are overlain and underlain by fossiliferous sediments, the former, and perhaps the latter also, being of Wenlock age. These igneous outcrops represent some of the very few signs of Silurian vulcanicity in the British Isles, the others being in south Pembrokeshire and south-west Ireland. It is noticeable, moreover, that all these places lie along a line that is not Caledonian in trend, but which extends into Belgium, and later becomes a belt of Hercynian folding.

It is possible that Lower Palaeozoic rocks appear even farther south, in the Hercynian thrust zone of south Cornwall (Fig. 32). Quartzite blocks with Ordovician trilobites have been found, but the fossils in slates and thin limestones, once thought to be Silurian, are not well preserved or diagnostic enough to make their ascription certain; they may well all be Devonian.

Apart from the small thickness of Tremadoc that succeeds the Cambrian shales at Nuneaton the only Midland coalfield in which Lower Palaeozoic rocks are exposed is that of South Staffordshire. Its westerly borders are only some twenty miles east of Wenlock Edge and a Shropshire-like facies is evident. The coalfield probably lies on a Silurian foundation, because beds of this age not only appear in several parts but are also found beneath Coal Measures underground. Although no Ordovician rocks have been found in the Midlands* there is indirect evidence

* Since this account was written, Llanvirn rocks have been found in a borehole at Great Paxton, Huntingdonshire.

of their deposition in some place unknown. Quite large boulders, up to 8 inches across, of an Ordovician fossiliferous quartzite are not uncommon in the New Red Sandstone pebble beds. Neither rock-type nor fossils resemble those of the Welsh Borders and are much more like certain quartzites now exposed in Normandy. The Midland boulders are too large however, to be easily reconciled with such a distant source and others are also found in the pebble beds of Devonshire. It is possible that somewhere now concealed, or eroded since Triassic times, there were Ordovician rocks of which we have no other knowledge.

The small Llandovery outcrops of south Staffordshire consist of fossiliferous sandstones and shales overlying Cambrian quartzites in the Lickey Hills, at the south end of the coalfield. Wenlock beds include representatives of the Woolhope Limestone, Wenlock Shales and Wenlock Limestone, and the calcareous rocks are particularly well developed. The Woolhope Limestone amounts to 30 feet and contains trilobites, tabulate corals and brachiopods. The Wenlock Limestone is particularly well exposed in three sharp, small anticlines at Dudley, north-west of Birmingham (Fig. 43), where it has been much quarried. It is thicker than on Wenlock Edge (nearly 200 feet), and is particularly rich in fossils, both in the bedded and reef-knoll facies, here called 'crog balls'. Trilobites, which are rare in Shropshire, are beautifully preserved here, including the typical *Calymene blumenbachi*. The Ludlow Series also resembles that of the type area though the uppermost beds are not well exposed. The calcareous siltstones and mudstones develop bands of nodular and impure limestone in part of the sequence to form the Sedgeley Limestone, which resembles the Aymestry Limestone facies in containing *Conchidium* valves.

It thus appears that in the western Midlands the limestones of the shelf facies are not only maintained, but accentuated compared with the main outcrop of Shropshire and Herefordshire, particularly in the thickness of the Wenlock Limestone. Nevertheless, there was probably a further limit to this facies somewhere in the Midlands, for it is not known from deep borings in eastern and southern England. Tremadoc shales have been encountered at Calvert, in north Buckinghamshire, and Wyboston in Bedfordshire; Silurian beds (some at least recognizable as Wenlock or Ludlow, and chiefly shales) are known from Calvert, Little Missenden (south Buckinghamshire), Ware (Hertfordshire) and from a number of boreholes near Harwich and Bury St Edmunds. There are also Silurian rocks found at depth in several places on the south of the Thames Estuary, in north Kent.

Structurally these records of these beds show that Silurian rocks are an important component of the London Platform, particularly on its northern and southern margins. Stratigraphically their dominantly shaly character is a slight indication of muddy deposits over much of south-eastern England, with the corollary that the Shropshire-Staffordshire calcareous facies was limited in distribution, and particularly characteristic of the shelf areas of the west Midlands and Welsh Borders.

NORTHERN ENGLAND AND THE ISLE OF MAN

The several Lower Palaeozoic outcrops of this region are at present separated by tracts of later rocks, by major faults or by the sea. Nevertheless, they have much in common, and during Ordovician and Silurian times were almost certainly a single area of sedimentation. No certain Cambrian rocks are known, nor is the relation of this region to those of Wales or southern Scotland at all clear. The sedimentary and volcanic history is decidedly different.

The central, and by far the largest and most important area, is the inlier that forms the main Lake District fells, together with the adjoining Howgill Fells on the south-east. Small inliers occur on the western and southern margins of the northern Pennines and the Manx Slates probably represent some form of south-westward extension from the northern Lake District.

THE LAKE DISTRICT

The major formations of the lakeland hills (Table 18) have a fairly simple areal distribution, but their detailed geological structures are complex, and their unravelling is hampered by a lack of recognizable horizons (often a lack of fossils altogether) through great thicknesses of rock. The oldest group, the Skiddaw Slates, occupies much of the northern section of the inlier; the succeeding Borrowdale Volcanic Series appears in a small area on the extreme north and a much larger one in the central fells. It is bordered on the south-east by a narrow strip of late Ordovician and early Silurian rocks, and the middle and upper Silurian compose the areas of the southern lakes and Howgill Fells.

The individual folds mostly have a Caledonian trend (west-north-west) but swing round to east–west or even south of east in the extreme south-east. The folding is tight and crumpled and accompanied by much faulting in the Skiddaw Slates, whereas the more competent Borrowdale rocks are less affected in detail but are thrown into broader folds, such as the syncline with its axis through Scafell. The massive late Silurian

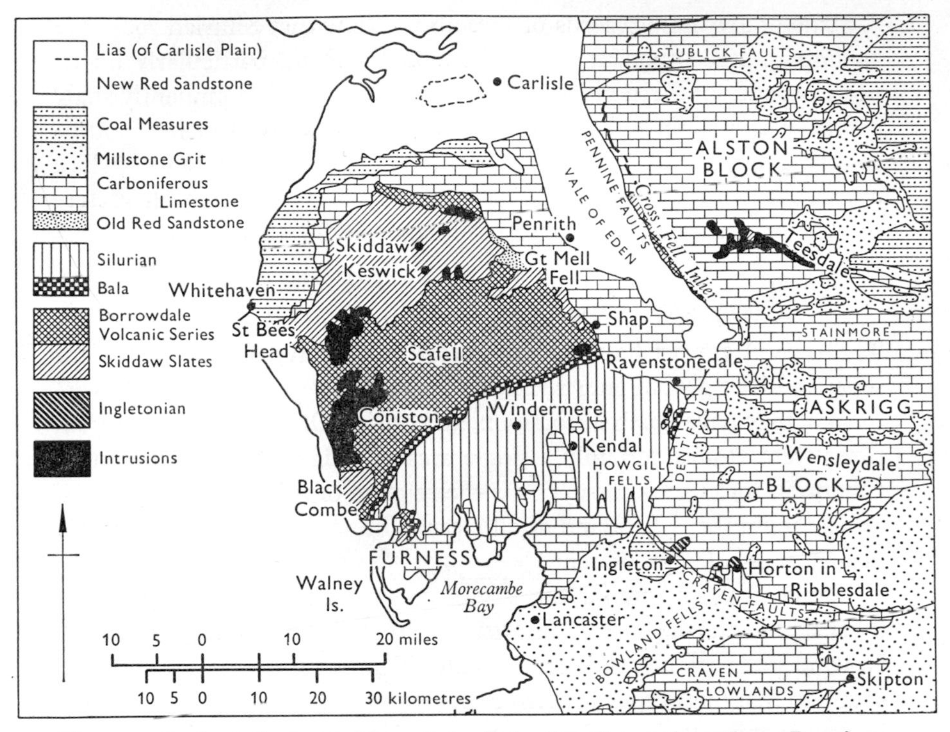

Fig. 24. Geological sketch map of the Lake District and northern Pennines. The intrusions of the Lake District are Caledonian; the Whin Sill (which includes all the igneous outcrops shown in the Pennines) is Hercynian.

rocks of the south exhibit similar structures, but the relatively incompetent strata between, especially Ashgill and Llandovery, have been the site of powerful thrusting.

The first phase, Skiddaw Slates and Borrowdale Series

The Skiddaw Slate group contains a variety of clastic rocks, only some of which are fine-grained enough to be called slates; others are greywackes and, more rarely, coarse pebbly sandstones. All fossils are uncommon; they include rare trilobites and brachiopods, but the graptolites are the most important, and the *extensus*, *hirundo* and *bifidus* zones have been identified. Earlier suggestions of Tremadoc faunas have not been substantiated though, since the base is not seen, such beds might exist at depth. The rarity of fossil bands and the complex local structures make a reliable estimate of thickness or succession difficult, and almost certainly there is a good deal of lateral facies-change. Over much of the central part a double grouping can be recognized:

Mosser and Kirkstile Slates	*extensus* and *hirundo* zones
Loweswater Flags	*extensus* zone

The coarser greywackes of the lower group show many turbidite characters. Towards the west, south of Ennerdale, an upper arenaceous group (the Latterbarrow Sandstone) appears above the Mosser and Kirkstile Slates and the Watch Hill Grit is intercalated among them. This is the coarsest of the Skiddaw Slates and contains igneous pebbles (granophyre and rhyolite) from some source unknown in direction or age.

In these varied rock types, and particularly in the abundance of greywackes, the Skiddaw Slates resemble the coarser massive beds of the Welsh trough, though there they were not much developed in Arenig or Llanvirn times. The junction with the volcanic rocks above has been much debated. In many places it is faulted or thrust, probably a reflexion of the differing competence of the two rock groups. Elsewhere there is considerable evidence for a conformable sequence, often with alternations of sediments, tuffs and lavas near the junction; in a few places there appears to be a local unconformity.

The intercalated sediments do not persist far up into the Borrowdale Volcanic Series and the remainder consists of a great thickness, unrival-

Table 18. *Ordovician and Silurian succession in the Lake District*

Series	Formations or groups	Approximate thickness (feet)
LUDLOW	Kirkby Moore Flags Bannisdale Slates Coniston Grits Upper and Middle Coldwell Beds	12,000
WENLOCK	Lower Coldwell Beds Brathay Flags	1,000
LLANDOVERY	Stockdale Shales	250
ASHGILL	Ashgill Shales Applethwaite Beds	c. 400
	~~ unconformity ~~~	
CARADOC	Stockdale Rhyolite Stile End Beds	0–450 c. 250
	~~ major unconformity ~~~	
LLANDEILO? LLANVIRN	Borrowdale Volcanic Series	12,000
LLANVIRN ARENIG	Skiddaw Slates	?7,000
	(base not seen)	

led even in Wales, of various types of lavas, tuffs and minor agglomerates. The age therefore can only be gauged by the underlying and overlying beds and the ascription to the Llanvirn, and possibly part of the Llandeilo series, follows from the conformable junction below and marked break above.

The Borrowdale lavas are mainly andesitic, though rhyolites are also abundant. Flow banding and flow brecciation are often seen and the thicknesses are spectacular. Flow groups of several hundred feet are not uncommon, and a sequence of steeply dipping andesites north of Skiddaw has been estimated at 10,000 feet. In many regions tuffs are nearly as important as the lavas; the coarser rocks approach agglomerates and the finest ones are well cleaved and form the typical green 'slates' of the Lake District. Some of the tuffs are well bedded and show many signs of deposition in water, including current-bedding and ripple marks; there must therefore have been local bodies of water, whether lakes or the fringes of the sea, but there is no lateral passage into non-volcanic sediments as in North Wales. Much of the vulcanicity is deduced to have been subaerial, and welded tuffs, or ignimbrites, occur in several of the tuff groups.

Any kind of uniform succession is not to be expected in such rocks, although definable sequences of lava and tuff groups, with their characteristic rock and mineral assemblages, can be recognized in one part of the outcrop or another. There is much lateral change and correlation is often uncertain, but a group of coarse tuffs in the upper part allows the broad structure of the Scafell syncline to be made out. On the whole the greatest thickness, 10,000–12,000 feet, appears to be in the south-west. Very little is known of the sources of the Borrowdale Volcanic Series. A few vents and plugs have been described, but not on a scale at all commensurate with the volcanic products. All the major intrusions of the Lake District are later in age, being late Caledonian.

The second phase, Bala to Ludlow

The extrusion of this great mass of lavas and tuffs was followed by an important earth movement. The Skiddaw Slates and Borrowdale Series were uplifted, folded and faulted, frequently along north-north-east lines, before the thin sequence of Bala beds was laid down. This is particularly evident from the way the latter transgress on to lower and lower levels in the volcanic rocks as they are traced along the outcrop from east to west. Finally, in the neighbourhood of the Duddon Estuary they lie for a short stretch on a small inlier of Skiddaw Slates. Even allowing for the irregular, impersistent distribution probable among the tuffs and lava flows, such a marked overstep, together with the truncation of the pre-Bala folds (for instance south-west of Coniston) indicates substantial uplift and erosion. It may be that this was related to a doming centred on the Skiddaw Slate outcrop of Black Combe, for the local strike in the Borrowdale rocks swings round this inlier.

The combination of subaerial vulcanism and pre-Bala uplift means that for a considerable part of Ordovician time (mid-Llanvirn to ? late Caradoc) the Lake District was part of a landmass among the geosynclinal complex of troughs and ridges. The Welsh history shows how a combination of lavas, tuffs and sediments in a shallow sea, the waxing and waning volcanic islands, is typical of Ordovician conditions. The volcanic landmass of the Lake District was larger and more long lasting than the Welsh islands, but it was a cognate feature.

The second phase of Lake District sedimentation begins with the submergence of the denuded volcanic mass. The earliest sediments include a miscellaneous array of limestones (often impure), shelly mudstones, ashy beds and a local rhyolite flow. These, which were earlier grouped together as the 'Coniston Limestone', contain a shelly fauna of fairly late Caradoc (Actonian) age. The Ashgill Shales above contain the zonal indices *Dicellograptus anceps* and *complanatus*. Since the outcrop is essentially linear we can form no clear idea of the areal invasion of the Lake District in Bala times. The small faulted Caradoc outcrop (the Drygill Shales) on the north side, near Carrock Fell, is slightly older so that the seas were extensive and invasion not simultaneous all over the region.

The Stockdale Shales succeed the Ashgill conformably, and all through the Llandovery period very slow deposition of muds prevailed. The lowest 50 feet, of black carbonaceous shales and occasional calcareous nodules, with graptolites preserved in pyrites, represent the whole of the Lower and Middle Llandovery beds. Quiet conditions, possibly stagnant, seem to have been the governing factor rather than any great depth of water.

There is a gradual upward transition to the grey, Upper Llandovery shales and an addition of slightly coarser detritus, the Wenlock and lowest Ludlow beds (Brathay Flags and Coldwell Beds) being grey mudstones and siltstones with sandy bands. The remainder of the Ludlow is a thick group of poorly fossiliferous strata in which greywackes are dominant, for instance the Coniston Grits; the Bannisdale Slates are finely banded siltstones and mudstones in which *Monograptus leintwardinensis* (the highest graptolite zone in Britain) is one of the rare fossils; the Kirkby Moor Flags, fine-grained flaggy sandstones and greywackes, contain *Dayia* and *Protochonetes ludloviensis*, which are characteristic of Upper Ludlow beds in the Welsh Borders. Sedimentation therefore continued until late in Silurian times and late Caledonian folding in this basin was presumably largely or entirely Devonian. Its completion is uncertain, however, as the age of the local Old Red Sandstone (? Lower) near Ullswater is not precisely determined (p. 144).

Isle of Man

The impressive central mountains of this island, which rise to nearly 2,000 feet, have long presented something of a geological enigma. They are composed of the Manx Slate Series, together with some small Caledonian granitic intrusions and dykes of several ages. A thickness of 25,000 feet has been estimated for the slates, but they are much folded. There is some lithological resemblance to the Skiddaw Slates in the greywackes, siltstones, slates and certain thick slumped beds. The only fossils known, apart from tracks and burrows, are some dendroid graptolites found in slates that on structural grounds are thought to be near the top of the succession, and microfossils in a group of flags lower down. The palaeontological evidence is not wholly conclusive, but on balance an early Arenig or late Tremadoc age is probable; it is unlikely that the group is as old as Cambrian.

Probably the Manx Slates originally continued north-eastwards and passed under the Skiddaw Slates of the northern Lake District. In the opposite direction there is less similarity and the thick group of slates and greywackes of the Cumberland–Manx trough does not fit easily into the pattern of Ordovician outcrops on the Irish coast north of Dublin. South of that city, however, there is some lithological resemblance to the Clara Group (or upper part of the Bray Series) in northern Wicklow (p. 65). It is possible that there was originally a Caledonian trough, north-west of the 'Irish Sea landmass', from north Leinster to the Isle of Man and Cumberland. The upland which it later formed falls in stratigraphical level north-eastwards from Cambrian(?) to Tremadoc and Arenig. At a still later stage much of it foundered beneath the Irish Sea. There is some geophysical evidence that at present the wedge of the Manx Slates is an uplifted mass or horst flanked by basins of lighter rocks—perhaps New Red Sandstone—after the model of the Vale of Eden, between the Lake District and the Pennines, or the basins of the Southern Uplands.

The Pennine Inliers

The surface rocks of the northern Pennines (Fig. 24) are largely Carboniferous, but beneath this cover there is a varied basement of Lower Palaeozoic and Ingletonian rocks and, part of the Alston Block, the unexposed Weardale Granite. Exposures are few and nearly all are restricted to the western and southern fault-scarp edges.

The Cross Fell Inlier lies at the foot of the western scarp, on the borders of Cumberland and Westmorland, and is itself much faulted. Its rocks repeat in miniature those of the central Lake District from Skiddaw Slates to Coniston Grits. In addition the Caradoc Series is better developed and its diminutive, faulted exposures somewhat resemble the contemporaneous beds of Shropshire and North Wales, in their calcareous mudstones with an abundant shelly fauna; they also mark a slightly earlier marine invasion than do the basal Bala beds of the Lake District. The Ashgill of Cross Fell also has its peculiar facies, the addition of the Keisley Limestone, noted for its trilobite fauna with affinities to those of Scotland and Ireland. Indeed, there is a 'shelf-like' aspect to the Ashgill facies in the northern Pennines, though certainly not to the beds below. How far the Ordovician and Silurian rocks extend eastwards is uncertain, but there is a small outcrop of Skiddaw Slates in Teesdale, and they were probably encountered in a deep borehole in the west of Co. Durham; presumably it is they that were the surrounding country rock of the Caledonian Weardale Granite. It is doubtful whether the Borrowdale rocks extend east of Cross Fell and thus uncertain whether the volcanic episode did not reach so far or whether its products were later removed, as they have been over the present Skiddaw outcrops of the Lake District.

The small exposures in the sub-Carboniferous basement of the Askrigg Block in west Yorkshire appear in the floors of valleys running southward off the Pennine hills; they comprise the inliers of Ingleton, Austwick and Horton in Ribblesdale. Neither Skiddaw Slates nor Borrowdale Series is present; the oldest Ordovician rocks have been compared to the Coniston Limestone Series but are probably all Ashgill. They are mudstones with a shelly fauna, and only slightly exceptional in being interrupted at one level by a minor volcanic episode and at another by a minor unconformity. Above there succeed Llandovery mudstones, and a sequence of greywackes and mudstones that does not extend above the lower Ludlow, equivalent to Brathay and Coldwell Beds of the Lake District and locally rather coarser in grade.

CHRIST'S COLLEGE
LIBRARY

In the Yorkshire inliers shales are less abundant, and in so far as the faulted exposure allows estimates to be made, the rocks are rather thinner than in the Lake District; it is possible that they were farther from the centre of the trough. To that extent a 'marginal facies' is detectable from Cross Fell to Shropshire, but it is not possible to deduce an easterly margin of the geosyncline in northern England in the same way that one can be drawn, in a general fashion, in the Welsh Borderland.

SOUTHERN SCOTLAND, NORTHERN AND WESTERN IRELAND

The narrowest part of the Irish Sea is that between Stranraer and Larne where the Lower Palaeozoic massifs on either side approach one another. They have common features in their strong Caledonian folds and faults, the tendency for Ordovician rocks to be exposed on the north-west and Silurian on the south-east, and the fact that both form some kind of upland (though slighter on the Irish side) which is overlapped on certain margins by Upper Palaeozoic rocks. In the Midland Valley of Scotland a few inliers expose Silurian beds beneath Old Red Sandstone and in the Irish continuation there are minor inliers in an analogous position. The Highland Border Series of Scotland includes slices of Ordovician rocks which have some important resemblances to those of Ireland and the Scottish Southern Uplands. In west Connaught (south Mayo and north Galway) a major Lower Palaeozoic trough was developed which is also part of the Midland Valley in so far as it lies south of the Highland Border Fault.

THE SOUTHERN UPLANDS OF SCOTLAND

This is slightly the largest, and the most studied, of the regions just enumerated, but one which still retains several stratigraphical enigmas. Many of them arise from the complex structures, which need good outcrops for their interpretation. Inland the hills of the Border Country are often rounded and grass covered and their streams cut through sharply folded and steeply dipping sequences between which correlation is not easy. The restricted cliff sections on the north-east coast expose vigorously folded sequences that are probably Silurian, but in which fossils are very rare or absent. Only in the south-west has a fairly satisfactory succession been compiled.

For many years, since the late nineteenth century, it was thought that the essential key to the Southern Upland structure was isoclinal folding: that the steep dips found in the field generally represented the tightly

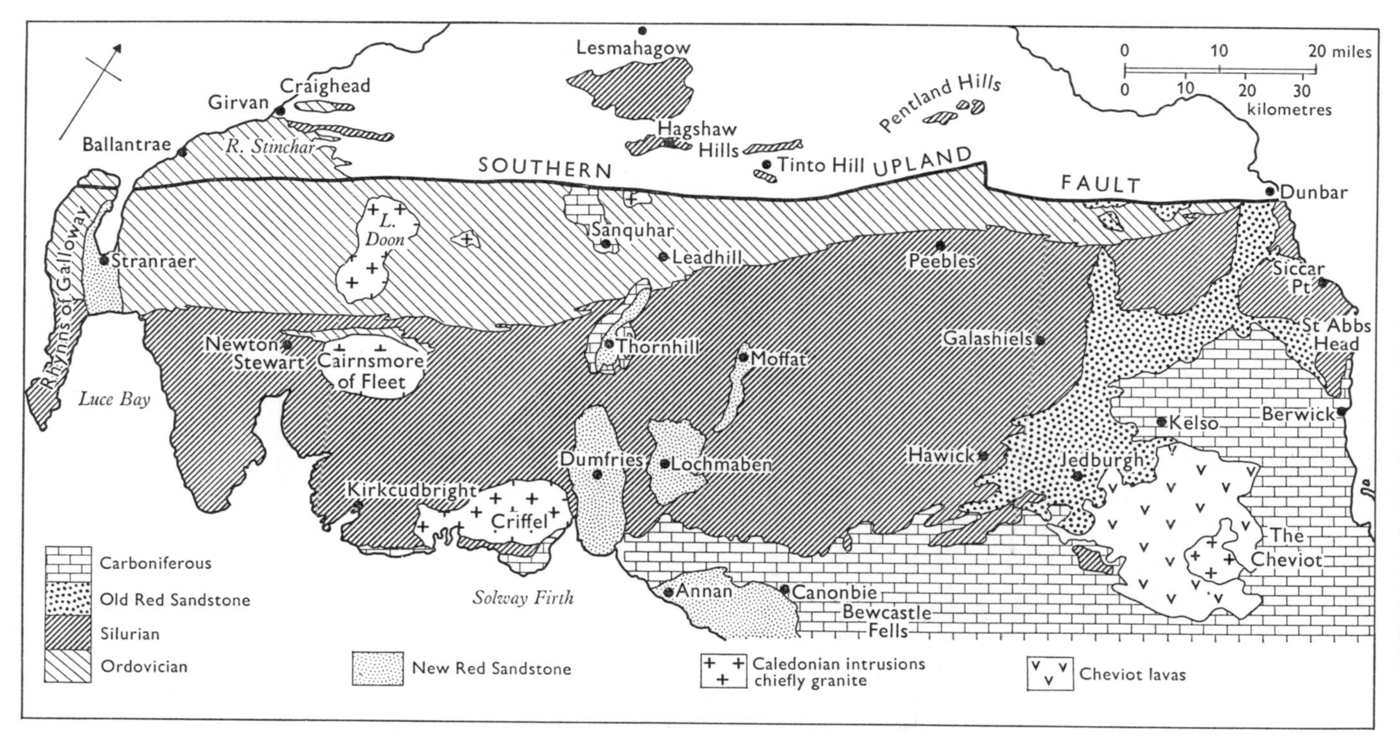

Fig. 25. Geological sketch map of the Southern Uplands of Scotland and neighbouring regions. The only outcrops shown north-west of the Southern Upland Fault are those of the Lower Palaeozoic rocks, including the Silurian inliers of the Midland Valley of Scotland.

packed limbs of relatively small folds, in which certain marker beds, usually graptolite shales, were repeated many times. The known dominance of Ordovician beds to the north-west and Silurian to the south-east was related to (respectively) an anticlinorium and a synclinorium.

This last concept, however, has been shown increasingly not to conform with many of the structures observed in the field. Isoclinal, or tight-packed, folds do occur, especially in the incompetent shales and mudstones, and the repetition of certain shale bands is substantiated. But much of the succession comprises coarser beds, particularly greywackes; though these are usually poor in fossils, their texture and structures allow their orientation, or 'way up', to be determined, and so they become a further tool in the identification of anticlines and synclines.

In some areas these later studies on orientation have radically altered the stratigraphical interpretations. In particular it appears in sequences dominated by massive greywackes the most important structures are some form of monocline—that is, a belt of steeply dipping rocks facing north-westwards is followed on the south-east by another belt of relatively flat-lying beds, and the same combination is then repeated. This arrangement should, however, result in the younger beds generally appearing to the north-west, whereas the opposite in fact occurs. The answer is probably that these sequences are also broken by large strike faults throwing down to the south-east which more than counteract the north-westerly down-dropping monoclines. Such a structural combination (see Fig. 13, p. 59) has been described from the south-west coasts (Galloway and Kirkcudbrightshire) and may be widespread. It is chiefly from this region and south-west Ayrshire that the following description is compiled.

The southern boundary of the Southern Uplands is a long, irregular, unconformable junction with Upper Palaeozoic rocks from Berwickshire to the Solway Firth. On the northern side the Southern Upland Fault for most of its course throws down the Old Red Sandstone and Carboniferous rocks of the Midland Valley. At the south-west end, however, the Lower Palaeozoic area of Girvan lies north of the fault, as do the Ordovician rocks at the mouth of Loch Ryan in Galloway. Here, moreover, there is a marked downthrow to the south.

Arenig: the Ballantrae Volcanic Series

This series is chiefly known in relatively restricted outcrops, on and near the coast for a dozen miles near Ballantrae (Fig. 26). It consists of basic lavas, with breccias and tuffs, agglomerates, radiolarian cherts and thin seams of black shales with graptolites and horny brachiopods. The lavas predominate towards the base, which is not seen, and are overlain by the breccias, tuffs and shales; the graptolites of the last, species of *Dichograptus*, *Didymograptus* and *Tetragraptus*, belong to the *extensus* zone. The lower lavas are spilitic with pillow structures, so that much of the extrusion was submarine, on to a sea floor of fine black muds. Numerous intrusions cut the volcanic series, of acid to basic and especially ultrabasic rocks.

In the cores of anticlines farther north-east, in the northern belt of the Southern Uplands, more volcanic rocks have been found associated with graptolite shales. The faunas in the Leadhills area, some 60 miles along the strike, suggest that there the lavas are Arenig and probably an extension of the Ballantrae Series. Where there is no palaeontological proof of age it may be that the lavas are intercalated with the later, Bala, sediments.

No Llanvirn rocks or fossils have been found in the Southern Uplands and around Ballantrae and Girvan there is no doubt about the large, pre-Caradoc unconformity. This has usually been considered as a major break in the region as a whole, but below basal Caradoc beds in western Galloway there are mudstones with conodonts (phosphatic 'teeth' of uncertain affinity) that have been taken to be Llandeilo in age. Since the base of the Caradoc is normally unexposed inland the presence or absence of Llandeilo there is not determinable.

Caradoc and Ashgill: Glenkiln and Hartfell Shales

The use of graptolites as zonal fossils dates from the pioneer work of Lapworth in the last quarter of the nineteenth century and was based on the faunas in the shale successions exposed near Moffat in the central belt of the uplands. Three main faunas were established there—Glenkiln, Hartfell and Birkhill; slightly amended they fit into the Anglo-Welsh series thus:

England and Wales	South Scotland (Moffat)
Upper Llandovery (approximately)	Gala Group
Middle and Lower Llandovery	Birkhill Shales
Ashgill	Upper Hartfell Shales
Caradoc, upper part	Lower Hartfell Shales
Caradoc, lower part	Glenkiln Shales

Above the Birkhill Shales most of the Upper Llandovery corresponds to a group called in Scotland Gala, or Gala-Tarannon, in which shales are no longer dominant.

In the Moffat country the shale groups are visible in good type sections; they are conspicuous for being extremely thin, the Glenkiln being about 20 feet and Hartfell 100 feet, of black shales, dark-grey mudstones (locally tuffaceous) and cherts. The shales are mostly fossiliferous and some of the mudstones barren. Such rocks represent the extremely quiet slow sedimentation already noted in the lowest Stockdale Shales of the Lake District, and there was little life besides the floating graptolites. In the south-west of Galloway somewhat similar rocks are found, but they are rather thicker and coarser; much of the Glenkiln and Lower Hartfell shales are still black, with cherts, but siltstones and greywackes become increasingly important in the Upper Hartfell.

On the older interpretation this central or axial belt was thought to be beyond the reach of the coarser arenaceous sediments found farther north-west; to that extent the central part of the trough represented the deeper water. If, however, the greywackes of the north were brought down the slopes and spread out on the sea floor by turbidity currents, which is broadly the more recent view, they themselves would come to lie in the lowest portion of the trough. In that case the Moffat shales perhaps rather represent a central 'rise' out of the reach of the turbidity currents, at least till later Hartfell times.

The facies-change across the strike, north-west from the central belt is extremely impressive. In the Girvan region the rocks are much thicker and coarser; the graptolite-bearing shales only appear rarely and, though useful when they do, much of the correlation has been based on the shelly faunas. The series here are named Barr and Ardmillan, and some 16,000 feet has been estimated as a maximum thickness for the lower portion, up to the top of the Lower Ardmillan.

Ashgill Series	Upper Ardmillan Group
Caradoc Series	Lower Ardmillan Group
	Barr Group

This thick mass of sediments was piled up against the steep, eroded edge of the Ballantrae Series and gradually overwhelmed it. So steep are some of the junctions that they are considered to be fault margins, the faults continuing to move spasmodically during Caradoc times. From the scarp edges there slid, periodically, great wedges of conglomerate; these tail away south-eastward into mudstones and greywackes and their pebbles reflect the erosion of the Ballantrae igneous rocks. The thickest and most extensive is the Benan Conglomerate (2,000 feet) but there are others above and below. At three periods limestones were formed on local shelves near the shore—the Auchensoul and Stinchar limestones low in the Barr Series (*gracilis* zone), and the Craighead Limestone, now only seen in an inlier north of Girvan, in the Lower Ardmillan Group. Their fossils include trilobites, brachiopods, molluscs, algae and occasional corals; they show affinities with American faunas, and to some extent Scandinavian and Irish, rather than with those of Wales.

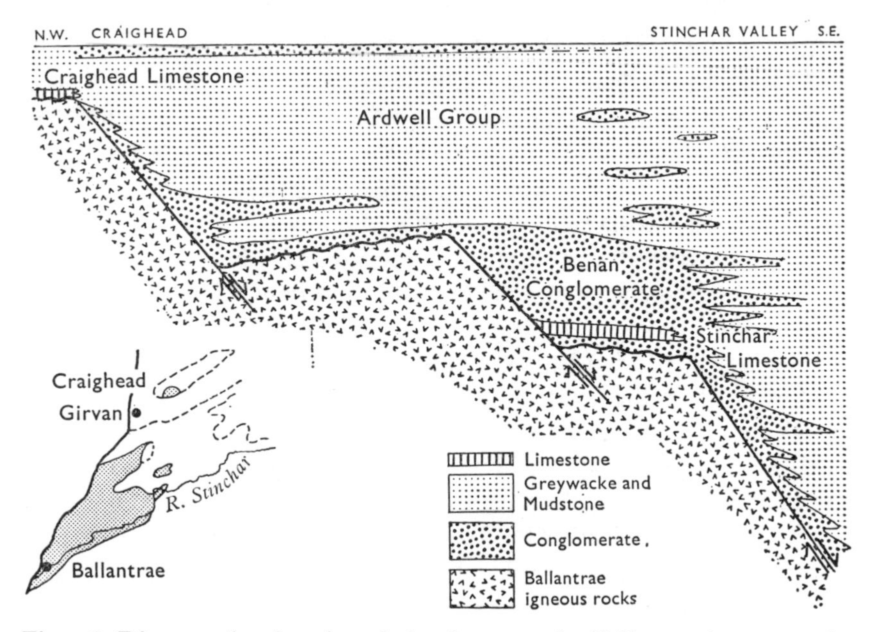

Fig. 26. Diagram showing the relation between the Ballantrae igneous rocks and the Caradoc sediments that are banked up against them. The coarse detritus, which includes the Benan Conglomerate, diminishes away from the Ballantrae mass and passes laterally into greywackes and mudstones. (After Williams 1962, p. 67.)

The massive conglomeratic facies appears to be rather closely restricted to the periphery of the submerged Ballantrae massif, but the Barr and Ardmillan greywackes spread farther into the trough, probably from more than one direction, and brought detritus from farther afield. In particular they include, in addition to Ballantrae-like components, mineral assemblages which can be matched in the Highland schists; in western Galloway there are further indications of a north-westerly derivation in a mineral suite comparable to that of the Dalradian rocks.

The Silurian succession

The stratigraphical groups of the central and southern belts are as follows:

?Hawick Beds	(age uncertain)
Riccarton Beds	Wenlock
Raeberry Castle Beds	?Wenlock
Gala Group	Upper Llandovery
Birkhill Shales	Middle and Lower Llandovery

In the Moffat region the quiet mud sedimentation of the Glenkiln and Hartfell periods continued with deposition of the Birkhill Shales; they are well exposed in the classic section of Dobbs Linn as a hundred feet of grey, graptolitic mudstones. Above these there is an abrupt change to the thick Gala Group—4,000 feet of massive greywackes, even conglomerates, with flaggy sandstones and shales. The quiet, central mud belt (or central 'rise' in the trough) had finally been invaded by the greywacke sediments that already dominated deposition to the north-west.

The higher Silurian beds are only seen in the southern belt and are best exposed on the coast of Kirkcudbrightshire. Of the three groups, which together may amount to 16,000 feet or more, only the Riccarton Beds bear reliable evidence of age. Among thin greywackes, conglomerates and sandstones, shale bands yield *Monograptus riccartonensis* and *Cyrtograptus murchisoni*. The Raeberry Castle Beds are thought to be older on structural evidence. They are finer in grade and consist of shales, thin greywackes and nodular limestones; gastropods, cephalopods and rare tabulate corals occur in the last, but are long-ranging forms. The position of the Hawick Beds is doubtful, as the boundaries appear to be faulted. They are a very thick group of greywackes and purple and red siltstones and mudstones, with ripple marks and current-bedding. If they really are latest in the sequence, they may indicate a gradual shallowing of the Southern Uplands trough. Even so there is no positive evidence of Ludlow sediments in the region.

Silurian history in the Girvan area is even less complete, for the beds are restricted to certain inliers, bordered by Old Red Sandstone or Carboniferous rocks; they have received many local names and nearly all belong to some part of the Llandovery Series. The facies-contrast between the northern and central belts persisted into the lower Silurian times and the Birkhill Shales are represented in the Girvan outcrops by a great variety of shelly facies, from sandstones with conglomerate and breccia bands to siltstones, calcareous mudstones, and thin bands of grey and black graptolitic shales. The Gala Group, however, is less coarse and possibly somewhat thinner than at Moffat, being fine, flaggy sandstones and mudstones with bands of graptolite shales. Wenlock beds are only found in a small faulted outcrop of grey shales with sandstones and conglomerate bands and a fauna that suggests correlation with the Riccarton Beds of the southern belt.

LONGFORD–DOWN AND BALBRIGGAN MASSIFS

In north-eastern Ireland, as in the Southern Uplands, the largest Ordovician outcrops lie in the northern belt bordering the 'Midland Valley', but here they are very badly exposed apart from the coastal section in Co. Down. Rather more is known of the inliers in the Silurian belt to the south.

As a whole the Ordovician rocks appear to be rather finer than in the analogous Scottish outcrops. There are some greywackes and sandstones but the coarser beds of Galloway and Ayrshire have no Irish counterparts; nor is there any equivalent of the Ballantrae Series unless it be represented by some little-known pillow lavas near Longford, in the west; most of the fossils recorded are Caradoc or Ashgill, all the graptolite zones being known at one place or another. More variation and coarser rocks appear in the Silurian beds, which include greywackes, conglomerates and sandstones. As in the Southern Uplands most of the recognizable horizons are Llandovery but some Wenlock beds are known.

Along the coast north of Dublin (Fig. 27) Lower Palaeozoic inliers protrude through a cover of Carboniferous Limestone. The largest is at Balbriggan where Llanvirn graptolitic mudstones are associated with volcanic rocks. Beds of this age are rare in Ireland and not recognized in southern Scotland. Above lie Caradoc slates and greywackes, 1,500 feet thick, succeeded by 5,000 feet of augite andesites, lavas and tuffs. The small inlier of Portrane and the island of Lambay, between Balbriggan and Dublin, expose rocks of the same volcanic group, interbedded with Ashgill slates and limestones. The brachiopods, trilobites and tabulate corals have affinities with the faunas of both Cross Fell and Girvan, and the sediments on Lambay appear to have been laid down on the flanks of a volcanic island around which also accumulated the tuffs and agglomerates.

In the north and east of Ireland there are thus some links in Lower Palaeozoic times with southern Scotland and northern England, but also

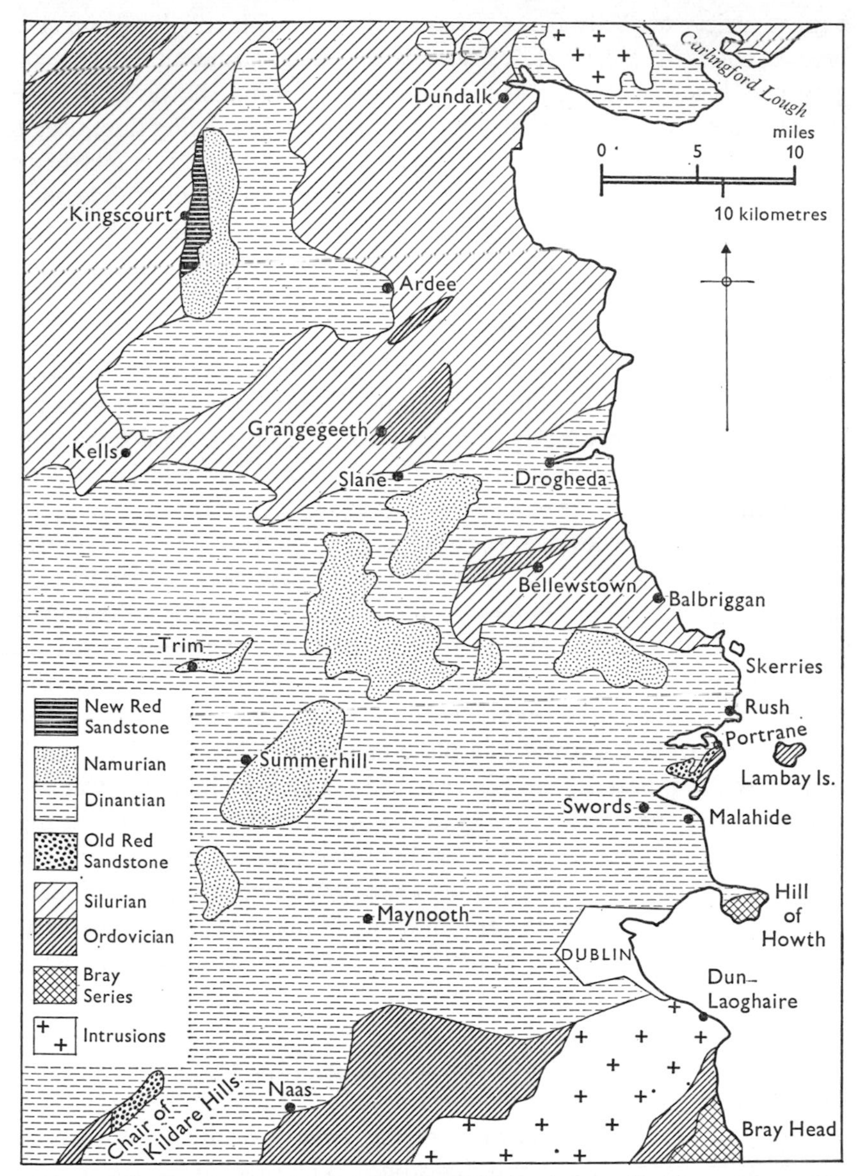

Fig. 27. Geological sketch map of eastern Ireland, from Dublin northwards to the southern part of the Longford–Down Massif and the Kingscourt Outlier.

some dissimilar characters, such as the much more pronounced Ashgill vulcanicity. All the outcrops lie north of the Bray-Manx-Skiddaw belt, in which, it has been suggested, there was a thick trough sequence in early Ordovician and perhaps Cambrian times.

THE MIDLAND VALLEY OF SCOTLAND

At present this is a down-faulted trough, between the Scottish Highlands on one side and the Southern Uplands on the other; most of it is occupied by Devonian and Carboniferous rocks but enough is seen of earlier beds to suggest that sedimentation comparable to that of the Southern Uplands extended into the Midland Valley, perhaps during the Ordovician, and more certainly during the Silurian period.

There are three principal groups of pre-Devonian outcrops: in Lanarkshire and the borders of Ayrshire in the south, among the Pentland Hills near Edinburgh (Fig. 25), and the Highland Border Series. The Downtonian rocks of Stonehaven in the extreme north-east are lowest Devonian and mentioned in the next chapter.

The largest of the southern inliers is at Lesmahagow, but it is supplemented by those of the Hagshaw Hills and near Tinto Hill. Common to all is the apparently conformable junction between Silurian and Old Red Sandstone, a succession only seen elsewhere in Britain in the Welsh Borders. In the Midland Valley, however, fossils are much rarer than in south Shropshire. The beds below the Old Red Sandstone are dominantly sandy and may be as thick as 5,000 feet; they contain some shelly faunas but these are not easily correlated with those of the Welsh Borders and graptolites are rare. The 'fish-beds' contain the scales and bones of early vertebrates (*Thelodus*, *Birkenia* and *Lasanius*), not the armoured types typical of the Downtonian of Herefordshire but more resembling the fragments from Ludlovian bone beds. In these circumstances it is not surprising that in the past various ages have been attributed to these inliers.

More recently the marine faunas near the base of the succession in the Hagshaw Hills have indicated that there is little older than Upper Llandovery; above this there may be a complete Silurian sequence in beds showing a gradual transition from greywackes to sandstones, the latter resembling the finer types of Old Red Sandstone; here and elsewhere a fish bed occurs near the top. A somewhat similar succession is known from Lesmahagow; 1,300 feet above the base in a succession of greywackes, shales and mudstones there is a shelly fauna, together with vertebrate and eurypterid remains. Again the age is probably Llandovery and a fish bed occurs higher up.

The three small inliers in the Pentland Hills have some resemblance to the larger ones just described, but the beds were sharply folded and eroded before the Old Red Sandstone was laid down. Their stratigraphical position is also better defined in that a few graptolites have been found, though the fauna is chiefly shelly, with starfish and eurypterids. As in Lanarkshire most of the faunas are Upper Llandovery but some Wenlock beds may be present near the top. The thickness is uncertain, but some thousands of feet are present, including mudstones, sandstones and conglomerates in the lower part, and grey-green sandstones, mudstones and conglomerates towards the top.

It is concluded that seas of the Southern Upland trough extended into the south-eastern part of the Midland Valley during Llandovery and Wenlock times at least. In Lanarkshire sedimentation continued without a major break, from the marine greywackes to the shallower sandy beds and eventually to the coastal flats and fluviatile conditions typical of the Old Red Sandstone. Such a transition is exceptional, however, and most of southern Scotland was affected by a major phase of folding and erosion. One result is that although Ludlow rocks may be present, Ludlow faunas have not been identified in Scotland. In the Southern Uplands they may have been deposited and later eroded, or they may possibly be represented by part of the unfossiliferous greywackes, purple and red beds of the southern belt. In Lanarkshire they are presumably present but lack diagnostic fossils, and if the facies change was earlier here than in the Welsh Borders, as is quite possible, they may have been partly non-marine.

At many places along the line of the Highland Border Fault there are small outcrops of Lower Palaeozoic rocks (the Highland Border Series) from Stonehaven on the north-east, through Perthshire and the Trossachs to Arran on the south-west. Their structural relations vary; commonly they are faulted against Dalradian rocks on the highland side and Old Red Sandstone on the other, but the latter junction at Stonehaven is an unconformity and in Arran the Lower Palaeozoic rocks are probably unconformable on the Dalradian schists. The commonest components are serpentines—altered basic igneous rocks which now form a vertical or steeply inclined screen between the parallel fractures of the fault belt. In several areas, such as Stonehaven, Aberfoyle in Perthshire, Loch Lomond and Arran, sedimentary and volcanic rocks are also present. They resemble the spilitic lavas, cherts and black shales of the Ballantrae Volcanic Series. The shales have yielded poor graptolites and horny brachiopods but neither group is well enough preserved to be diagnostic, and the usual assumption of an Arenig age is based on the strong lithological similarity. In a few places an upper more sandy group of sediments is present the fossils of which have been taken as Caradoc.

There is an obvious resemblance to the Girvan area, and although we cannot know how far to the north-west the Ballantrae-like assemblage may have continued, a broadly similar history from the Solway at least as far as the Scottish Highlands is a reasonable supposition.

THE MIDLAND VALLEY IN IRELAND

Inliers in the north and centre

The much greater spread of Lower Carboniferous rocks in the Irish continuation of the Midland Valley reduces the outcrops of the earlier formations and in many places obscures the faulted margins themselves. The most easterly inlier is at Pomeroy, in County Tyrone, where both igneous and sedimentary rocks appear. The Tyrone Igneous Series is overthrust on the north-west by Dalradian schists and consists of basaltic and spilitic lavas overlain by black shales, cherts and sandstones. There is some similarity to the Ballantrae Series but the age may be materially different, since the very few, poorly preserved graptolites from the tuffs have been tentatively identified as Caradoc.

The transgressive, sedimentary series above also contains a doubtful Caradoc fauna near the base, and above that ranges from Ashgill to Upper Llandovery. Succeeding a basal conglomerate there are black shales, calcareous mudstones and sandstones. The abundant fauna is largely shelly with some resemblance to that of Girvan, but the beds are much thinner as well as finer, amounting only to some 600 feet.

Farther west there is very little information to fill in the picture between Tyrone with its partial Scottish resemblances and the markedly different succession of south Mayo. In the two small inliers of Lisbellaw (Fermanagh) and Charlestown (eastern Mayo) the main group consists of Llandovery sediments, often coarse. There is, however, some hint of westerly thickening, for the Silurian beds at Charlestown reach 7,000 feet; they are underlain by felsites, tuffs and agglomerates associated with Arenig graptolitic shales.

The Trough of South Mayo and North Galway

In the south-west of C. Mayo the Highland Border Fault reappears along the south side of Clew Bay, as a fault belt associated with serpentine. To the north lies the Dalradian country of north Mayo and to the south the geosynclinal trough of Ordovician and Silurian rocks, beyond which emerge the Dalradian mountains of Connemara. The Lower Palaeozoic outcrops fall structurally into three east–west belts (Fig. 28). In the northernmost Silurian beds form a deep syncline, slightly over-

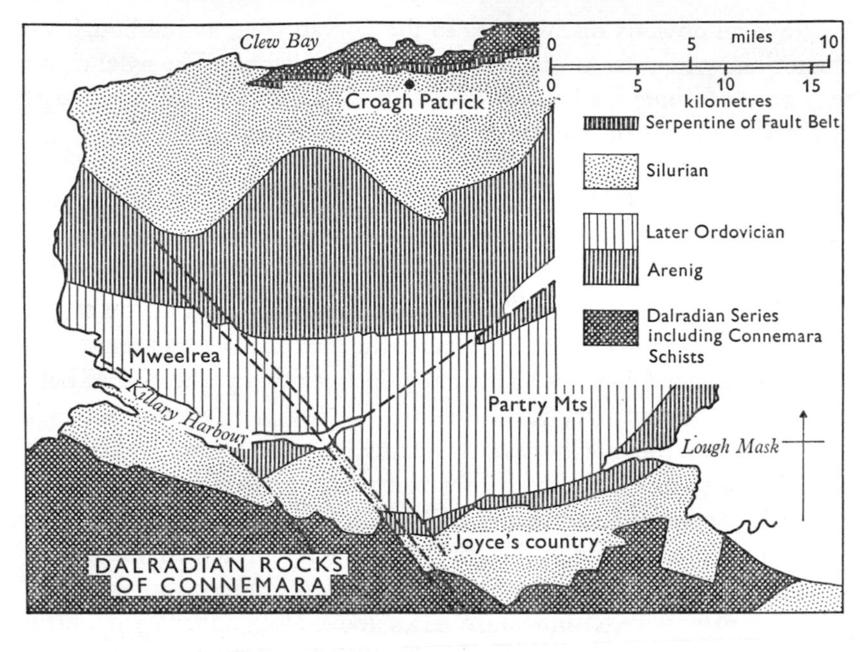

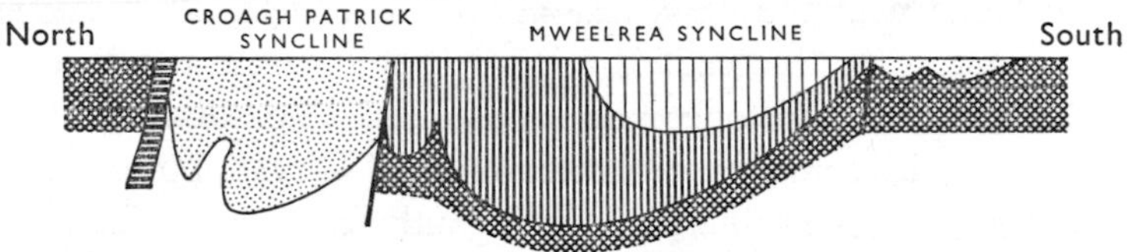

Fig. 28. Geological sketch map of the Lower Palaeozoic rocks of the South Mayo Trough and their Dalradian borderlands. The section below is on the same scale as the map and to true scale. (Simplified from Dewey, 1963, p. 315.)

turned on its northern flank. The junction with the Ordovician outcrop to the south is partly faulted and partly unconformable; this belt is also a complex syncline. Still farther south there are a number of faulted, and again partly synclinal, Silurian outcrops, which overlap on to the Connemara schists. The whole tract is not more than 20 miles across and the dips are commonly steep. In spite of its isolation, relatively small area and the tantalizing way the outcrops disappear westwards under the Atlantic, this region, which has been called the 'South Mayo Trough', is an important part of the geosynclinal complex. The succession is notable for its great thickness, few fossils, and a major unconformity between the Ordovician and Silurian portions. The succession in Table 19 is composite; in particular the Silurian beds differ somewhat on the north and south.

The lowest Arenig beds consist of black shales, spilitic lavas and acid tuffs; their base is not seen owing to the Silurian cover on the southern margin, but almost certainly they abutted on the Connemara uplands, which formed the shore of the growing trough. Above lie slates, greywackes with typical turbidite characters, bands of tuffs and boulder beds, the whole mounting up to an exceptional thickness. Fossils are very rare but faunas of the *extensus* and *hirundo* zones have been found. Some of the tuffs are graded and probably represent ash showers that fell into the sea. The individual bands differ enough in composition and texture to be recognizable in the field, and so form marker horizons—a most valuable feature in rocks so poor in fossils. In the easterly outcrops, along the slopes of Lough Mask, the Arenig rocks include a few limestones and limestone breccias, whose trilobites and brachiopods have American and Scandinavian affinities.

Table 19. *Ordovician and Silurian succession of south Mayo*

Series	Southern outcrops		Northern outcrops
WENLOCK	Salrock Group Upper Owenduff Group	6,600	Croagh Patrick Series 5,600
	~~ local unconformity ~~~		
UPPER LLANDOVERY	Lower Owenduff Group	1,900	
	~~ major unconformity ~~~		
	Central outcrops		
CARADOC?	Partry Series	15,000	
LLANVIRN?	Glenummera Series	4,000	
ARENIG	Murrisk Series	15,500	
	Owenmore Series	2,000	
	(presumably unconformable on Dalradian schists)		

The table is composite and the figures represent maximum thicknesses in feet.

The remaining Ordovician beds have yielded only very sparse and poor faunas, though enough to suggest that they do not rise above Caradoc. In facies they are like the main Arenig succession—slates and greywackes with tuffs and conglomeratic bands; current-bedded sandstones appear near the top. From the lateral changes in thickness and the way in which the sediments between identifiable tuff bands expand to the north-west, it appears that the greatest thickness and subsidence of the Ordovician trough was in this direction. The Dalradian uplands on either side supplied much recognizable detritus, though in fluctuating proportions, at one period the southern source being the more influential and at another the northern. After Caradoc times there was a major phase of uplift, folding and erosion; thousands of feet of strata were removed, especially on the northern side, before the Upper Llandovery submergence. The area of sedimentation, however, continued to be much the same, though the overlap of the Lower and Upper Owenduff groups on the south side suggests a slight extension over the Connemara shore.

The Lower Owenduff Group is largely confined to the south-eastern outcrops, where, overlying a basal conglomerate, the beds decrease in grade through sandstones to fossiliferous limestones and shales. The group is cut out progressively westward by a sub-Wenlock unconformity, and in the south-west (south of Killary Harbour) the Upper Owenduff Group rests on Connemara Schists. The Wenlock groups on the south side revert again to the greywacke facies, together with slates and shales. The fossils include very rare graptolites, and two horizons with shelly faunas.

In the northern belt the pre-Wenlock transgression was also effective, and no more than 20 feet of Llandovery beds are known, the base of the Croagh Patrick Series lying either on these or on Arenig. On this side the Wenlock beds lack the greywacke-turbidite facies and show some signs of belonging to a littoral facies belt, in which much of the detritus was derived from the northern Dalradian uplands. Current-bedded sandstones and quartzites are important, together with finer beds, which include some tuff bands, and there is much small-scale variation in facies. No Ludlow strata are known in this region, though they are possibly represented in the uppermost unfossiliferous beds, and the Caledonian movements responsible for the major folding are not closely dated, being only later than Wenlock beds and earlier than the Lower Carboniferous that overlies the eroded folds on the east.

Compared with other Lower Palaeozoic troughs, that of south Mayo and north Galway is in some ways simpler in its stratigraphy and better defined. In no other are the marginal landmasses so clearly seen. Some may be fairly safely deduced, and fragments of others, such as Anglesey, still survive; but the picture presented in western Ireland of a gulf of deeply subsiding sediments, hemmed in on either side by Dalradian uplands, is particularly graphic. Arenaceous rocks are conspicuous, particularly greywackes, which are exceptionally thick and, like most greywackes, markedly unfossiliferous. Relative simplicity in this aspect should not, however, be equated with normality: it is more 'normal' for the Caledonian geosynclinal sequences to be more varied and complicated, and among Ordovician rocks the complications are commonly related to vulcanicity. The tuff bands, especially the Arenig examples, have their own interest and importance, but their source is unknown, and they make up but a small proportion of the whole.

SOUTHERN IRELAND

The Leinster Massif is the largest Lower Palaeozoic block in Ireland but not one of the most informative. Apart from the Bray Series (p. 64) most, if not all, of the sedimentary rocks known are Ordovician, separated into two outcrops by the 70-mile stretch of the Leinster Granite, the largest Caledonian intrusion in the British Isles. Formerly they probably arched over the plutonic mass, as is shown by the attitude of the beds on either side and by the few patches that remain high on the granite surface; these are considered to be the remains of roof pendants, and include the metamorphosed sediments of Lugnaquillia, the highest point in Leinster. In addition to the main arch, the Ordovician sediments are affected by a complex series of north-easterly Caledonian structures.

The smaller outcrop, north-west of the granite, is relatively little known, but includes slates with greywackes and quartzite bands that tend to become coarser towards the west. On the east side there are fine coastal exposures but inland little is visible. The most coherent sequences come from the south, at Tramore Bay, Co. Waterford, and the north-east, on the borders of Wicklow and Wexford. In the former unfossiliferous slates and mudstones are overlain by the Tramore limestones, a variable group of calcareous rocks rich in trilobites, colonies of polyzoa and nodular masses of the coral *Monticulipora*. As in other limestones, the fauna is Scottish and Baltic rather than Welsh. The succeeding shales contain *Nemagraptus gracilis* and other Caradoc forms, brachiopods and trilobites.

Neither in the south nor elsewhere in the Leinster Massif have Llanvirn, Llandeilo or Ashgill beds been found. The north-easterly outcrops include Arenig banded slates but more outstanding is the thick Caradoc volcanic group. The lavas, tuffs and agglomerates amount to 8,000 feet at their maximum, of which rhyolitic pyroclastic rocks are the most important, but which also contain rhyolitic and basic lavas. The volcanic rocks are interbedded with the sediments and also pass into them laterally; much of the extrusion was submarine, but the lava pile shows signs of emergence late in its history.

There are certain parallels here with the Welsh sediments and lavas, particularly in the dominance of the Caradoc vulcanicity; the Caradoc shales of Tramore have also been compared with the *Dicranograptus* Shales of South Wales. On the other hand, the faunas have their differences, no Silurian rocks remain in south-east Leinster (or at least none have been described) and there is a lesser proportion of coarse detrital deposits. Moreover, somewhere between Leinster and Ireland there probably rose the 'Irish Sea landmass', and on that hypothesis no very close similarity is to be expected.

Some 10 miles from the north-west margin of the Leinster Massif and 30 miles south-west of Dublin, the small inliers of the Chair of Kildare protrude through a cover of Carboniferous Limestone and drift. Their steeply dipping succession includes Llanvirn (*bifidus*) shales at the base, which are overlain by tuffs and lavas and these again by the Chair of Kildare Limestones (Ashgill). The trilobites and brachiopods of the limestones have considerable resemblances to those of Cross Fell and the vulcanicity is reminiscent of that on Lambay island north of Dublin. This small group of inliers thus links the Dublin outcrops with those of central Ireland, and slightly reinforces a more distant comparison with northern England.

Several mountains or hilly regions arise from the south-central Irish plain and are chiefly formed from anticlinal outcrops of Old Red Sandstone. Some of these enclose central inliers of Lower Palaeozoic rocks, the most important being in the Silvermines Mountains, Galtee Mountains, Cratloe Hills, Slieve Bernagh and Slieve Aughty (in counties Limerick, Clare and Tipperary). The last two are unusual in that Caradoc and Ashgill beds are present but elsewhere these pre-Devonian rocks are nearly all Silurian. Ludlow and Wenlock are well represented, with both shelly and graptolitic faunas in mudstones and siltstones, greywackes and slates; some Llandovery beds are also found. At Slieve Bernagh the Silurian succession has several geosynclinal characters, being exceptionally thick (estimated at 20,000 feet) and dominated by greywackes and banded mudstones. The rather sparse faunas here and in the neighbouring Cratloe Hills indicate various horizons from Upper Llandovery to Ludlow.

The Dingle Peninsula, the most westerly point of County Kerry, is chiefly formed of Old Red Sandstone, but it includes two small inliers. The 5,000 feet of Silurian slates and calcareous sandstones range from Upper Llandovery to Ludlow and contain a shelly fauna of brachiopods and some corals. Thus far they fit into the pattern of inliers farther east, but the Wenlock beds are interrupted by several rhyolite flows and there are tuffs interspersed among the associated sediments. This volcanic phase has already been aligned with those of Pembrokeshire and the Mendips (p. 107) though there the rocks were basalts and andesites.

The Silurian beds in Kerry are overlain with marked unconformity by the Dingle Beds, which rest at some point on each of the three earlier divisions. There was thus a phase of considerable upheaval and erosion after the Ludlow rocks were deposited, and though the Dingle Beds are unfossiliferous they probably belong to the lowest part of the Old Red Sandstone.

SUMMARY OF CALEDONIAN BELT

Igneous activity

In its widest sense Caledonian igneous activity ranges from the lavas intercalated with the early Caledonian sediments, such as the Dalradian, to the youngest granites and dyke swarms, which cut Lower Old Red Sandstone. Since the extrusive rocks are part of the surface history that is the main theme of this book, they have appeared in the two preceding chapters. By far the greater part of the intrusions of the British Isles are also Caledonian; a few references to them (as a starting-point) are given at the end of this chapter and they are only very briefly summarized here.

There is a convenient distinction in sedimentary, tectonic, metamorphic and igneous history between the 'metamorphic' Highland Caledonian belt—the Moine and Dalradian areas of Scotland and Ireland—and the 'non-metamorphic' (or much less altered) belt that comprises the remaining Lower Palaeozoic outcrops south and east of the Highland Border. The intrusions of the former are much more varied in age and type than those of the latter, and include five major groups:

(1) Large regional migmatites of the Scottish Highlands. These have also been called the 'Older Granites' and were intimately related to the folding and metamorphism of this belt. They cover extensive areas of the Moine country in Sutherland and Inverness-shire and also in the northern and north-eastern parts of the Grampian Highlands, for instance in Aberdeenshire. Comparable areas of migmatization are known from Co. Mayo and Connemara.

(2) The alkaline intrusions of Assynt, in the Northern Highlands—a group without clear affinities in Caledonian history, though they are clearly post-Cambrian.

(3) The gabbros and associated basic intrusions of north-east Scotland. These intervene between the 'Older Granites' and 'Newer Granites' and are later than the deformation of the surrounding Dalradian rocks. Although seen at present in separate outcrops, all these gabbroic masses probably belong to one very large basic sheet, from a single source.

(4) The 'Newer Granites'. These are the most extensive of all the

Caledonian intrusions, and the only group well represented in the non-metamorphic belt. In the Scottish Highlands they include such well-known examples as the granites of the Cairngorms and Aberdeenshire and in Ireland the Galway and Main Donegal granites. In the non-metamorphic belt there are the intrusions of the Southern Uplands (and the analogous Newry Granite of Co. Down), the Lake District, the Leinster and the (subsurface) Weardale granites. In some outcrops they cut folded Silurian rocks, and a late Caledonian age is also supported by isotopic determinations. In addition to true granites, granodiorites and allied rocks are common.

(5) Ring complexes. This group, wholly post-tectonic, is largely known from the south-west Scottish Highlands, in the Ben Nevis, Glencoe and Etive intrusions, where Lower Old Red Sandstone lavas are also preserved. It may be that the Cheviot Granite should be associated with the group, and there are two ring complexes in Donegal.

The Caledonian Troughs

So much of the history of the British Isles from late Pre-Cambrian to late Silurian times lies within the framework of the Caledonian geosyncline that a summary of its component troughs is useful at this point. Within the long time-span (*c*. 350 million years) it appears that the separate, or partly separate troughs developed at different times, passed through somewhat different histories and were principally deformed and uplifted in different phases of earth movement. The sedimentary and volcanic histories can be interpreted with some confidence, but the beginning and end in many provinces are much less clear: the former because the earliest strata, and any Pre-Cambrian floor, are so often not exposed, and the latter because subsequent erosion tends to remove the later rocks (normally upper Silurian) the most effectively.

However, bearing these provisos in mind, the primary distinction can also be made in this context between the Dalradian outcrops (the 'Dalradian trough') and those farther to the south-east, which belong essentially to Cambrian and later periods. Thus four major areas can be reasonably defined:

(1) The Dalradian trough of Scotland and Ireland. Sedimentary history from late Pre-Cambrian to some period after Lower Cambrian. At least one Lower Palaeozoic phase of folding and metamorphism. Margins very uncertain, but the foreland lay to the north-west.

(2) The Welsh trough. ? Early Cambrian to late Silurian; major vulcanicity during the Ordovician period, and several phases of earth-

movement in different sectors—pre-Arenig, middle and late Ordovician, early and late Silurian. Bounded by the eastern shelf on one side and perhaps by the Irish Sea landmass on the other.

(3) The Lake District trough, with extensions to the Isle of Man and northern Pennines. ?Cambrian to late Silurian; major vulcanicity especially Llanvirn. Pre-Bala and late Silurian or early Devonian folding. Margins uncertain, but the trough probably lay north of the postulated Irish Sea landmass.

(4) South Scotland, north and west Ireland. Arenig to Silurian; vulcanicity chiefly Arenig. Mid-Ordovician and early Silurian deformation in some regions; late Silurian or early Devonian effects widespread. Margins not clear except for Dalradian uplands of the south Mayo sector—which is a somewhat separate entity.

This enumeration does not indicate that there is any formal equivalence in size or importance between the four regions, nor is their physical separation or definition necessarily very clear, as for instance in the outcrops north and west of Dublin.

If a fifth category were added for 'Southern Ireland' it would be even more heterogeneous, and brought together by little more than a geographical label. It might be, for instance, that Leinster would be better grouped with Wales, than linked, through a series of imperfectly known and isolated exposures, with Dingle. Accordingly, though Southern Ireland seems to have been the site of the same sort of sedimentation as the other sections of the Caledonian belt, no trough can be defined and so far little cognate history detected.

REFERENCES

Wales. Basset, 1963; Beavon, Fitch and Rast, 1961; Cox and Wells, 1927; Cox *et al.* 1930; Cummins, 1957, 1962; Jones, 1938, 1956; Jones and Pugh, 1949; Rast, Beavon and Fitch, 1958; Shackleton, 1954*b*; Thomas and Thomas, 1956; Williams, 1953; Wood and Smith, 1959.

Welsh Borderland, etc. Dean, 1958, 1964; George, 1963*a*; Holland, 1959; Holland and Lawson, 1963; Lawson, 1955; Pocock *et al.* 1938; Squirrel, 1958; Stubblefield, 1967; Whitaker, 1962; Whittard, 1928, 1932, 1952.

Northern England, Isle of Man. Dean, 1959; Mitchell, 1956; Simpson, 1963.

Scotland. Anderson, 1947; Craig and Walton, 1959; Flinn, 1959; Holtedahl, 1952; Kelling, 1961; Rolfe, 1960, 1961; Walton, 1963, 1965*a*, *b*; Williams, 1962.

Ireland. Charlesworth, 1963, chs. 4 and 5; Dewey, 1963; Harper, 1948; Stanton, 1960

Igneous activity. Read, 1961; Mercy, 1965.

General. George, 1962*b*, 1963*a*, *b*; Sutton, 1963; Whittard, 1960, 1961.

British Regional Geology. South of Scotland, Northern England, North Wales, South Wales, Welsh Borderland, Bristol and Gloucester.

5

THE DEVONIAN SYSTEM

One of the best known facies contrasts is that between the marine Devonian strata of the Hercynian geosyncline and the continental Old Red Sandstone of northern Europe and other regions bordering the North Atlantic. The latter facies was derived from the Caledonian mountains and consequently occurs among their relics or on their borders.

Fossils are only occasionally plentiful in the Old Red Sandstone and there are great thicknesses of rock which have so far yielded none. The most common and the most useful stratigraphically are the vertebrates. Although these are sometimes referred to indiscriminately as 'fishes', representatives of modern groups (the ancestors of the bony fishes, lung fishes and coelacanths) did not appear until Middle Devonian times. The others are chiefly primitive armoured vertebrates with bony plates protecting the head and anterior part of the trunk. They are divided into the Agnatha (ancestors of the modern lamprey) which lack true jaws, and the Placoderms; the latter, a heterogeneous group, possess jaws and some form of lateral fin. The Agnatha comprise the cephalaspids, the pteraspids and a number of smaller groups; the pteraspids in particular are employed in the subdivision of the Lower Old Red Sandstone of the Welsh Borders and Wales.

Although there is occasional interdigitation, where continental beds with vertebrates temporarily invaded marine areas, such cases are few and direct stratigraphical correlation between the two facies is very rarely possible. It is somewhat a matter of convenient chance that the three traditional divisions—Lower, Middle and Upper Old Red Sandstone—do correspond approximately with Lower, Middle and Upper Devonian, as employed in the marine successions of Belgium, northern France and Germany. In Britain the marine facies is almost confined to Devon and Cornwall, and although the system was first named from these outcrops they do not form so good a standard succession as some of those on the continent, especially in the Ardennes of southern Belgium. On the other hand the Old Red Sandstone, although not complete in any one outcrop, is widespread and typical in this country and it will be described first. The correlation between the two successions, their stages and zones, is given on p. 412.

As the Caledonian troughs gradually silted up and the Caledonian hills and mountains arose, there was a shift in the main centres of deposition. In a few places on the fringes of the uplands or within the mountain basins sedimentation was continuous, and there was a gradual transition from shallow marine, through littoral, intertidal and coastal conditions to fluviatile or flood-plain. Elsewhere new basins were established and in these the red sandstones, conglomerates and siltstones lie with marked unconformity on Lower Palaeozoic or Pre-Cambrian rocks. In the few conformable areas, of which the Welsh Borders is much the most important, the changing conditions were naturally accompanied by a changing, and for the most part, a diminishing fauna, so that it is not easy to find a satisfactory sequence of fossils to correlate and subdivide the rocks over the facies-change. A similar problem arises over all formal boundaries that were originally based on a conspicuous change in lithology and conditions, but that between the Silurian System and the Old Red Sandstone is an outstanding example.

THE ANGLO-WELSH LOWLANDS

The southernmost of the Old Red Sandstone outcrops represents a varied expanse of lowlands, shores and coastal flats between the newly developed Caledonian uplands and the Hercynian geosyncline to the south. The rocks are exposed in a triangular area of Herefordshire, south-east Shropshire and Monmouthshire, together with a south-westward extension along the northern boundary, or 'north-crop', of the South Wales Coalfield and into Pembrokeshire; there are also small strips among the folded Lower Carboniferous outcrops of the Gower Peninsula of south-west Glamorgan. In the north the junction with the Ludlow beds is conformable, but south-westward, on the flank of the Towy Anticline, a major unconformity develops at the base of the Old Red Sandstone, till that formation itself is overstepped by Carboniferous Limestone. Old Red Sandstone reappears in Pembrokeshire but there is no clear conformity here with the small amount of Ludlow beds present. (Figs. 19 and 20.)

The type areas are thus around Ludlow and the Clee Hills and on the Brecon Beacons and Black Mountain; there are also useful supplementary sections on the borders of the Usk and Woolhope inliers and in Monmouthshire. The Middle Old Red Sandstone appears to be missing throughout (at least no faunas are found) and the Upper Old Red Sandstone is thin and most probably incomplete.

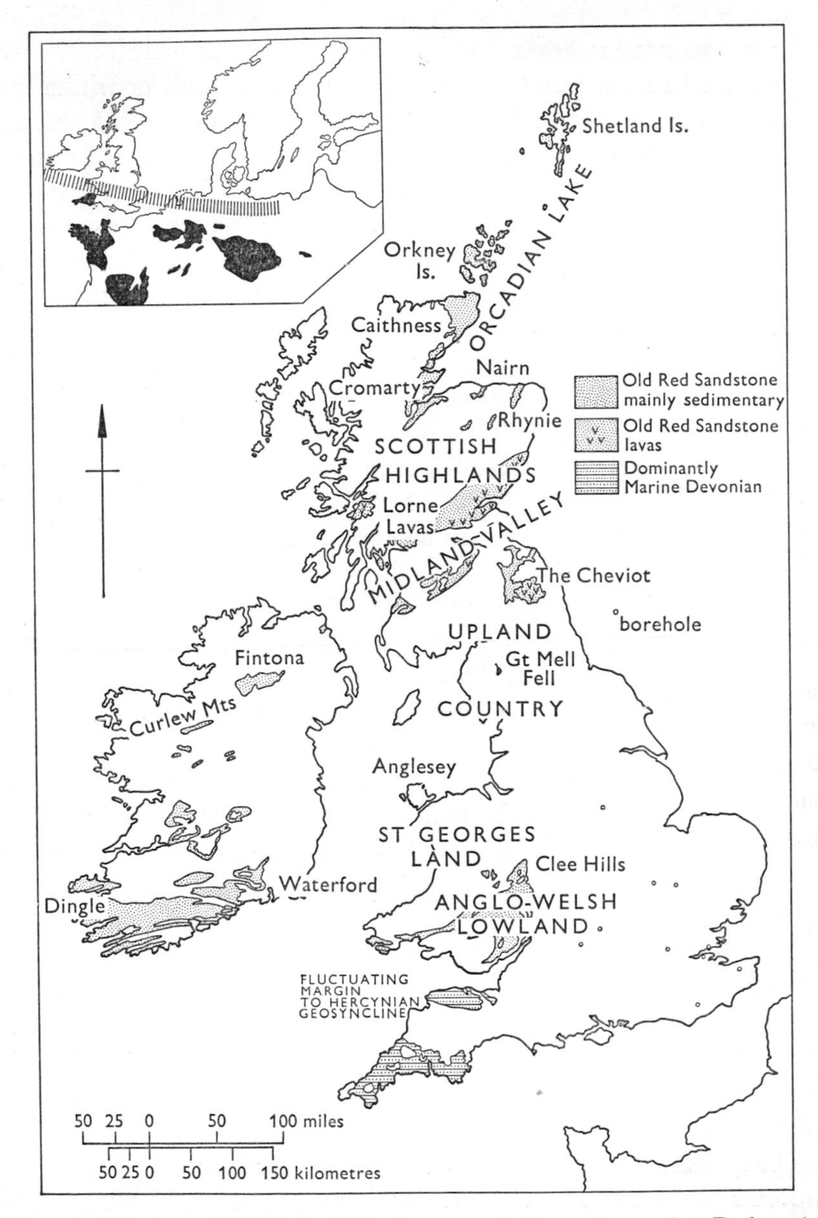

Fig. 29. Outcrops of Old Red Sandstone and Devonian rocks. Red rocks, chiefly Old Red Sandstone, are known from several boreholes. Inset: approximate southern margin of Old Red Sandstone continent; Middle European horsts, black.

Lower Old Red Sandstone

The series and their subdivisions shown in Table 20 are predominantly lithological but the lower two can be equated fairly well with Downtonian and Dittonian stages. The Dittonian vertebrate faunas contain a number of distinctive pteraspid species; the Downtonian forms are less useful stratigraphically and carry on a gradual change in vertebrates and invertebrates from Ludlow faunas. The Breconian beds of Shropshire have so far proved unfossiliferous but the Senni Beds have yielded *Rhinopteraspis dunensis* and a characteristic flora.

Table 20. *Lower Old Red Sandstone, Brecon Beacons to Clee Hills*

Stage	Series or Groups	Maximum thickness (feet)
	(Upper Old Red Sandstone)	
BRECONIAN	Brownstones	1,250
	Senni Beds (Clee Series of Shropshire)	2,850
DITTONIAN	DITTON SERIES	
	Red siltstones with subordinate limestones and thin sandstones	1,450
	Main 'Psammosteus' Limestone	0–16
DOWNTONIAN	DOWNTON SERIES:	
	Upper Red Group, dominantly red siltstones	450
	Holdgate Sandstone Group, coarse micaceous sandstones	400
	Lower Red Group, red siltstones	400
	Temeside Group, grey-green siltstones	250
	Downton Castle Sandstone Group, yellow sandstone, grey-green siltstones, bone beds	100
	Ludlow Bone Bed	
	(Ludlow Series)	

There is some lateral passage from the grey-green Temeside siltstones into the red siltstones of the Lower Red Group.

There is more than one type of upward transition in this succession. At base, above the Ludlow Bone Bed, the sediments are dominantly grey, greenish grey or yellow; these pass upwards into red beds (earlier called Red Marls, but which are mainly siltstones), with some lateral transition at the horizon of the Temeside Shales. Farther to the west, for instance in Carmarthenshire and Breconshire, these beds are represented by the 'Tilestones', micaceous flaggy sandstones once used for roofing. Thereafter the sandy and silty beds are reddish or brown, with thin grey-green layers. There is similarly no sharp change-over in the fauna: bone beds are found in the fully marine Silurian below and they continue up into the Downton Castle group. Molluscs persist as high as the lower Dittonian and with *Lingula* are often abundant in the lower Downtonian beds. Drifted plant remains are found in both the Downtonian and Dittonian beds.

The Ludlow Bone Bed is the most famous of these thin remanié deposits. It consists of worn vertebrate debris (bones, scales and spines) together with bits of brachiopod and molluscan shells, eurypterids, ostracods and phosphatic fragments, the whole being held together with a calcareous cement. This heterogeneous assemblage marks a period of very shallow seas, when the debris accumulated and was rolled and worn by currents. The bed has been traced over many miles in the Welsh Borders from Ludlow to Malvern and Usk, and in the absence of a full zonal sequence it forms a useful, though inevitably regional, base to the Old Red Sandstone. Above the bone bed there is a gradual transition, through the various Downtonian facies from littoral sands and silts with current-bedding, ripple marks and abundant shells, to intertidal flats and shoals, where the shells decrease but animal tracks and burrows abound; finally in the Dittonian facies fluviatile conditions become dominant.

A particular type of repetitive or cyclic sequence has been described from the Dittonian beds in particular, both from the Welsh Borders and Pembrokeshire; the components are:

(3) Siltstones, in places with calcareous concretions
(2) Sandstones, commonly current-bedded
(1) Conglomerate, resting on a local erosion surface.

These three make up a cyclic unit, or cyclothem, which may be from 10 to 50 feet thick; unlike some of the Carboniferous cyclothems these are not persistent laterally, but are lenticular in distribution. The pebbles of the conglomerates are chiefly local and many consist of hardened siltstones similar to those of the upper part of the cyclothem. Bony plates, as detrital fragments, are occasional components and the cement of the sandstones and conglomerates is calcareous. Other features include ripple marks, desiccation cracks and occasional plant remains.

These cyclic beds are interpreted as the products of winding river channels, redistributing deposits formed nearer the source. The whole formed a broad alluvial plain on which silt, sand and pebbles were laid down, to be shifted onwards by the rivers which frequently changed their courses. Occasionally far-travelled pebbles were brought down, together with plants and vertebrate skeletons which were disarticulated during their passage.

The calcareous beds (limestones or cornstones) of the Lower Old Sandstone have received a good deal of attention, partly because of their problematic origin and partly because they are a rich source of vertebrate remains. The term 'cornstone' includes two distinct types. Concretionary cornstones are usually associated with the red siltstones and carry varying amounts of quartz, mica, and clay minerals; they probably represent carbonate deposition under quiet conditions of shallow water and strong evaporation. There is also some resemblance to the calcareous crusts or 'caliche' of arid and semi-arid climates. The second type, conglomeratic cornstones, result from the erosion and redeposition of the first, and it is these that include local concentrations of vertebrate plates, spines and scales. One outstanding group of calcareous beds, the '*Psammosteus*' Limestones, can be recognized from Shropshire to Pembrokeshire and forms a convenient basal unit to the Dittonian stage; although the limestones themselves are unfossiliferous the associated beds contain the diagnostic pteraspid *Traquairaspis symondsi*, once thought to be a species of *Psammosteus*.

The picture drawn so far of littoral deposits, passing upwards into more varied, and often coarser, fluviatile deposits, applies to much of the Anglo-Welsh region. But in addition there are lithological additions and changes as the beds are traced westwards into Carmarthenshire and Pembrokeshire, particularly towards the top. As the basal unconformity increases the grey Downtonian beds tend to be reduced, and in places in the west are absent altogether; here much of the Downtonian and Dittonian stages are represented by a great thickness of red siltstones with subordinate sandstones, the 'Red Marl Group'.

At the Brecon Beacons and Carmarthen Vans the Senni Beds (equivalent to the Clee Series of Shropshire) form a distinctive group in the lower half of the Breconian stage—grey, green and dull red sandstones with more plant remains than are usually found. Normal Breconian brown sandstones (Brownstones) follow and the top of the Brecon Beacons is formed by the Plateau Beds, which belong to the Upper Old Red Sandstone. Farther west, in the more northerly Pembrokeshire outcrops, rocks presumed to be Breconian in age are represented by the Cosheston Beds, with conglomerates and breccias, the latter including volcanic debris. In the south-west of the county none of these upper facies is known and on top of the 'Red Marls' (? Upper Dittonian) there rests the Ridgeway Conglomerate, 700 feet thick with massive banks of cobbles and boulders. The latter include a quartzite with very poorly preserved fossils, most nearly matched in the Ordovician of Normandy, but not in Wales or the Welsh Borders. It is the same kind of problem that is raised by the Ordovician fossils in the Midland quartzites and the source is likewise obscure.

There is also a regional variation in thickness, but this may be due chiefly to post-Breconian erosion—that is to the uplift and denudation which is reflected in the absence of Middle Old Red Sandstone. Without a reliable system of zones it is not possible to be certain whether a thinner sequence in a certain area really represents persistently less accumulation, series by series, or whether some or all of the boundaries are diachronous and certain facies expand, or contract, at the expense of others. However, there is some suggestion that the red Downtonian siltstones do not vary so much as the overlying beds, and that those regions where the Breconian is thin, or deduced to be absent, are those least in total thickness.

Thus on the whole the thicker successions are in the north, such as those in Breconshire and Carmarthenshire where the Lower Old Red Sandstone exceeds 5,000 feet; it is thinner in south Pembrokeshire and at Cardiff. A similar north–south variation can be seen in Monmouthshire; from the north, on the edge of the Forest of Dean, where it is locally as much as 4,000 feet, the Breconian is progressively reduced southwards by the overstepping, unconformable Upper Old Red Sandstone and south of Monmouth is cut out altogether. Probably for the same reason no Breconian beds are known from the few outcrops south-east of the Severn Estuary.

Upper Old Red Sandstone

Throughout the region the upper division is feebly developed compared with the lower. In spite of the major gap between the two, little tilting accompanied the Middle Devonian uplift, and apart from one place in east Pembrokeshire (Pendyne) there is no significant difference in dip. The Upper Old Red Sandstone is not divisible on faunal grounds, but both the characteristic vertebrate genera, *Bothriolepis* and *Holoptychius* (chiefly the latter), have been found in the Anglo-Welsh region; there are also occasional plants and the large mussel, *Archanodon*.

At the Clee Hills the division comprises the Yellow Farlow Sandstone below and the Grey Farlow Sandstone above, the former being locally fossiliferous; together the two amount to some 500 feet. Along the north crop and at Monmouth the most conspicuous component is the Quartz

3. The Black Mountain, Brecknockshire. The north-facing scarp is formed of some thousand feet of Lower Old Red Sandstone, the Brownstones.

4. Siccar Point, Berwickshire. 'Huttons unconformity' where almost horizontal Upper Old Red Sandstone overlies steeply dipping Silurian greywackes and shales. The upper figure shows a general view and the lower one detail from another part of the outcrop.

Conglomerate—a grey, yellow and red formation with large pebbles, some of which can be matched in the Welsh uplands to the north. A thicker and more varied group is known from south Pembrokeshire, the Skrinkle Sandstone. This consists of 1,000 feet or so of interbedded sandstones, silts and mudstones, variegated in colour and including bands of breccia, especially near the base. Towards the top, but still among red and yellow beds, there is a local influx of marine fossils. They include spirifers, rhynchonellids, fish teeth and crinoids—a fully marine assemblage; nevertheless, they are interbedded with plant-bearing silts and even poor coal laminae, so that this invasion was on a small scale before the wide transgression of Lower Carboniferous seas.

In most of the Welsh outcrops and at Monmouth (but not at Clee), there is a conformable passage from Upper Old Red Sandstone into marine Lower Carboniferous shales and limestones—the converse of the Siluro-Devonian transition. There is also a like difficulty in establishing a satisfactory boundary, but in southern Britain at least the incoming marine faunas are more prolific and more helpful than were the relics of the late Silurian seas and the sparse Downtonian vertebrates.

A lesser marine fauna occurs in the Plateau Beds, the lowest part of the Upper Old Red Sandstone on the Brecon Beacons. It has been equated with the Skrinkle Sandstone invasion; or, since there is evidence of a stratigraphical break above as well as below, the fossils here may very well represent an earlier marine episode (?Frasnian), succeeded by widespread erosion. In either case the Upper Old Red Sandstone of the Anglo-Welsh lowlands is a thin and incomplete residuum formed in fluctuating conditions compared, say, with the 8000 feet of Frasnian and Famennian sediments of North Devon (p. 158).

CALEDONIAN UPLANDS, WALES TO SCOTLAND

Over the long stretch of irregular uplands between the Welsh Borders and the Midland Valley of Scotland little Old Red Sandstone remains. This does not mean that none was deposited, but rather that the sands, silts and gravel beds were rarely permanent, but were shifted onward by the streams and rivers to lower ground. Moreover, there may have been further erosion in subsequent periods, from Lower Carboniferous to Tertiary; in North Wales, for instance, the Carboniferous submergence was much later than in South Wales.

The principal uplands are the central Lake District, the Southern Uplands of Scotland and St George's Land. The last is a name given to the Lower Palaeozoic massif of Wales and adjacent regions. It was probably connected in some degree with the Leinster Massif in south-east Ireland, and a rather ill-defined ridge (the Midland Barrier or Mercian Highlands) extended eastwards into the English Midlands. The southernmost of the Caledonian land masses, St George's Land was a persistent geographical element and an intermittent source of detritus throughout Upper Palaeozoic times. In northern England the Pennine blocks were probably minor upland relics as well.

With the exception of the outcrop that laps round the easterly edge of the Southern Uplands and the Cheviot, and which is best regarded as an extension of the Scottish Midland Valley, the red rocks among these heterogeneous uplands are considered to be Old Red Sandstone on lithological grounds and stratigraphical position, for no fossils have been found. The most important relic is on Anglesey, where though the area is small, the beds are over 1,600 feet thick. The basal conglomerate overlies Pre-Cambrian and Ordovician rocks and contains recognizable pebbles of local origin; the succeeding strata include red and brown sandstones, red siltstones and cornstones—fluviatile sediments with many similarities to the Dittonian and Breconian beds farther south. On this lithological basis, and because the red rocks have been folded and faulted before the Carboniferous sandstones and limestones were laid down, they are thought to be Lower Old Red Sandstone.

On the western, scarp, edge of the Alston Block and around the eastern and northern edge of the Lake District, the Lower Carboniferous is locally underlain by patches of reddish rocks, often coarse, of various ages. Most of them are considered to be the Carboniferous Basement Beds, but two—the Polygenetic Conglomerate of the Cross Fell Inlier and the more extensive Great Mell Fell Conglomerate of Ullswater—are generally taken as Old Red Sandstone. The second group consists of unfossiliferous red conglomerates and thin sandy beds and is estimated to be 5,000 feet thick. The pebbles were derived from rocks similar to those of the central inlier of the Lake District, especially the Silurian greywackes of the southern belt. The succeeding Lower Carboniferous limestones overlie the conglomerates with marked unconformity, after a period of minor folding; on this account, together with their thickness and coarseness, they are ascribed to the Lower rather than the Upper Old Red Sandstone.

THE MIDLAND VALLEY OF SCOTLAND

The Devonian stratigraphy of this region differs in several respects from that of the Anglo-Welsh Lowlands: first, where the base of the Old Red

Sandstone is seen it is usually an unconformity; second, there was much vulcanicity; third, along the north-eastern margins in particular there are very thick, coarse conglomerates; and fourth, palaeontological correlation is rarely possible, the pteraspids in particular being very scarce. As in the Anglo-Welsh outcrops the Middle Old Red Sandstone is missing.

The thick sequence of dominantly red rocks accumulated in a broad, mountain-girt lowland with approximately the same trend, but not quite the same margins, as the present Midland Valley, the latter being largely due to the differential erosion of the upper Palaeozoic rocks, which are less resistant than those on either side. Within the rift the disposition is synclinal, so that a central Carboniferous tract is flanked on either side by Devonian belts; there are many subsidiary folds and faults, however, especially east–west Hercynian structures, and along the south-eastern belt in particular the outcrops are discontinuous (for map, see Fig. 48, p. 232). At present the Lower Old Red Sandstone is almost, but not quite confined between the bordering faults of the rift; the upper division is more extensive and there is some evidence of fault movement in Middle Old Red Sandstone times.

Lower Old Red Sandstone

Downtonian faunas have been found only in the extreme north-east of the northern belt, where a group of mudstones, sandstones, tuffs and lavas is brought up by an anticline that comes out on the coast at Stonehaven, in Kincardineshire; these beds, which lie unconformably on the Highland Border Series, have yielded eurypterids, crustacea and a species of the pteraspid *Traquairaspis* that allows correlation with the Welsh Borders. The remainder of the Lower Old Red Sandstone, up to 18,000 feet of sediments and lavas, follows conformably.

The great masses of coarse conglomerates are most abundant in the lowest (Dunnotar) Group, where there are over 6,000 feet with only minor sandy bands. The pebbles and cobbles, including much quartzite, were derived from the Dalradian uplands to the north, carried down by torrential streams and dumped at the mountain foot. Higher up there are groups of lavas, chiefly basalts and andesites, sandstones, red shales and more conglomerates; the lava flows were broken up by contemporary denudation and contribute to the coarser beds. Although much of the finer sediment is water-laid there is little evidence of much standing water and no pillow structures have been found in the volcanic rocks.

Rocks of the same age can be traced southwards into Forfarshire, though the lower beds are not exposed and the sequence is somewhat thinner; grey siltstones and sandstones occur in addition to the red beds, and from these, near the base, rich vertebrate faunas have been collected. Since pteraspids are so rare the assemblage is not closely comparable with the Anglo-Welsh, but it includes many acanthodians (placoderms) as well as cephalaspids. Plants are found towards the top. Despite the difficulties of correlation it seems probable that in these north-eastern areas most of the Lower Old Red Sandstone is present.

The lavas of the north-east and centre contribute to the upstanding ridges of the Sidlaw and Ochil Hills, but farther to the south-west they fail, though there are spreads of andesite pebbles. In the Trossachs the Lower Old Red Sandstone consists of conglomerates and dark brown or chocolate-coloured sandstones, as yet found to be barren. Along the outcrop here the north boundary is formed by the Highland Border Fault, but as this structure sweeps round to the south, to cut across the Isle of Arran and then skirt the Mull of Kintyre, small patches are found on the westerly or upthrow side.

In the south-eastern belt minor outcrops are found bordering the Silurian inliers of Lanarkshire and the Pentland Hills. The base of the Devonian in the Hagshaw Hills is taken arbitrarily at the 'Greywacke Conglomerate', where there may also bc a minor break or disconformity. It was seen in the last chapter that the beds below were most probably Silurian, although they included vertebrate horizons and were once called Downtonian; this term in Scotland has been used rather widely, to include beds transitional in facies below the more typical, brown or red, Old Red Sandstone. Above the conglomerate with its pebbles from the nearby Southern Uplands, there are some thousands of feet of brown sandstones and conglomerates, lavas and tuffs that are presumably Lower Old Red Sandstone, though no diagnostic fossils are known; if true Downtonian beds are present they are unfossiliferous. As these rocks are traced south-westwards, towards the Southern Uplands and the Girvan district they pass out of the circumscribed areas of Silurian and Devonian conformity (or near conformity), and thick Old Red Sandstone conglomerates lie discordantly on a floor of folded Ordovician rocks.

In the Pentland Hills, south of Edinburgh (Fig. 25), the stratigraphical relations are clearer than in Lanarkshire because the Lower Old Red Sandstone rests with an angular unconformity on the Silurian rocks of the inliers. The lower part consists of some 2,000 feet of sandstones and conglomerates; here also Silurian detritus is identifiable, both from local sources and the more distant Southern Uplands. Above there is a thick pile of lavas, pyroclastic rocks and sediments (6,000 feet), in which no fossils have been found.

There is only one Lower Old Red Sandstone outcrop of any extent south of the Southern Upland Border Fault, and that is a patch extending a few miles inland from the coast of Berwickshire, between Eyemouth and St Abbs Head. The red sandstones, siltstones and conglomerates overlie the upturned edges of the Silurian greywackes and are covered by the more extensive Upper Old Red Sandstone. This relation, together with a species of *Pterygotus*, is their chief reason for their stratigraphical position.

Substantial bands of lavas, tuffs and small vents, are found in these sediments and they are probably of the same age as the Cheviot lavas. The Cheviot is the only Caledonian igneous mass of the Southern Uplands in which the extrusive phase is preserved. The lavas overlie folded Silurian rocks and after an early phase of agglomerates and rhyolites there followed the main flows of increasingly basic pyroxene-andesites, amounting to several thousand feet. There are virtually no associated sediments and the eruption was subaerial, with only occasional ash showers. This phase was followed by the central granite intrusion and later by numerous radiating porphyrite dykes. The lavas are overlain on the west by Upper Old Red Sandstone, and boulders of the granite are found in early Carboniferous conglomerates, so that in addition to being a classic example of an ancient volcano, the age of the Cheviot mass is well established.

Upper Old Red Sandstone

This division is thinner, normally finer in grade than the lower one, and the lava flows are few and scattered. The base is everywhere unconformable, sometimes on Lower Old Red Sandstone, sometimes overlapping on to older rocks. The upper junction may also be an unconformity, but elsewhere has been taken at some arbitrary lithological or colour change because diagnostic fossils on either side are rare or absent. Unlike the comparable boundary in South Wales, the transition here is not to a marine Carboniferous facies but to freshwater beds.

Within the Midland Valley the outcrops are much less extensive than those of the lower group and the characters of the northern and southern belts may be summarized together. The principal rocks are red, pink and buff and yellow sandstones, lighter in colour and more quartzose than the Lower Old Red Sandstone; they are often current-bedded and pebbly and though conglomerates occur, especially near the base, massive, coarse conglomeratic beds are rare. A thickness of nearly 3,000 feet is known north of Glasgow, but in most regions it is a good deal less. There are also red, brown and grey-green siltstones and shales and the concretionary type of cornstone. Desiccation cracks, mud-flake breccias and animal tracks are known, and in a few places the scales of *Holoptychius*; *Bothriolepis* is recorded from Ayrshire. On the whole, however, these rocks are not fossiliferous and the sandstones of Dura Den, in Fife, thickly strewn with *Holoptychius*, or the articulated plates of *Bothriolepis* at Duns, in Berwickshire, are rare exceptions.

The conditions suggested by these lithological characters are those of a very shallow, large, inland body of water, subject to desiccation and evaporation, between uplands which, though considerably reduced, still supplied substantial amounts of sandy detritus and occasional spreads of pebbles. Gradually the beds become finer, with more carbonate, and the grey sediments result, with rare plants and shells; these are usually called Lower Carboniferous, but it is unlikely that the transition is simultaneous throughout the lowland, or that sedimentation was necessarily continuous. In places there is a clear unconformity and deposition was renewed much later.

Beyond the boundary faults on the west minor areas of overlapping Upper Old Red Sandstone are seen around Loch Lomond and the Clyde Estuary, where red sandstones overlie the Highland Border and Dalradian rocks; there are similar smaller outcrops in Bute, Arran and the Mull of Kintyre. In the opposite direction the overlap is much more extensive, and the outcrops that stretch from East Lothian into Berwickshire, and thence along the eastern edge of the Southern Uplands, form one of the major Upper Old Red Sandstone areas in Britain. The rocks lie with gentle dips on the Cheviot lavas and fill in hollows and valleys in the folded Silurian greywackes and shales. In these areas there is much locally derived conglomerate and thicknesses up to 2,000 feet have been recorded. An outlier at 1,345 feet O.D. on the Lammermuir Hills, together with other, lower patches, is an indication of a much more extensive, earlier covering.

This is also a classic region in the history of stratigraphy: the striking junction of gently tilted Old Red Sandstone overlying steeply dipping greywackes (seen at Siccar Point in Berwickshire, Pl. 4, and near Jedburgh) was used by Hutton to illustrate an unconformity, the forces that produced it, and to support his theories of geological processes.

CALEDONIAN UPLANDS, THE SCOTTISH HIGHLANDS

North of the Highland border the main sedimentary basin is the Orcadian lake, the outlying southern portions of which are found in Cromarty, along the shores of the Moray Firth and in the northern section of the Great Glen. In addition, however, there are small but significant outliers on the Grampian Highlands and larger outcrops to the south-west, in Argyll. The last-named include the major volcanic group of the Lorne Plateau Lavas, stretching from Oban north-eastwards to Loch Etive. They are principally andesites, with agglomerates and tuffs and a variety of sediments, from breccias and conglomerates to sandstones, shales and cornstones. The sediments are best seen in the west of the outcrop and have yielded a number of vertebrates including *Cephalaspis*, eurypterids, crustaceans and plants—a Lower Old Red Sandstone assemblage.

Small but significant masses of sediments are also associated with the volcanic centres of Ben Nevis and Glencoe. In the latter they form part of the central block of the cauldron subsidence; most of the 4,000 feet so preserved are andesites and rhyolites but there are minor sedimentary intercalations. Essentially similar relations are present on Ben Nevis, where the summit is formed of a mass of lava and agglomerates; for the most part these rest on Dalradian schists, but on one side there is a small thickness of shale and conglomerate.

In Aberdeenshire there are two small outliers, at Tomintoul and Rhynie, south of Huntly. The latter is best known for its plant-bearing cherts, which contain members of the Psilophytales—small leafless plants, exquisitely preserved, some apparently in position of growth. The silicified plant bed also encloses small terrestrial invertebrates and a microflora. Apart from its profound palaeobotanical interest the Rhynie bed also represents one of the few examples of pre-Carboniferous fossiliferous, terrestrial sediments, in which plants are preserved where they grew, as distinct from drifted remains in streams or lakes.

The full sequence at Rhynie, which includes sandstones, shales, conglomerates and a volcanic horizon, has been estimated at 1,500 feet and it lies at about 1,000 feet O.D. At Tomintoul, 20 miles to the west, 500 feet of conglomerates, some with large boulders, reaches a height of 1,950 O.D., the boulders and pebbles having accumulated on an irregular Dalradian surface. The age of the Rhynie plants is usually taken to be Middle Old Red Sandstone, but they may be older. What is particularly impressive about these remnants, however, together with those in

Argyll, is that they represent more than a thin transitory veneer or minor valley gravels over the Grampian Highlands; during much of Devonian time it is probable that variable sediments, often coarse, with local lavas, accumulated over much of this upland, to be later removed almost in their entirety.

THE ORCADIAN LAKE

Many of the remnants of this region are now far scattered, separated by the waters of the North Sea and North Atlantic, but once the sediments must have extended at least from the northern part of the Great Glen to the Shetland Islands, a distance of about 150 miles. This province differs from all others in the British Isles in the thick, fossiliferous sequence of Middle Old Red Sandstone, much of which is not red, or even brown, but grey, fairly fine in grade and carbonate-bearing.

Moray Firth, Cromarty and Caithness

In these outcrops the red facies comprises a great variety of rocks, such as sandstones and shales, conglomerates, arkoses and breccias, which are locally interbedded with cornstones and brown shales. In the south such beds are distributed rather irregularly from Cromarty to Inverness, and along the northern edge of the Grampian Highlands in Nairn and as far east as Gamrie in Banffshire.

Table 21. *Old Red Sandstone of the Orcadian region*

UPPER OLD RED SANDSTONE	Fossiliferous sandstones of Elgin and Nairn Hoy Sandstone of Orkney Dunnet Sandstone of Caithness
MIDDLE OLD RED SANDSTONE	John o' Groats Sandstone of Caithness Caithness Flagstone Series Red fossiliferous beds of Moray Firth
? LOWER ? MIDDLE OLD RED SANDSTONE	Barren Red Measures of Moray Firth, Cromarty and Caithness

The various formations are not necessarily superposed in any one region.

In some places, such as Cromarty, the higher formations include red or grey limestones and shales with a rich assortment of vertebrates (acanthodians, placoderms and fishes); these are equivalent to part of the grey Caithness Flagstones to the north and the fauna is similar to that of the Achanarras Limestone (Table 22). Part of this facies is thus Middle Old Red Sandstone. Occasionally, however, there is evidence for an unconformable break below the vertebrate horizons, and so the age of the lower red beds may not be Middle, but Lower Old Red Sandstone, though there are no fossils to prove it. Conversely the break may be only a small one in a variable series of continental beds.

The Upper Old Red Sandstone of the Nairn and Elgin districts is poorly exposed but nevertheless includes a number of valuable fossiliferous horizons. Five successive faunas have been found, characterized especially by Psammosteids, as well as the longer ranging *Holoptychius*, *Bothriolepis* and *Asterolepis*. Comparable successions are known in north-west Russia and the Baltic, and through these a broad correlation is possible with the marine facies; the Scottish beds probably correspond to stages from upper Givetian to Famennian.

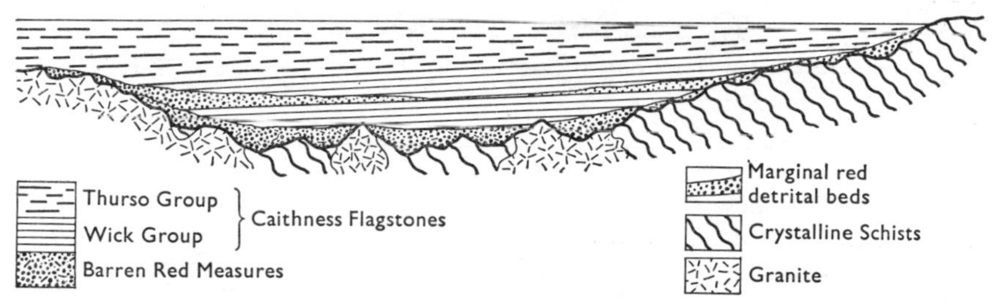

Fig. 30. Diagrammatic section to show the infilling of the Orcadian lake basin in Caithness. The irregularities of the land surface (Moine Schists and granites) are filled up with breccias and other components of the basal Barren Red Measures. The succeeding flagstone facies has a greater extent and uniformity but a coarser detrital marginal facies develops locally. (Redrawn from Crampton and Carruthers, 1914, p. 85.)

In Caithness and eastern Sutherland the Middle Old Red Sandstone is the thickest and most striking division. Again there is a basal barren red group; locally it is separated by an angular unconformity from overlying breccias which themselves pass up conformably into the grey flagstone series. These basal beds are highly variable, both laterally and vertically, as they filled up pockets and valleys in the floor and lower slopes of the basin, here formed of the Moine Schists and their associated intrusions. The irregularities were largely evened out, however, by the time the grey flagstones were laid down and these extend with little facies-change over much of Caithness and nearly all the Orkney Islands.

The principal rock types of the Caithness Flagstone Series are the grey calcareous 'flags' (finely laminated carbonate and quartz siltstones, dark with carbonaceous matter), mudstones with desiccation cracks, and fine sandstones which usually show slumping. The flakes splitting off the mudstones were often redeposited soon after and form layers of mud-flake breccias. These rock types are repeated over and over again, and at some levels the repetition is orderly enough to be called cyclic.

The fauna is extremely restricted in variety and only abundant in the calcareous flags. Vertebrates (fishes and placoderms) are by far the most important, especially the Dipnoans or lung fishes; there are also small crustacea, very rare eurypterid remains and some drifted plants. No shells and no marine forms at all are known. The richest and most famous fossiliferous layer is the Achanarras Limestone of central Caithness, somewhere in the middle of the series, which is the chief marker horizon and the basis of correlation with the Orkneys, Cromarty and Nairn.

Table 22. *Middle Old Red Sandstone*

Marine equivalents		Caithness	Orkney
? GIVETIAN		John o'Groats Sandstone	Eday Sandstone
approx.	Caithness	Thurso Group	Rousay Flags
....	Flagstone	.........Achanarras and	Sandwick Fish Beds ....
? EIFELIAN	Series	Wick Group	Stromness Beds (local
approx.		Barren Red Measures	basement beds only)
		Moine Schists and intrusions	

These grey, carbonate-bearing rocks are interpreted as the slow accumulation in a large, shallow, playa-type of lake that frequently dried up; the coarsest detritus was fine sand, and during the deposition of the laminated beds conditions were exceedingly tranquil. Some 15,000 feet of lacustrine sediments accumulated in this way, accompanied by gentle, intermittent subsidence. The margins of the lake basin have largely been removed by erosion, for it is the Barren Red Measures that abut on the older rocks to the south and west. But along the north coast into Sutherland the flagstones in a few places approach their contemporary shore line, and here they pass into buff or red calcareous sandstones or develop spreads of breccia and local tufa screens. Towards the top of the Caithness Series red sandstones begin to appear, sometimes cutting shallow channels in the underlying flags; similar red beds, often current-bedded, are dominant in the John o'Groats Sandstone, which bears considerable resemblances to the fluviatile facies of the Welsh Borders and Wales.

Orkney and Shetland Islands

Geologically the archipelago of the Orkneys is a straightforward northward continuation of Caithness; the same grey facies is present under different stratigraphical names and the Achanarras Limestone is equated with the Sandwick Fish Bed. No thick, basal red beds appear, though small knobs of granite and other intrusions have produced their own marginal facies of breccias and sandstones. The rock types, repetitive sequences and sedimentary environments of Caithness are repeated here, but the upper beds equivalent to the John o'Groats Sandstone have a wider distribution.

On the south-western island of Hoy there are some 3,500 feet of red and yellow sandstones, often current-bedded, which are usually taken to be Upper Old Red Sandstone. Neither they, nor their counterparts on Dunnet Head, south of the Pentland Firth, contain any fossils, however, and the age is deduced from an unconformable contact between the basal lavas and the flagstone facies at the north-west corner of the island.

The Shetland Islands are largely composed of schists, which are part of the Caledonian belt from north Scotland to Scandinavia. Their age is not easily determined, but there are some similarities to highly altered Ordovician rocks of Norway. In a few areas these older rocks are faulted against, or overlain unconformably by, Old Red Sandstone. The principal outcrops are a long strip from Lerwick south to Sumburgh Head and that forming the Walls Peninsula on the west. The rocks consist of conglomerates, grey and brown sandstones and grey flagstones, but calcareous and fine-grained types are less characteristic than in Orkney and Caithness. More than one 'fish band' has been found, and the Melby Fish Bed has a fauna similar to that of Achanarras; the others are probably higher in the succession and possibly equivalent to the Eday faunas of Orkney. Both Devonian rocks and faunas of Shetland lack a modern description, however, and it is possible that there are remnants of several divisions in the Old Red Sandstone; the outcrops of Walls are also exceptional in this region in being interbedded with lavas and ashes.

THE OLD RED SANDSTONE IN IRELAND

No marine Devonian rocks have been found in Ireland, and so few fossils are known that comparisons with Wales and Scotland depend chiefly on the relations of the various outcrops to the tectonic framework —Caledonian structures in the north, Hercynian in the south—and on some broad lithological similarities. The greater cover of later rocks in the north and centre, especially of Antrim Basalts and Carboniferous Limestone, means that there is much less Old Red Sandstone exposed in the Irish equivalent of the Midland Valley.

The largest outcrop is the Fintona block, in Fermanagh and Tyrone, of some 300 square miles. On the north the Devonian beds are brought by the Highland Border Fault against Dalradian schists; to the east they overlie the Ordovician Tyrone Igneous Series and to the south are themselves overlain by Carboniferous beds. All the older rocks, and especially the Dalradian, have contributed to the conglomerates and mineral assemblage. A single pteraspid fragment is the only indigenous fossil found, but at least it confirms a Lower Old Red Sandstone age, which would also be deduced from the coarse components and volcanic rocks, similar to those in Scotland. Farther to the south-west, along both the strike and the line of the 'Midland Valley', the anticline of the Curlew Mountains brings up a further elongate outcrop. The lithology of the 7,500 feet is very similar to that of Fintona and probably these beds also belonged to the lower division.

North of the Highland Border Fault there are several remnants. At Cushendall (Fig. 53), on the eastern Antrim coast a small outcrop lies immediately north of the fault, but the beds none the less amount to 3,000 feet—conglomerates with large boulders of andesite and quartzite, brown and purple sandstones and mudstones, tuffs and agglomerates. A smaller outcrop near by with a lesser dip, which incorporates pebbles of an already lithified conglomerate, might be Upper Old Red Sandstone, but there are no confirmatory fossils. There is a strong resemblance to the outcrops of Kintyre, only 20 miles across the sea to the north-east, and they can be considered as part of the same depositional area. Much farther to the north there is a square mile or so of brown sandstone, siltstones and conglomerates, local in origin, lying on Dalradian quartzites at Ballymastocker, on Lough Swilly. These are trivial except as an indication of a previous cover over the highlands. In County Mayo, however, there are larger areas north of the Boundary Fault—such as those forming a composite sequence of some 5,000 feet from Clare Island to Castlebar, at the western end of the Ox Mountains.

These northern outcrops of the Irish Old Red Sandstone extend the picture drawn from the Scottish Midland Valley, but have not as yet added much that is new; the covering outcrops of the Dalradian highlands are a little broader, but that is all.

Southwards from the Curlew Mountains, across the central plain of Ireland, there are many inliers of Old Red Sandstone, from a few square miles to substantial upstanding tracts such as the rims of the Lower Palaeozoic inliers of the south and centre, the Galtee Mountains of Co. Tipperary and the Comeraghs of Co. Waterford. Little detailed information is available on the former group; the latter are on the northern margin of the Hercynian folded belt. This belt contains some of the thickest and most extensive Old Red Sandstone developments in the British Isles, and some of the most tantalizing. Apart from the Kiltorcan Beds at the top, there is no reliable fossil record to determine the age of the beds; where the base is seen it is a pronounced unconformity and the upper limit is usually a conformable passage into grey Carboniferous rocks.

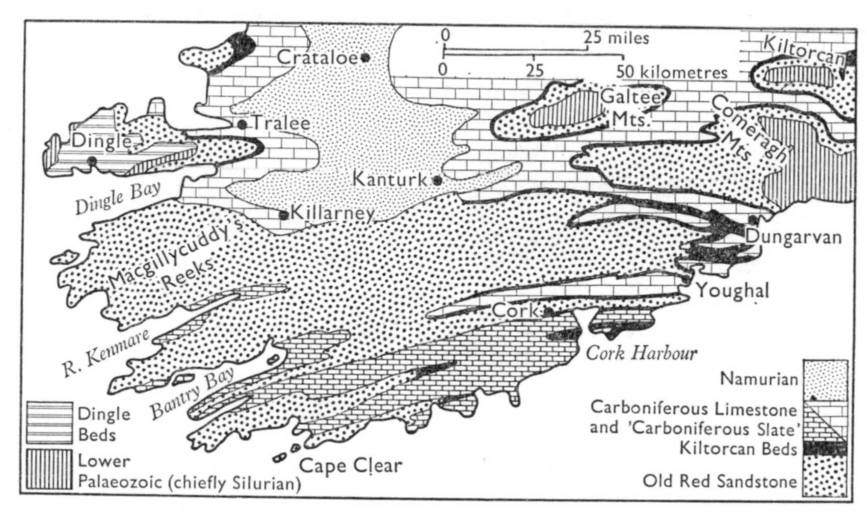

Fig. 31. Geological sketch map of southern Ireland. The Lower Palaeozoic outcrops are almost all Silurian but Ordovician beds are present in the extreme south-east, south of the Comeragh Mountains. Coal Measures are present at Crataloe and Kanturk. (Redrawn from Charlesworth, 1963, p. 185.)

The dominant lithological types are red, purple and grey-green sandstones, often fine and interbedded with siltstones and mudstones; the finer rocks locally exhibit slaty cleavage. There are also conglomerates, especially in the east, where in the Comeraghs they have a recognizable source in the Leinster Massif; cornstones, on the other hand, do not seem to be abundant. Because these rocks, particularly the sandstones, are more resistant than the succeeding Carboniferous limestones and shales, the synclinal tracts of the latter have been eroded into the long inlets of west Cork and Kerry (Bantry Bay, Kenmare River and Dingle Bay) while the Old Red Sandstone forms anticlinal mountain tracts between; the highest range of Ireland, Macgillycuddy's Reeks (3,414 O.D.), is among them.

It is difficult to generalize satisfactorily over this area, 120 miles from east to west, but as in some other outcrops the sediments tend to become finer upwards, and there is probably a gradual, but not necessarily regular, thickening to the south and west. About 3,000 feet of Old Red Sandstone is present in north Kerry and something of the same order in the Galtee Mountains, 10,000 feet in the Comeraghs, and 17,000 feet in south Kerry.

The Dingle Peninsula is exceptional not only for its Silurian inliers, but also for the overlying Dingle Beds. These are a thick group (?10,000 feet) of sandstones and flaggy beds, with current-bedding, ripple marks and desiccation cracks, which have unconformable junctions with the Ludlow beds beneath and the normal Old Red Sandstone above; moreover they are grey, and have suffered slightly more cleavage and folding than the latter. Because of this and their position the Dingle Beds have been called Downtonian, but there is no palaeontological evidence to support this—or for that matter to fix the age of the Old Red Sandstone above. It is possible that certain conglomeratic sandstones, siltstones, tuffs and agglomerates in the Comeragh district occupy a similar position, in that they are similarly more deformed than the nearby, normal Old Red Sandstone.

Over much of the western outcrops the upper boundary of the Old Red Sandstone is marked by a gradual colour change to the grey, slightly calcareous, sandy Coomhoola Beds. In most places these have an early Carboniferous fauna, but occasionally it is mixed with a late Devonian assemblage comparable to that in North Devon. Because of this conformable passage much of the red and purple rocks below has been considered as Upper Old Red Sandstone, but again there is no proof.

Over a wide stretch of the eastern and northern outcrops the uppermost Devonian is represented by Kiltorcan Beds, or their presumed equivalents. At the type area in Co. Kilkenny there are yellow sandstones, siltstones and marls with a fauna of arthropods, *Archanodon*, fish and a particularly abundant flora; the last includes lycopods and fern-like forms, especially the large *Archaeopteris hibernica*. Plants in this facies have also been found in several places, as far west as north Kerry. The combination of species is not known outside southern Ireland; it is generally taken to indicate a lacustrine assemblage of very late Devonian age and is in striking contrast to the great barren mass of rocks below.

In the unconformable base, gradual transition above and position in the Hercynian folded belt there is some similarity between the southern Irish outcrops and those of south-west Pembrokeshire. The conspicuous lack in Ireland is any proof of age, or suggestion of the Middle Old Red Sandstone break; it will be recollected, however, that in Wales and England this is very rarely a conspicuous or angular unconformity.

THE MARINE DEVONIAN FACIES IN DEVON AND CORNWALL

During Devonian times there was presumably a shoreline taking an east–west course across southern Britain, separating the Old Red Sandstone continent from the sea that lay to the south. The two facies are easily distinguished in their typical areas, but any sharp demarcation is another matter. Over the great span of Devonian time the shoreline clearly did not remain in one place but shifted to and fro. It is unfortunate that for much of the period the western section lay somewhere in the present site of the Bristol Channel and that the eastern is hidden under the Mesozoic cover of southern England. But it is also probable that we are here dealing with a very gently shelving surface, and that there was a gradual belt of transition, perhaps over miles, from the lower reaches of the flood plains to broad coastal and intertidal flats and finally to true marine conditions. The concept of a shore-line thus remains, but even were we provided with a complete section from north to south it might be very difficult to determine its position unequivocally.

The central structure of Devon is a complex syncline, or much faulted synclinorium, of Carboniferous rocks and the Devonian strata emerge from beneath these on the north and south. The northern outcrops are rather less complex in structure than those of the south, where there are also the local effects of metamorphism surrounding the Dartmoor and other Hercynian granites; in addition the more competent beds, such as the thick limestones or coarser sandstones, have suffered less deformation than the argillaceous rocks, which are now often in the form of slates and much folded and thrust. The greatest difficulties in stratigraphy thus tend to arise in the slate belts, particularly where these are poor in fossils, such as those of Lower Devonian age.

In these circumstances the stratigraphy is based on whatever criteria are available. Originally the names given in the Tables 23 and 24 were largely lithological, based on type areas, but as palaeontological knowledge has increased they have been fitted into a wider frame. The international stages are mostly Belgian or German in origin; in recent years these stages have been applied successfully to most of the divisions in Devon and some of those in north Cornwall. The most valuable fossils are goniatites, supplemented by brachiopods, trilobites, ostracods and other groups. The Lower Devonian is especially difficult to deal with, though brachiopod and vertebrate faunas have proved useful. Middle and Upper Devonian rocks are, however, often rich in fossils and the remaining uncertainties in their stratigraphy are largely caused by imperfect exposure (away from the coastal sections) or by severe tectonic complications.

North Devon

The Devonian rocks of North Devon have a regional dip to the south or south-west, so that there is a gradual descent in the succession along the fine cliff exposures from the Barnstaple area on the edge of the Carboniferous outcrops round the western and northern coasts as far as Lynmouth, where the lowest beds appear (Table 23). From here an anticlinal axis trends nearly eastward and on the northern side the Foreland Grits, long considered to be the lowest formation exposed, are probably a repetition of the Hangman Grits between Lynton and Ilfracombe. Further outcrops in eastern Exmoor, the Brendon Hills and the Quantocks result from another anticline bringing up Old Red in the midst of New Red Sandstone.

Table 23. *Devonian succession of North Devon*

Stages	Lithological groups	Approximate thickness (feet)
	(Upper part of Pilton Beds, Carboniferous)	
FAMENNIAN	Lower part of Pilton Beds	1,600
	Baggy and Upcott Beds	2,200
	Pickwell Down Sandstone	4,000
	Morte Slates	?
FRASNIAN GIVETIAN	Ilfracombe Beds	2,000
? EIFELIAN	Hangman Grits	4,000
? UPPER EMSIAN	Lynton Beds	1,300+
	(base not seen)	

Thicknesses can only be given in round figures and some of the correlations are inferred.

Goniatites are virtually unknown in this succession and the fauna is chiefly shells, found in the sands, shales (often transformed into slates) and impure limestones. They are sometimes accompanied by trails, tracks and burrows of organisms that have left no other remains. Such trace fossils add to the deductions of very shallow water, local erosion and current-swept shell concentrations; at times the conditions may have been intertidal. The least muddy seas are represented by the limestone bands of the Ilfracombe Beds which, in addition to brachiopods, contain simple corals. The Lynton Beds have an impoverished fauna of brachiopods and lamellibranchs and polyzoa: their age is probably Emsian but may possibly be younger. The lowest Devonian stages are thus not exposed in North Devon.

Two incursions of Old Red Sandstone type interrupt these off-shore sediments on the northern edge of the geosyncline. The Hangman Grits comprise a great variety of sandy beds, some with marine fossils, but there are also fragmentary plants and fish scales. Drifted plants alone are no firm indication of non-marine conditions, being known elsewhere in fully marine rocks, but these Old Red Sandstone incursions are reasonably similar to the northern continental facies in lithology as well. The Pickwell Down Sandstone is better provided and contains the remains of *Bothriolepis*, *Holoptychius* and other fishes. This is a valuable intercalation among established Famennian beds and one of the few faunal confirmations of the relative ages of Upper Old Red Sandstone and Upper Devonian in Britain.

One curious anomaly in the North Devon sequence is the apparent lack of any reflexion of the very large stratigraphical gap in South Wales. The absence of all the Middle Old Red Sandstone (and small thickness of Upper Old Red Sandstone) strongly suggests uplift during much of the Eifelian, Givetian and Frasnian stages—yet where are the products of this long erosion period? Some part of this time-span is represented by the least detrital rocks of the succession, the Ilfracombe Beds. As yet there is no straightforward answer to this problem.

South Devon and North Cornwall

A full sequence of marine Devonian rocks, down to the Siegenian, can be made out south-east of Dartmoor, from Newton Abbott to Torquay and Dartmouth, although the outcrops are partly separated by faulting or by New Red Sandstone and Tertiary deposits. There is also a good correlation westwards towards Plymouth, the Tamar valley and Tavistock; the outcrop belt can then be traced across to the Cornish coast, from Boscastle on the north to Newquay on the south.

In south Cornwall our knowledge is much less complete but there are broad outcrops of Lower and Middle Devonian rocks (Mylor and Gramscatho Beds) including much slate and some grits. Along the southern coast, from the Lizard to Dodman Point, there is a very important belt of thrusting that is not wholly understood. Its southern border is only seen in the thrust margins of those headlands, bounding the Lizard complex and the Dodman phyllites, but between these the belt can be traced across the Falmouth Estuary and along the intervening cliffs. Included in the thrust slices there are conglomeratic and sandstone wedges of the Gramscatho Beds that contain fragments of wood (*Dadoxylon*, probably Middle Devonian), and shales with cherts, tuffs and spilitic lavas that may be Upper Devonian. Much discussion has centred on fossiliferous blocks—chiefly quartzite and limestone—that are involved in this thrust belt, because although some of the fragmentary

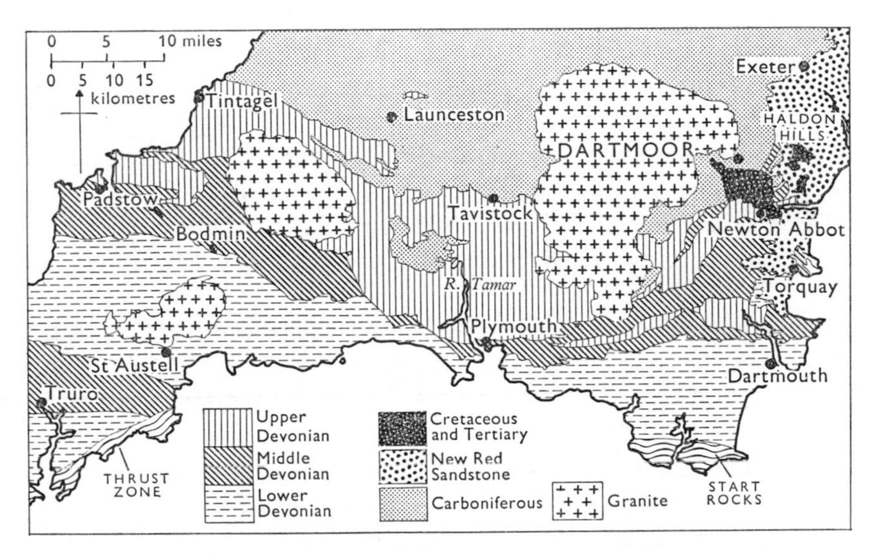

Fig. 32. Geological sketch map of south Devon and north Cornwall. (Redrawn after House, 1963, p. 3.)

fossils are probably Devonian, others have been called Ordovician or, more doubtfully, Silurian (p. 107). The blocks have been interpreted as exotic masses within the thrust belt, but whether they are part of a thrust breccia, a conglomerate, or a submarine slumped assemblage in the Devonian sequence is uncertain. In any case they represent beds of an age and facies not found elsewhere in the south-west peninsula. It is possible that the schists of Start Point, in South Devon, are to be linked with the Dodman phyllites and Lizard complex, lying south of the same line of thrusting that continues in a broad arc under the waters of the English Channel, north of the Eddystone Lighthouse; the lighthouse itself is built on a resistant reef of metamorphic rocks.

The sequence in the Torquay region is shown in Table 24. The incomplete exposures and structural complexities make it very difficult to estimate thicknesses reliably in these beds, but the succession in South

Table 24. *Devonian succession in the Torquay district*

Stages	Lithological groups
FAMENNIAN	Ostracod Slates
UPPER AND MIDDLE FRASNIAN	Saltern Cove Beds
LOWER FRASNIAN, GIVETIAN AND UPPER EIFELIAN	Torquay Limestone
LOWER EIFELIAN	Calceola Shales
EMSIAN AND UPPER SIEGENIAN	Meadfoot and Staddon Beds
?BRECONIAN AND DITTONIAN	Dartmouth Beds

Devon is probably a good deal thinner than that of North Devon, with its generally coarser grained rocks and sandstone intercalations.

The Dartmouth Beds at the base stand out as the only continental formation in the south. The 3,000 feet or so consist of grey-green and purple sandstones and siltstones with some pyroclastic rocks. Fossils are poorly preserved, but molluscs, plants and fragments of pteraspids are recognizable; the last suggest an equivalence with the Upper Dittonian and Breconian of Wales, or approximately Lower and Middle Siegenian. There is no indication of the source of the detritus, so we do not know whether it was derived from the south, or represents an early Devonian, southward extension from Wales.

Whatever the surroundings of this basal continental facies, it was succeeded in Upper Siegenian times by fully marine conditions. The Meadfoot Beds of Torquay are richly fossiliferous, with a Lower Emsian brachiopod fauna, and the Staddon Grits of Plymouth are approximately their more quartzose equivalent. Farther west the Looe Grits of east Cornwall contain a fauna like that of the Meadfoot Beds. On the north coast of Cornwall this belt of Lower Devonian slates and grits is represented in the cliffs around Newquay.

Above the lowest shales, with rare *Calceola* (also called 'Couvinian Shales'), the Middle Devonian is almost wholly calcareous and as the Torquay Limestone this facies continues up into the Lower Frasnian; stromatoporoids and corals are particularly conspicuous in the limestones that crop out in the neighbourhood of Torquay itself. The Plymouth Limestone, of the same age, has been shown to contain a full series of Middle Devonian coral zones, as originally established in Germany. West of the Tamar this calcareous facies gives place to a more muddy one and near Padstow the Middle Devonian is represented by grey slates with Eifelian and Givetian goniatites. This is not the only east–west facies-change in the southern Devonian outcrop, but it is the most conspicuous and important. There are also abundant lavas of this age and near Totnes the whole of the Torquay Limestone is replaced by spilitic lavas and tuffs—the Ashprington Series.

At Torquay the Middle and Upper Frasnian is represented by the reddish, but still partly calcareous, Saltern Cove Beds, which have long been known for their goniatite faunas, concentrated in a band at the top. Above this the Upper Devonian rocks in general can be roughly divided into two facies. The 'ostracod slate' is an argillaceous rock with a rather sparse fauna, the ostracods affording a valuable link with similar facies in Germany. It forms all the Famennian in the Torquay district and can be picked up again as part of that stage in north Cornwall. Near Camelford the Delabole Slates represent a similar facies and contain the conspicuous *Cyrtospirifer verneuili*—an Upper Devonian index fossil, though not so short-ranged as the goniatite species.

The second facies comprises the cephalopod-rich, impure or nodular limestones and shales. This type occurs east of Dartmoor, at Chudleigh, where the total Famennian thickness is about 170 feet; it is also known to the west of Tavistock, and in the inliers at Launceston. The ostracod slate facies has been interpreted as argillaceous sediments, accumulating more thickly in 'basins', contrasted with the thinner, more calcareous, deposits on intervening 'swells' on the sea floor. On the Cornish coast, for instance at Pentire Head and Tintagel, volcanic activity continued into this period in the form of tuffs and great submarine flows, producing spilitic lavas.

Southern and south-eastern England

From Devon and Somerset the Devonian rocks disappear eastwards under a Triassic cover but they are known from several borings in the east and south-east. Most of them are on the London Platform, particularly on the north-west margin (Cambridge to Oxfordshire) and in an east–west belt through London itself. The latter appears to be flanked by earlier rocks and may be a syncline near the southern margin of the platform. In most places the facies is Old Red Sandstone and thus far resembles an eastward extension of the Anglo-Welsh area, but there are some marine records. These include the Willesden boring in north-west London which penetrated over 1,000 feet of fossiliferous grey and red mudstones with a fairly abundant Upper Devonian fauna, including *Cyrtospirifer verneuili*. At Witney in south-west Oxfordshire the rocks contained a marine fauna that may be Lower or Middle Devonian; this is surprisingly far north for such a facies, and though the records as a whole are too sparse to encourage much palaeogeographic deduction, the picture appears to be more complicated than a simple extension of the sedimentary belts as they are seen in South Wales and Devon.

Marine beds, possibly Lower Devonian, were also met at depth in the central Weald (Brightling, Sussex) but here they are to be expected, as they are on the Hercynian belt from Devonshire to northern France and Belgium, lying south of the London–Brabant Massif.

Any synthesis of marine conditions in Britain during the Devonian period depends largely on the outcrops of Devon itself and even here some matters are uncertain. Gedinnian history is unknown and any possible northern counterparts of the continental Dartmouth Slates of the south are probably unexposed. With the late Siegenian and Emsian marine transgression (Meadfoot Beds and Linton Beds) the picture becomes plainer. In the south the seas gradually cleared so that by Givetian times colonies of stromatoporoids and corals contributed to the Torquay Limestone. It follows that the arenaceous invasion of the Hangman Grits did not reach beyond mid-Devon; in the north marine conditions were subsequently re-established, even to the deposition of the thin limestones of the Ilfracombe Beds.

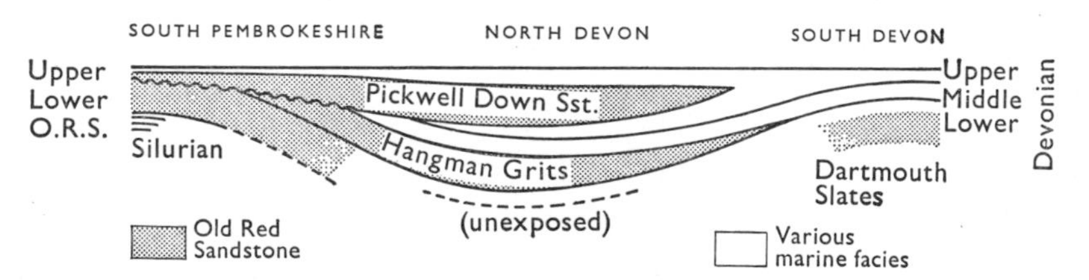

Fig. 33. Diagrammatic section to show the possible relations between the Devonian rocks of South and North Devon and the Old Red Sandstone of Pembrokeshire, the base of the Lower Carboniferous beds being taken as horizontal. Two continental incursions swell the thickness in North Devon where, however, no beds of Siegenian or Gedinnian age are recognized. Any northward continuation of the Dartmouth Slates is unknown. Not to scale.

A second invasion—of the Pickwell Down Sandstone—was likewise restricted and during the Famennian stage there was a very marked contrast between the heavy detrital sedimentation of the north (whether of inshore seas or Old Red Sandstone in type) and the much thinner shales and limestones or ostracod slate facies that are found east and west of Dartmoor. The Middle and Upper Devonian vulcanicity of the south emphasizes the dissimilarity.

On review it appears that in some ways the commonly drawn distinction between the Old Red Sandstone continent and the seas beyond its margin is too simple. The former comprises such diverse facies as the conglomerates and lavas of central Scotland, the red fluviatile Dittonian of the Welsh Borders and the lacustrine carbonate silts of Caithness. In the south there was indeed a fluctuating, transitional belt in the region of the Bristol Channel but the facies-change hidden under the Carboniferous rocks of mid-Devon is nearly as striking.

Any southern margin to this section of the Hercynian geosyncline is almost wholly obscure, but there may be some hint of it in the coarse sandstones and conglomerates of the Gramscatho Beds where these contain plant debris in south Cornwall.

REFERENCES

Old Red Sandstone. Allen, 1960, 1962, 1963, 1965; Allen and Tarlo, 1963; Ball, Dineley and White, 1961; Crampton and Carruthers, 1914; Charlesworth, 1963, ch. 8; George, 1958*a*, 1960*a*, 1963*a*; Hickling, 1908; Mitchell and Mykura, 1962; Shackleton, 1940; Welch and Trotter, 1961; Waterston, 1965; Westoll, 1951.

Marine Devonian. George, 1962*a*, *b*; Goldring, 1962; Hendriks, 1937, 1959; House, 1963; House and Selwood, 1966; Lloyd, 1933; Simpson, 1951, 1962.

General. Simpson, 1959*b*.

British Regional Geology. Northern Highlands, Grampian Highlands, Midland Valley of Scotland, South Wales, Bristol and Gloucester, South-West England.

6

THE CARBONIFEROUS SYSTEM

LOWER CARBONIFEROUS OR DINANTIAN

The scope of the Carboniferous System in the British Isles is immense. In duration it is exceeded only by the Cambrian and Ordovician; in outcrop it covers a larger area than any other system; the variation of facies is unrivalled and as a result the faunal and floral range is outstanding. The Upper Carboniferous Coal Measures are the most valuable of the country's natural resources, and for this reason, if for no other, the volume of research on British Carboniferous rocks and their problems is prodigious.

Table 25. *Principal divisions of the Carboniferous System in Great Britain*

	Stages		Lithological groups
UPPER CARBONIFEROUS	STEPHANIAN*		Coal Measures
	MORGANIAN		
	AMMANIAN		
	NAMURIAN		Millstone Grit
LOWER CARBONIFEROUS	DINANTIAN	VISÉAN	Carboniferous Limestone
		TOURNAISIAN	

* Only the lowest Stephanian beds are found in this country.

In North America there is a convenient stratigraphical division of this time-span into two systems, Mississippian and Pennsylvanian, but the dividing line between them does not fit conveniently into traditional European practice or faunal boundaries, the Lower Carboniferous of Europe being shorter than the Mississippian. The main divisions employed in north-western Europe are given in Table 25, together with the traditional lithological grouping that is used in much of England and Wales; in Scotland, however, this grouping is only partly applicable, and in Devon and Cornwall not at all. Upper Carboniferous rocks are poorly represented in Ireland, particularly the Coal Measures. Even in England and Wales the traditional terms on the right hand side of the table are only generalized labels for great groups of strata, and are not simple lithological descriptions. Each group may contain a great variety of rock-types.

Many of the areas of outcrop belong to the sedimentary provinces that were established in Devonian times; these include Devon and Cornwall (which continue to be part of the Middle European or Hercynian geosyncline), southern Ireland, the Middle Valley of Scotland, and the Anglo-Welsh Lowlands, the last being known in Carboniferous stratigraphy as the South-west Province. On the other hand in the Midlands and north of England the invading Carboniferous seas covered several areas where Devonian erosion was active, so that Old Red Sandstone is unimportant or absent and the base of the Carboniferous rocks, where visible, is often unconformable on Lower Palaeozoic strata. The Pennines, Northumberland and the borders of the Lake District are in this category, as is North Wales, and possibly parts of Northern Ireland.

In the British Isles there was a general progression through Carboniferous times from dominantly marine conditions to dominantly fresh water and terrestrial, and the faunas and floras follow suit. It might be thought that, in the earlier part of the period at least, a fairly straightforward system of zonal fossils could be used to subdivide the marine rocks, but unfortunately this aim is only partly fulfilled. The marine rocks are commonly limestones, and the goniatites, continuing up from the Devonian period, are hardly ever found in this facies. The most abundant groups are brachiopods and corals. These are more closely governed by local conditions and individual species are less widespread; they were probably also more slowly evolving. Goniatites appear in numbers where black or dark-grey marine shales come in and disappear again at the top of the Middle Coal Measures. They are occasionally found in the reef limestones but very rarely in other calcareous rocks.

The principal freshwater group (or more correctly, non-marine) is the lamellibranchs, or 'mussels' of the Coal Measures. These are locally abundant in the mudstones and ironstones, and there are also bands containing fish remains and small crustacea. The succession of plants forms a zonal system which is used side by side with that based on the lamellibranchs. Macroscopic plants are being increasingly supplemented by spores, which have the advantage of being obtainable from the coal seams themselves as well as the intervening shales; they form a third series of zones in the Coal Measures (p. 411).

The term 'band' (fossil, faunal or marine band) is not confined to Carboniferous stratigraphy but it is one much used in that system. It

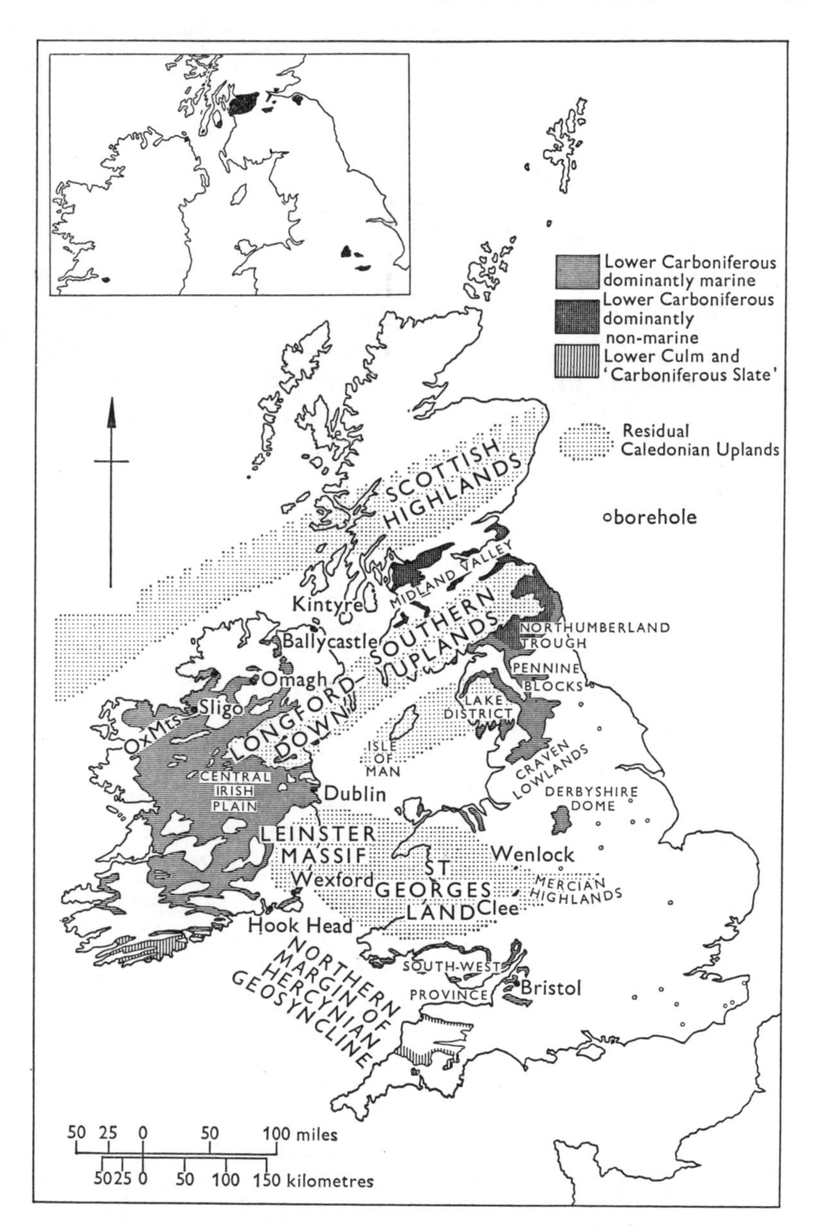

Fig. 34. Lower Carboniferous outcrops and residual Caledonian uplands. Boreholes reaching these rocks are numerous in the east Midlands and are also found in the north-east and south-east. Inset: principal areas of Lower Carboniferous lavas.

refers to a small thickness of strata, usually not more than a few feet, rich in one or a few species. Bands are useful mapping, or 'marker', horizons and as such have a severely practical basis. The marine bands of the Coal Measures are extremely widespread, and the most important, with their distinctive faunas, are found on the continent as well as in Britain. The bands of the Dinantian limestones are slightly different and act as marker horizons in, say, a single sedimentary province. They are characterized by species of brachiopods, corals, algae or foraminifera and some locally form biostromes; even distinctive rock-types may occasionally serve. Were good zonal fossils abundant they could be used instead, and in some degree the use of faunal bands in Carboniferous rocks is an indication of the difficulties experienced in Carboniferous zonal stratigraphy.

The remainder of this chapter is concerned with the Lower Carboniferous or Dinantian. The latter name is taken from the Belgian town of Dinant, where there are good sections in the limestone cliffs of the river Meuse, or Maas; the English name, Avonian, from the Avon gorge at Bristol, has approximately the same limits.

THE SOUTH-WEST PROVINCE

The most extensive Dinantian outcrops in this province are those north and south of the South Wales Coalfield, in the Gower Peninsula and westwards along the Hercynian strike into the folded belts of South Pembrokeshire (Figs. 19 and 20, pp. 90–1). East of the Usk Anticline the Forest of Dean Coalfield is also bordered by Lower Carboniferous rocks and to the north the Clee Hills bear small outliers of this age. South-east of the Severn Estuary the Mesozoic cover is more extensive but isolated limestone outcrops protrude through it. They include those associated with the Bristol Coalfield in the north and Radstock Coalfield in the south; the last is bordered by the complex ridge of the Mendips, consisting of four short anticlines (periclines) *en echelon*, each with a core of Old Red Sandstone (Fig. 23, p. 106).

There are also a number of outcrops west and south-west of Bristol and one of them is cut through by the River Avon on its way to the Bristol Channel. The gorge so produced was the original type section of Carboniferous Limestone in this province, where the coral-brachiopod zones (p. 411) were first established in the early years of this century; fossiliferous sequences are also well exposed in the steep valleys or gorges of the Mendips. The K zone in the Bristol region is also known as the Lower Limestone Shales; the remainder is the Main Limestone. The coral-brachiopod zones have proved to be useful in some types of limestone, especially in the south-west, but in other rock-types, or in more distant areas, they are less reliable. The Viséan zones are more generally applicable than the Tournaisian, which in Britain is also a more restricted stratigraphical division.

Tournaisian facies

The Lower Limestone Shales do not differ much in lithology in the various outcrops of the South-West Province and, as at Bristol, they comprise a variety of calcareous shales and mudstones, with subordinate crinoidal, oolitic and polyzoan limestones. They represent the first major invasion of marine waters over the coastal flats of the Old Red Sandstone; the colour change at the base is usually abrupt, but there is no conspicuous break, except possibly in the northernmost outcrop, on the Clee Hills. In thickness the group varies significantly, from 100 feet along the north crop of the South Wales Coalfield to 200 feet in the Forest of Dean, and 500–600 feet at Bristol, the Mendips, Gower and south Pembrokeshire. This tendency to a southerly thickening, which is repeated so frequently in many of the Carboniferous divisions in this province, is part of a general downwarping of the sedimentary basin, and thicker accumulation, away from the upland of St George's Land.

Table 26. *Carboniferous Limestone in the Avon Gorge, Bristol*

Zones	Lithological formations	Average thickness (feet)
D (largely D_1)	Massive, oolitic and rubbly limestones with corals and brachiopods	450+
S_2	Fine-grained limestones with '*Seminula* Oolite' in the middle and algal limestones at the top	760
C_2S_1	Fine-grained limestones and calcilutites, partly dolomitized	190
C_1	'*Caninia* Oolite' above and '*Laminosa* Dolomite' (altered crinoidal limestone) below	180
Z	Dark-grey, crinoidal limestones with brachiopods and simple corals	400
K	Calcareous shales and impure limestones; basal division ('Km') of more detrital beds and red crinoidal, polyzoan limestones	480
	(Old Red Sandstone)	

The same influence continued during Z and C_1 times and produced not only conspicuous variations in thickness but also in facies. All areas show a decrease of detritus as the Z_1 beds succeed the shaly beds below. Away from the contemporary shore bioclastic limestones were formed; some contain brachiopods and simple corals, first zaphrentids and later caniniids, but there are other rock-types that are poor in fossils. Chief among these are the inshore, current-bedded oolites which occupy much of these zones to the north, but only appear at Bristol towards the top of C_1. The thickness changes are most striking in the west, where in Pembrokeshire some 1,300 feet of Z and C_1 beds become thin and finally disappear northwards in a distance of 8 miles. Part of this reduction is due to later removal (the Mid-Dinantian movements), but there is contemporary thinning of individual beds as well.

At the Clee Hills the Lower Limestone Shales are overlain by 50 feet of oolitic and crinoidal rocks, which is all that now remains of the Main Limestone; but again more may have been deposited and then eroded, since the sandstone lying unconformably above is Upper Carboniferous in age.

Viséan facies

The inshore type of C_2S_1 beds consists of calcite mudstones (or calcilutite) and algal limestones—products of broad, very shallow water flats (Fig. 35). The algae sometimes grow as sheet-like masses or incipient patch reefs. At some places on the southern edge of the lagoonal flats, for instance in west Gower, the mudstones show desiccation cracks, breccias and slumping. The deeper water equivalent in south Pembrokeshire, south Glamorgan or the Mendips is a continuation of the bioclastic, standard, limestones.

The S_2 beds in the north are often highly oolitic—the Seminula Oolite; but again to the south this facies is either restricted to the portion of the zone, as at Bristol, or absent, as in south Pembrokeshire. Above there follow more algal limestones which are very widespread. The return to the bioclastic, fossiliferous limestones, rich in corals, at the base of D_1 is one of the most striking facies changes in the province and this zone is more uniform than any of the earlier ones. It represents a general gentle depression of the basin as a whole, so that fragmental rocks and their varied faunas could be carried over areas previously held by more restricted facies. The uppermost limestones, D_2, are thin, and soon give place to the detrital facies that brings the Dinantian sedimentation to a close. At Bristol these are shales, and then massive sandstones, but over most of South Wales they take the form of dark limestones and shales, often with chert; they are without corals and characterized chiefly by brachiopods and very rare goniatites. Although thin, this facies (sometimes called the Upper Limestone Shales) can be traced from South Wales to Belgium; the age is uppermost Viséan (P_2, p. 411)

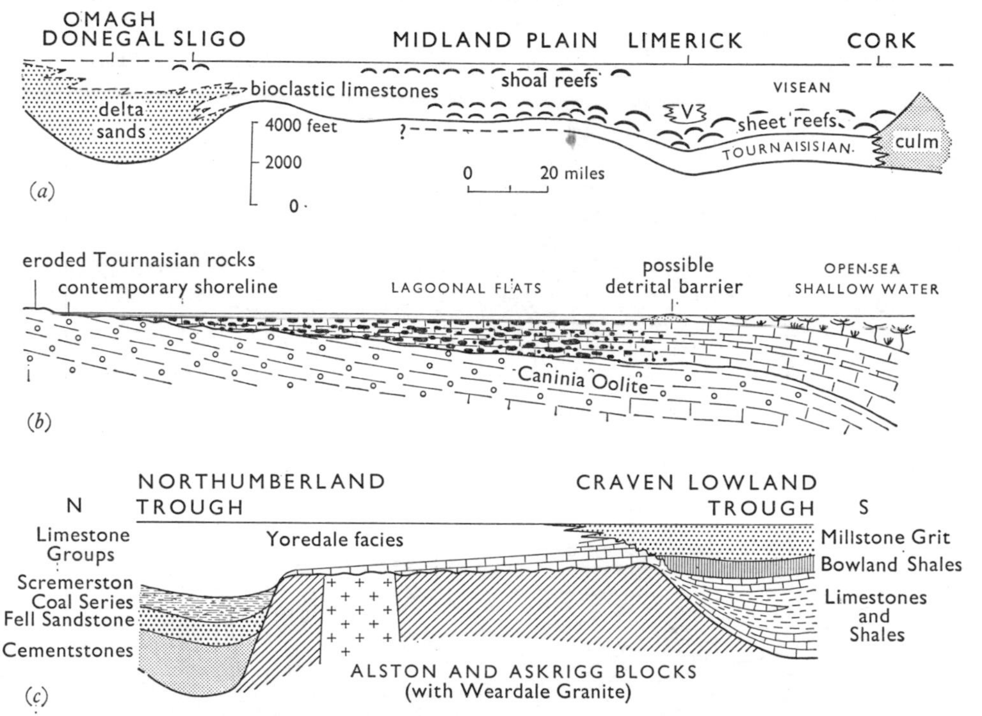

Fig. 35. Lower Carboniferous sedimentation. (*a*) Generalized section across Ireland to show the principal changes in facies (after George, 1958, p. 280). (*b*) The inferred conditions of sedimentation of $C_2 S_1$ limestones in South Wales (from George, 1958, p. 255). (*c*) Diagrammatic section from the Craven Lowlands to the Northumberland Trough, showing relation of the blocks and basins. The upper surface is approximately that of the E_2 (lower Namurian) zone. Not to scale.

Sandstones of the north-east

The limestones described in the last section are the normal marine Viséan facies and in South Wales show the same expansion in thickness to the south as do the Tournaisian limestones below. In the north-east they are partly or largely replaced by a local detrital incursion, which in the Forest of Dean is called the Drybrook Sandstone; in addition to current-bedded sandstones with drifted remains of plants there are siltstones, sandy limestones and shales. These beds are thickest and occupy the longest time-duration in the north of the Forest of Dean (S and D zones); farther south they are thinner and, moreover, are split by a limestone into Upper and Lower Sandstones, so that there was an oscillation between sandy influx and normal S_2 calcareous deposits. The sands did not reach the northern part of the Bristol Coalfield till D times, and very little reached the Mendips at all, nor spread westward over the Usk Anticline into Wales. Such a restricted accumulation, which may amount

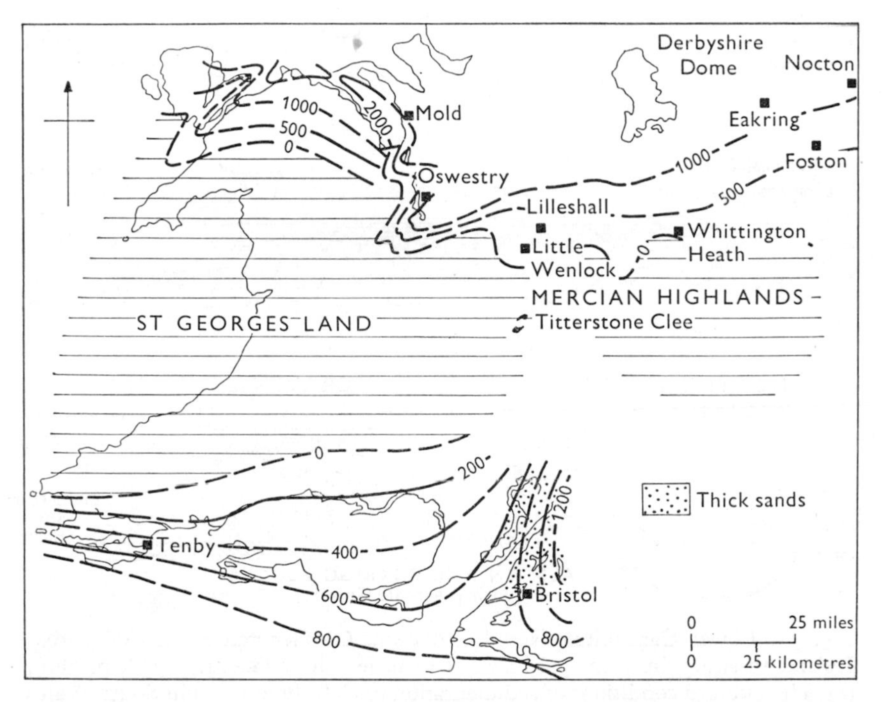

Fig. 36. Map illustrating the form of St George's Land and the Mercian Highlands as shown by the approximate isopachs of the D zone. The sandy influx (Drybrook Sandstone, etc.) is seen in the south. (After George, 1958, p. 284.)

to as much as 700 feet, is probably best interpreted as the deltaic deposits of a river, piling up locally on the shallow sea floor. The isolation of this facies is part of the general mystery that surrounds St George's Land, since otherwise it seems to have supplied virtually no detritus into the Dinantian seas, at least from Z times onwards. It is possible that we should not envisage it as an upland of substantial relief, but as a relatively low swell, that was only rejuvenated gently, and tilted, as the sea floor to the south sank beneath its weight of calcareous sediments.

The Mid-Dinantian movements in South Wales

These were not a vigorous orogenic phase and only rarely produced any angular discordance, but were rather an accentuation of the tilting movements just described, so that what was for most of the period only a northerly thinning of the limestones became for a time actual emergence and erosion. It is shown by such features as eroded, irregular limestone surfaces, local conglomerates, and the reduction or absence, in varying degree, of C_1 and C_2 beds over much of the northern outcrops. The succeeding, overstepping unit is usually the Seminula Oolite, and the effects of these Mid-Dinantian movements may be graphically demonstrated by observing the way in which the oolite transgresses the zones below; in particular westwards along the north crop of the South Wales Coalfield there is a transgression across all the underlying beds, C_2, C_1, Z, K and Old Red Sandstone. Finally for a short stretch, in mid-Pembrokeshire, the S_2 limestones lie on Silurian mudstones. Both tilting and emergence were thus most pronounced in a region near the north-west of the present outcrops.

SOUTHERN AND CENTRAL IRELAND

The various obstacles that hinder our reconstruction of Dinantian conditions in Wales and southern England—the blank areas of north Devon and the Bristol Channel, the positive barrier of St George's Land—are only partly reflected in southern Ireland. St George's Land is probably represented by the Leinster Massif, but west of this there seems to have been a continuous area of sea, much of it shallow, covering all the southern and western parts of the country (Fig. 35). The first signs of a major landmass are far away on the borders of the Dalradian highlands of the west and north. This impressive stretch of Dinantian rocks ought to provide us with a much more continuous picture of conditions, and particularly of facies-change from south to north, but unfortunately much of it is hidden under the drift and peat of the central Irish plain.

The tectonic setting has comparable differences. The main Hercynian folded belt in the south has a northern margin in the Hercynian front, a belt of high dips and thrusting (Fig. 6) extending from south of Dingle Bay eastwards to Dungarvan. Beyond this the Upper Palaeozoic rocks were folded, though more gently, by Hercynian movements that were not deflected by Caledonian and earlier structures as much as they were in Wales and the Midlands. The Carboniferous rocks of the southern belt are nearest to the Culm facies of Devon (p. 196), and 'Carboniferous Limestone' only appears north of a line from Kenmare to Cork Harbour—that is, parallel to the Hercynian front but well to the south of it. There appears to be no direct relationship between facies-change and the later structure. The former has an abrupt appearance, but this may partly be due to the fact that the most conspicuous limestones on the north are Viséan, and Viséan rocks are virtually absent on the south.

The closest comparison with the Carboniferous Limestone of south Pembrokeshire is in the small peninsula of Hook Head, east of Waterford Harbour, some 75 miles away along the westerly strike. 900 feet of beds are known here, probably all Tournaisian, and there are extensive sequences to the north and west, in Tipperary and Limerick, and across into Clare. Along this belt the lowest beds are again recognizable as the Lower Limestone Shales (usually conformable on the upper Old Red Sandstone, or Kiltorcan Beds), but to the north the shaly facies probably rises into the base of the Z zone, so that the lithological division is diachronic. The shales are followed by the upper Tournaisian limestones—the equivalent of the lower part of the Main Limestone—and these are again well displayed at Hook Head; in thickness and in the absence of oolites they are most like the offshore facies of south Pembrokeshire.

The lesser known and more isolated outcrops round the southern margin of the Leinster Massif do not give the extensive view of northward thinning against a gently rising landmass that is afforded in South Wales. Nevertheless, in the Wexford syncline, near Rosslare, the Main Limestone is exceptional in resting directly on Lower Palaeozoic rocks, with neither Lower Limestone Shales nor Old Red Sandstone intervening. Here at least we have some evidence of Dinantian sediments overlapping the edge of the Caledonian upland. The extent to which Tournaisian beds stretch northwards under the Viséan limestones (themselves all too often obscured by drift) is uncertain. They are still present near Dublin, and probably in patches at least farther north and west. On the other hand, there is much evidence of Viséan overstep and in many places along the edges of the various northern landmasses Tournaisian rocks are certainly absent. A gradually rising pre-Carboniferous floor may thus be deduced in this direction.

It might be expected that in the more uniform conditions of Ireland the gentle Mid-Dinantian earth movements would have little scope for marked effects such as emergence and unconformity, and this is in fact the case. The only results they may have occasioned are sudden changes in lithology or signs of shallowing. At Wexford the late C_1 oolites (perhaps a thin representative of the Caninia Oolite) are abruptly followed by algal limestones and calcite mudstones with suncracks, like those in the C_2 beds of Wales.

Over most of the southern Irish Viséan the outstanding feature is the great stretches of reef limestones. Typically these are C_2S_1, but near Cork, where they reach 2,000 feet in thickness, reef growth may have started in the late Tournaisian times. The total spread of reefs was of the order of 3,000 square miles, and the facies can be traced through Cork, to the west coast of Kerry and Clare, and north into Tipperary and Limerick. The most abundant reef-dwellers are algae and polyzoa, and there are pockets rich in shells. Reef facies of this type (sometimes called Waulsortian after the Belgian examples) are known in different topographic and growth forms elsewhere in Dinantian rocks, but the Irish spread of reefs is exceptional in its thickness and extent. It is noticeable that this facies appears abruptly along the line of junction between the southern 'Culm' and the Carboniferous Limestone, but if the reef did form some kind of barrier or fringe, it was not accompanied by any recognizable lagoon or back-reef facies on the northern side.

The S_2 and D beds that overlie the reef limestones are the fossiliferous, bioclastic or standard rocks that have already been noted as typical of the late Dinantian over much of the British Isles. There is no thick representative of the Seminula Oolite of Wales, though in Ireland some of the S_2 limestones are locally oolitic. In and around Limerick volcanic rocks occur in the limestones above the reef facies.

THE MIDLANDS AND NORTH OF ENGLAND, NORTH WALES AND DUBLIN

The Lower Carboniferous rocks and structure in this extensive region have not the coherence of those in the south-west, and there is no one province to which they are customarily assigned. Essentially they comprise the outcrops around the Irish Sea, north of St George's Land and south of the Southern Uplands; comparable beds also continue eastwards, under the Mesozoic cover of the eastern and north-eastern counties. The base of the Carboniferous rocks is often a marked unconformity and the underlying basement strongly diversified, both in structure and topography. This irregular foundation not only influenced the post-Carboniferous movements, which produced a wide range of strikes (including Caledonoid and Charnoid), but also the history of Dinantian sedimentation and subsidence. It is common in this midland and northern province to find a number of partly connected troughs or basins, with somewhat differing facies and thicknesses; when contrasted with intervening or bordering areas of lesser subsidence, this pattern has been called 'block and basin' sedimentation, but not all the positive elements are fault-bounded blocks.

One result of this lateral change is a marked variation in the associated faunas, bringing difficulties in correlation. The age of the lowest beds varies from place to place, according to the relief of the Caledonian uplands undergoing submergence. In general the sediments become less calcareous towards the north, bioclastic limestones almost vanishing in Northumberland; here 'Carboniferous Limestone' is a misleading title for rocks of Dinantian age. Where marine shales are abundant, goniatites come into their own as zone fossils, and they also are found locally in reef limestones; the coral-brachiopod zones have been applied in the more standard, bedded limestones, but the actual species used are not always those of the South-West Province and the correlation between north and south is sometimes a little dubious.

The southern margins: Shropshire, North Wales, Derbyshire and north Staffordshire

The most northerly outpost of the South-West Province is the small outcrop of Lower Limestone Shales and oolitic Main Limestone on Titterston Clee (Fig. 10). How much farther beds of this age once extended is uncertain; their present isolation is more probably due to erosion than to original limits of deposition. The next step northward presents a very different picture; under 20 miles away, south-east of the Wrekin, Carboniferous Limestone appears from under Coal Measures in a complex of outcrops at Little Wenlock. The basal sandstones rest with marked unconformity on Cambrian and Silurian rocks; they are succeeded by 180 feet of strata, including a basalt flow, and the limestones are wholly of D age. In the long preceding period of erosion (which almost certainly included the removal of much Old Red Sandstone, because there are 2,000 feet of it on Brown Clee nearby) it may be that Tournaisian limestones were deposited, and then removed. On the other hand there is no sign of them anywhere on this northern flank of St George's Land, where all the positive information points to a Viséan marine invasion, and in several places late Viséan.

This invasion is much more extensively shown in North Wales. Here there are patchy outcrops along the north coast from Anglesey to the Vale of Clwyd, where, with Upper Carboniferous and Triassic rocks, they are affected by post-Triassic faulting. Lower Carboniferous limestones form a westerly rim to the long strip of Coal Measures from

Flintshire south into Denbighshire, and share in the general easterly or north-easterly dip under the Cheshire Plain and Dee Estuary. The basal Carboniferous beds are strongly transgressive on the Caledonian structures and Lower Palaeozoic rocks below. They are a varied series, part detrital and part calcareous; in colour they locally resemble Old Red Sandstone, but the few fossils they contain show that they were deposited at the margins of a late Viséan sea.

The succeeding limestones are varied in the amount of detritus they contain and are very largely all D in age, with possibly a small amount of S_2 at the base. They are outstanding for their exceptional thickness, being nearly 3,000 feet in the extreme north-east. This is more than the total Carboniferous Limestone at Bristol, and there must have been a singularly deep subsiding basin off the edge of St George's Land; unlike most basins in the north it received little debris from the land since much of the limestone is a pure calcareous, fossiliferous rock. The absence of detritus from St George's Land is again apparent on this northern side.

At the southern end of the Pennines (Fig. 42, p. 217), Carboniferous Limestones, C_2 to P_2 in age, are brought up by a rather asymmetrical anticline known as the Derbyshire Dome. In the Lower Carboniferous terminology this is a 'block' area, or a 'massif'. Although the Carboniferous floor is not exposed, it is probably Pre-Cambrian, for rocks resembling the Uriconian Volcanic Series were met in a borehole near Buxton; the basal limestones were apparently of S age but C beds are known in the folded westerly outcrops. 1,500 feet of limestones are exposed and another 900 feet are known underground.

Most of the limestone in the centre is of massif facies—that is, similar to the fragmental, bioclastic limestones of the south-west; it consists of a well-bedded, light-grey to white bioclastic rock with a sparse coral-brachiopod fauna. Along the western margin there is a transition, first to a belt of reef facies and then to a series of thin, impure limestones with shales. The last by reason of its greater thickness and increase in mud is a fairly typical basin facies. C_2 and B zonal goniatites have been found in the reef limestones and fossils from C_2 to P_2 in the limestones and shales.

Other impressive, though local, reef facies are found on the north and north-east of the massif, for instance at Castleton. Here they lie with steep original dips against the bedded limestones which form the central plateau; there are small unconformities between horizontal bedded limestones (D_1–D_2) and the bounding 'apron reef', but probably also a partial transition from one to the other. The reef limestones contain B and P_1 goniatites and are again overlain unconformably by Namurian shales. In a boring put down to the north of the limestone plateau a thick shaly basin facies was encountered beneath this Namurian cover, so the present edge of the limestone approximates to its late Viséan edge and the reef limestones, often brecciated, accumulated against an arc of contemporary limestone cliffs.

The three types of facies, massif, reef and basin, have been compared with those of modern reefs. The massif represents the back-reef deposits, the reef limestones the actual barrier with its steep seaward-facing slope, and the more muddy rocks the basin on the seaward side of the reef. Over the considerable period of accumulation, however, the reefs grew at different horizons and in slightly different places and there were minor uplifts, erosion, and unconformities.

The southern edge of this the Viséan sea is not evident in the Derbyshire Dome itself, but farther south at Breedon Cloud, in Leicestershire, beds of the same age are reduced to 600 feet, and in the Whittington Heath borehole (through a thick Triassic cover in south-east Staffordshire) a mere 2½ feet of sandy and muddy limestones of the D zone is all that remains. This attenuation took place against an eastward extension of St George's Land, which divided the southern seas from those of the Midlands. It is sometimes called the Midland Barrier or Mercian Highlands (Fig. 36). Farther to the west the comparable, but rather thicker, D beds at Little Wenlock tell the same story. To the east a comparable state of affairs can be deduced from boreholes penetrating Carboniferous rocks in Nottinghamshire (p. 231). From this southern shore, the Derbyshire massif may have stood out as some promontory or island with a Pre-Cambrian core to be gradually submerged in C and S times, and developed local reef facies on the slopes facing the deeper seas on the north and west.

The Craven Lowlands

The long outcrop up the centre of northern England falls into four structural and stratigraphical units; these are, from south to north, the Craven Lowlands, the Askrigg and Alston Blocks, and the Northumbrian Trough. The two centres are the twin blocks of the northern Pennines, and they are flanked on the south and north by the two basin areas (Figs. 35 and 37).

The Lower Carboniferous sequence in the Craven Lowlands, on the Yorkshire–Lancashire borders, is probably a continuation of the basin type of facies that abutted on the reefs and promontory of Derbyshire, but there is little positive information in the region of thick Millstone Grit and Coal Measures that lies between. The dominant structures are

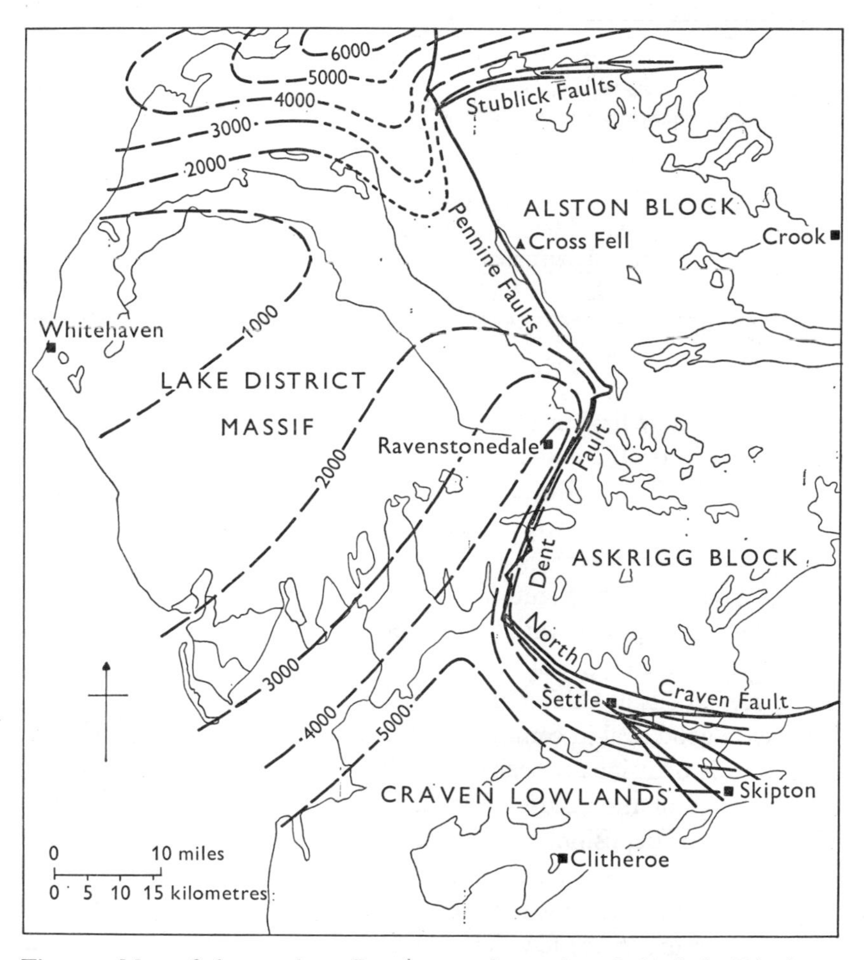

Fig. 37. Map of the northern Pennines and margins of the Lake District to show approximate and generalized isopachs of Lower Carboniferous rocks. Whether the centre of the Lake District was submerged is uncertain. (From George, 1958, p. 290.)

short, north-east trending anticlines (the Ribblesdale Folds), which are terminated by the Craven Faults; north of these rises the Askrigg Block, once also called the Craven Uplands. The lower Dinantian rocks are only exposed in the cores of a few folds, of which the Clitheroe Anticline may be taken as typical.

Table 27. *Lower Carboniferous, the Clitheroe district*

Zones	Groups	Maximum thickness (feet)
	(E_1—Upper Bowland Shales)	
P_1 and P_2	Lower Bowland Shales, including some sandy beds and limestones at the base	900
B_2 D_1, B_1 S, C_2	Worston Shale Group { Calcareous shales and limestones, reef knolls }	3,000
C_1	Chatburn Limestone Group, dark-grey, well-bedded limestones	2,750
	(base not seen)	

The lowest beds are assigned to the C_1 zone, and there is no knowledge of what may lie beneath them; it is probable, however, that this extensive basin on the east of the Irish Sea is one of the few areas of relatively early, Tournaisian, marine invasion in the north. Other important features are the inclusion of both coral-brachiopod and goniatite bearing rocks and the thick development of Bowland Shales, which straddle the Dinantian–Namurian boundary. The limestones are chiefly dark-grey and crinoidal with brachiopods and simple corals, not the light-grey, standard types of the Derbyshire massif or the Askrigg Block.

The reef knolls at Clitheroe are particularly impressive; in lithology they have a good deal in common with the reefs of Derbyshire (where a knoll form is occasionally seen) and with a later, D_1, series at Cracoe, near the edge of the Askrigg Block. Characteristic features are steep original dips of deposition and much poorly fossiliferous calcite mudstone; the last carries a number of obscure features, partly the result of recrystallization, and may have been largely algal in origin, though visible algal remains are few. The fauna in the reef facies is varied and local, but includes polyzoa, pockets of shells, occasional goniatites and crinoids; corals and brachiopods are rare. The origin of the facies and form of the knolls is a debated problem, but there is some evidence to suggest that they grew as hillocks on a shallow sea floor. Even the Cracoe group do not form a simple 'barrier' separating a back-reef of the Askrigg Block from a fore-reef and basin; there is little to suggest deep water in the Lower Carboniferous basins of the north, though it was probably a little deeper than that on the blocks. The principal difference is the greater amount of subsidence and thickness of sediments (Fig. 35).

Askrigg and Alston Blocks

Most of the faults bounding these units are post-Carboniferous, but their positions often correspond with earlier 'hinge-lines', or belts of marked thickening of the strata, where the sequence on the block passes over into that of the neighbouring basin.

The North and South Craven Faults separate the Craven Lowlands from the Askrigg Block, the northern fracture coinciding more particularly with the facies change. For a short distance there is also a Middle Craven Fault, which, unlike the others, was a fault scarp in late Viséan times; the Bowland Shales are seen banked up against it and locally override the scarp-top. The western boundary is the Dent Fault, and the more complex Pennine Faults (Outer and Inner) hold the same position for the Alston Block; the northern margin of the latter is formed by the Stublick Faults, *en echelon* (Fig. 37, p. 179). To the east the Lower Carboniferous cover dips gently under Upper Carboniferous and Permian strata, and the two blocks become less defined. The southern part of the Durham Coalfield, however, is some kind of continuation of the Alston Block, and the 'basin' type of Lower Carboniferous facies met in the Cleveland Hills borehole suggests that the eastern margin of the Askrigg Block is reached before this point.

The Dinantian succession on the Askrigg Block comprises the Great Scar Limestone below and the Yoredale rocks above. The former is not one of the really thick limestone formations in Britain despite its name, and does not approach those in North Wales or south Pembrokeshire; it consists of 700 feet or so of pure, white or pale-grey bioclastic limestones ranging in age from C_2 to D_2. The highest zone is rich in corals and brachiopods, but much of the Great Scar is rather barren, and correlation depends a good deal on faunal bands. These include, for instance, that of the algal nodules (*Osagia*) or '*Girvanella* Band' at the base of D_2 and the brachiopod *Davidsonina septosa* in the middle of D_1. The base of the limestone is decidedly irregular, the invading seas covering an uneven plateau, and C beds are found only locally.

The Great Scar Limestone does not maintain its thickness and purity to the north and in the Alston Block is only partly represented by the Melmerby Scar Limestone. The beds beneath the Yoredale facies here contain a larger proportion of detritus, and the marine invasion was later, most of the upland not being submerged till D times. Carboniferous Limestone in the sense of strictly Dinantian limestones, is thus much reduced in the extreme north and north-east of England.

Yoredale facies

The type area of the Yoredale succession is Wensleydale (valley of the Ure), in the centre of the Askrigg Block, but the facies has a wide extent and importance in the north of England and in Scotland. In essence it is a cyclic sequence, and in its simplest form the Yoredale cyclothem comprises the following:

coal, always thin and sometimes absent,
sandstone, often current-bedded and containing a rootlet bed at the top,
shales and silty shales, sparse in fossils,
shales, with marine fossils,
limestone, with marine fossils, sometimes abundant.

This sequence is often complicated by minor intercalations or subsidiary cycles.

Table 28. *Yoredale limestones and underlying beds*

Zones	Askrigg Block	Alston Block	
			BASE OF UPPER LIMESTONE GROUP
E_1	Main Limestone	Great Limestone	
P_2	Underset Limestone	Four Fathom Limestone	MIDDLE LIMESTONE GROUP
	Three Yard Limestone	Three Yard Limestone	
	Five Yard Limestone	Five Yard Limestone	
	Middle Limestone	Scar Limestone Cockle Shell Limestone Single Post Limestone	
	Simonstone Limestone	Tynebottom Limestone	
	Hardraw Limestone	Jew Limestone	
P_1, D_2 D_1 S C	Great Scar Limestone	Lower Little Limestone Smiddy Limestone Robinson Limestone Melmerby Scar Limestone Basement beds (variable and detrital)	LOWER LIMESTONE GROUP

Over much of the Pennines the correlation of the Yoredale beds is frankly lithological; the cyclothems are named after the basal limestone and on the Askrigg and Alston blocks there is a fairly certain scheme of equivalence (Table 28). The Main or Great Limestone is the thickest and most persistent and is the only one that can be traced as far as the Scottish border. Correlation farther afield is chiefly dependent on the rare goniatites in the marine shales, which are just sufficient to allow a broad zonal grouping, supplemented by the less precise limestone faunas.

It is fairly easy to interpret the local conditions of the Yoredale rocks —the limestones, shales and sandstones—but the ultimate origin of the cyclic facies as a whole is much more problematical; it extends not only into the three Limestone Groups of Durham and Northumberland, but into Scotland as well. The limestones, which are largely bioclastic in type (though algal rocks are known), represent a continuation of the shallow Dinantian seas; those seas were periodically invaded by muds, then silts and sands until the sediment rose above water-level, and the vegetation of the coal or rootlet beds was established. Then the supply of detritus ceased, and clear, limestone-forming seas were re-established. This is a satisfactory picture for any one area, with the minor retrogressive or repetitive phases implied by the minor cycle or anomalous bed that may be intercalated in the major cyclothems. The source of the sediment is uncertain but is deduced to be somewhere to the north or north-east. It fails, largely, in south Yorkshire and Lancashire and in west Cumberland, and forms an increasing proportion of the facies northwards; moreover a 'northern continent' is implicit in much of Carboniferous times—the major Caledonian range from Ireland and Scotland to Norway.

The intermittent supply of detritus is a puzzling problem for which many suggestions have been made. Chief among these are periodic diversions of the large river which was the transporting agent, periodic rejuvenation in the northern landmass (and hence in the carrying power of its rivers), or a combination of the latter with a eustatic rise and fall of sea-level, related to climatic fluctuations.

Erosion channels are not rare in the Yoredales of the Pennines, but the only extensive erosion resulted from an uplift in late Viséan times of the south-east corner of the Askrigg Block. Here there is marked transgression of the lowest Millstone Grit (E_1) on to beds as low as the Middle Limestone, presumably a local effect associated with the instability of the Craven Fault belt. At this southern extremity of the Yoredale facies its upper limit (i.e. the termination of the thick, persistent limestones) almost coincides with the Viséan-Namurian junction; farther north, however, this type of bed continues a good deal higher; fully marine limestones are known of E_2 age in Northumberland and Scotland and R_1 in County Durham.

Northumberland Trough

In some ways this is analogous to the Craven Lowland Trough but there is a marked difference in facies. The structures also have not the regularity of the Ribblesdale Folds though sharp flexures do exist, especially in north-east Cumberland, and there is a broad swing of the strike around the Cheviot volcanic mass. To the south-west the Carboniferous beds are covered by the New Red Sandstone of the Solway Firth and Carlisle plain. North-west of this cover, across the borders into south Scotland, the intermittent outcrops along the northern shores of the Solway continue the line of the sedimentary basin towards the more persistently marine area of the Irish Sea.

Table 29. *Dinantian and Lower Namurian succession in the Northumberland Trough*

Zones	Groups
E_1 and part of E_2	Upper Limestone Group
P approximately	Middle Limestone Group
D	Lower Limestone Group
?C_2 to ?S_2	Scremerston Coal Group Fell Sandstone Cementstones

The succession in this trough, which is some 6,000 feet thick, is shown in Table 29. The three lower divisions come in north of the Stublick Faults (Figs. 24 and 35); in the centre and north of the county they are almost entirely non-marine and their grouping is lithological. It may be that Tournaisian beds are lacking, and if so the junction with the Old Red Sandstone on either side of the Cheviot is a major unconformity.

There are various, local, littoral facies at the base of the Cementstones (for instance where they overlie the Cheviot lavas), but above this the beds are dominantly argillaceous, with some sandstones and the thin, impure limestones from which the name is derived. The rare fossils include small invertebrates and plants and towards the centre of the trough there are intercalations of algal limestones. The Fell Sandstone is a group of variegated, massive and current-bedded rocks. The general south-westerly thinning and direction of foreset dips suggest a north-easterly derivation; in this, and the deltaic type of deposition, they resemble the later, much more extensive, influx of the Millstone Grit. More varied conditions, with intermittent extensive vegetation followed, to form the coals, shales, sandstones and impure limestones of the Scremerston Coal Group.

These three lower divisions bear witness to the steady accumulation in the Northumberland Trough, but for the most part sedimentation was non-marine. The succeeding Lower Limestone Group represents a major marine invasion, and thereafter cyclic Yoredale conditions persisted over all north-east England. The conspicuous difference between block and basin is the increased thickness in the detrital beds (the sandstones and shales) in the latter, the line of the Stublick Faults acting as a hinge at the edge of the Alston Block.

Towards the west of the Northumberland Trough, for instance in the Bewcastle Anticline of north-east Cumberland (Fig. 25), the lower divisions include an increasing number of marine limestones, often algal, and only the Fell Sandstone maintains its identity as a recognizable unit; the faunas also are more varied, and include brachiopods in particular. These are not ideal stratigraphical fossils, but they suggest that the lowest visible beds in the anticline are C_2. The scattered outcrops westwards along the shores of the Solway Firth, from Dumfriesshire to Kirkcudbrightshire, continue this picture of increasingly marine conditions. The rocks include richly fossiliferous limestones, some with corals; to the north they lie with conspicuous unconformity, and sometimes local pebble beds and sandy detritus, on the Silurian rocks and Caledonian granites of the Southern Uplands. Despite the local variation and incomplete exposure, the main picture from Northumberland to the Solway shows clearly how the non-marine facies of the north-east gives way to a greater proportion of marine rocks in the south-west.

Lake District and Isle of Man

The Lower Palaeozoic inlier of the Lake District was an upland in Dinantian times, probably continuous with the Manx hills. At present the Lower Carboniferous rocks form a discontinuous ring round the central inlier (Fig. 24); for the most part they dip gently away from it, except where locally deflected among the faulted southern outcrops north of Morecambe Bay. This Lake District fringe is now separated from the Pennines by the Dent and Pennine Faults and the New Red Sandstone of the Vale of Eden, but in late Viséan times at least there was some continuity from one province to the other.

The largest single outcrop of the fringe is that of Ravenstonedale and

Shap on the east, stretching from the Dent Fault to the margins of Ullswater. It was here that the coral-brachiopod zones of the south-west were first applied to the north; only the D zone, however, contains a sufficiently far-reaching fauna to have a number of species common to north and south and more local forms have had to be employed in lower beds.

Tournaisian rocks are present only in Ravenstonedale where they consist of dolomitic shales and mudstones (the Pinskey Gill Beds); they are overlain by red and green sandstones, locally conglomeratic, with abundant debris from the Shap Granite—one of the several Caledonian intrusions of the Lake District. The succeeding dolomitic and algal limestones of the C zone are also local, but in S times there was a widespread submergence around the flanks of the Lake District as a whole and pure white limestones were laid down in Ravenstonedale. They are interrupted by the incursion of the Ashfell Sandstone; this is an arenaceous wedge, diachronous in both upper and lower boundaries, that is probably a remnant of the much more extensive Fell Sandstone in the north-east. The D_1 limestones are standard in type, and although D_2 beds are largely missing through later erosion, there is some suggestion that the cyclic Yoredale sedimentation, with its detrital invasions, reached the eastern and northern borders of the Lake District. The Ravenstonedale gulf received about 4,000 feet of Dinantian sediments (Fig. 37), and is an important area of early marine invasion on this side of the Irish Sea. It lay between the upland areas of the Lake District and Askrigg Block, and perhaps was continuous with a Tournaisian basin in the Craven Lowlands.

On the north-western side of the Lake District a rather different picture is revealed by the Lower Carboniferous rocks that crop out between the Skiddaw Slates and the Coal Measures of the Whitehaven Coalfield. At the base there is a local detrital facies, and above that a series of limestones from the S to E_1 zones, interrupted only by relatively thin beds of shale and one sandstone. Neither the Ashfell Sandstone nor the Yoredale cyclothems reached West Cumberland. The 700 feet of beds are also relatively thin, compared with the sediments of Ravenstonedale or Bewcastle. This dominantly calcareous sequence, in an area of lesser subsidence, has some parallel with the limestones on the north side of the Solway Firth; they are a further indication of a sea area that lay between the Southern Uplands and the Cumbrian-Manx ridge.

The outcrops on the Isle of Man are distinctly similar to those of the Lake District and the Pennines. On the north a limestone succession, hidden under Trias and drift, forms an extension to that at Whitehaven; on the south there are reef and bedded limestones, together with dark grey limestones and shales (C_2 to D_1). These latter rocks and fossils have a certain amount in common with those of the Craven Lowlands and probably form a small remnant in an east–west facies belt that extended across from Yorkshire to Ireland.

Dublin

Whether the northern margin of St George's Land took a direct course across the Irish Sea is uncertain, but the remnants, in North Wales and the Leinster Massif, lie on the same latitude. The Longford–Down and Southern Uplands mass, however, is Caledonian in plan as well as in age, and trends south-westwards from Scotland into northern Ireland. Thus whereas the two major Upper Palaeozoic landmasses are something like 100 miles apart on the east, with all the complex units of north-western England in between, in Ireland the distance narrows to a strait of less than 20 miles—the lowlands of County Dublin.

Table 30. *Dinantian succession in north County Dublin*

Zones	Lithological groups	Average thickness (feet)
P_2 and P_1	*Posidonia* Shales and Limestones	350
B or S_2D_1	*Cyathaxonia* Beds, dark-grey cherty limestones	200
S_1 and C_2	Limestones, including pebbly and oolitic rocks	?400
C_2	Rush Conglomerate Group, conglomerates, sandstones, shales and limestones	500
C_2 and ?below	Rush Slates, with shales and occasional limestones	1,500+
	(base not seen)	

The most striking exposures in County Dublin are on the coast north of the city and Dublin Bay, as far as the Balbriggan Massif (Fig. 27, p. 124). Most of the Dinantian rocks (Table 30) consist of dark-grey, impure limestones. They are diversified inland by abundant reef knolls, chiefly C in age, and the upper beds tend to be shaly and contain chert. There is no palaeontological evidence of Tournaisian rocks in the Rush section (though there is a great thickness below the lowest Viséan faunas) but they are present farther north, at Swords. Goniatites appear in the C_2 zone and the P_1–P_2 beds at the top resemble a rather more calcareous version of the Lower Bowland Shales of the Craven Lowlands.

On the south side of the Dublin strait the *Cyathaxonia* Beds contain blocks of granite and metamorphic rocks that have slid seawards off the edge of the Leinster upland, and on the north there is some indication, in the highest beds, of overlap on to the Balbriggan Massif. The Rush Conglomerate, which crops out on the coast 15 miles north of Dublin, is a striking local example of a submarine landslip, in which angular boulders have ploughed into and contorted the bottom muds; the Rush Slates also include boulders from the Leinster Granite. This coarse detritus diminishes westwards, passing into a more normal calcareous sequence.

Most of the Dublin rocks have clear resemblances to those of the Craven Lowlands. The Irish version is, however, much thinner, even including the local increment of the Rush detrital groups. The Dublin strait was thus in communication with a Dinantian 'Irish Sea', but it was not a subsiding basin comparable with those of northern England.

SCOTLAND AND THE NORTH OF IRELAND

It has already been shown how in Devonian times the Midland Valley of Scotland carried across into north-central Ireland, but was less easily defined there owing to the great spread of Lower Carboniferous rocks. In a study of the latter the situation is reversed and the Irish picture is more continuous than the Scottish. The Dinantian geology is also more coherent in this westerly region, because, apart from the small Ballycastle Coalfield on the Antrim coast, the rocks are more thoroughly marine. There was thus a gradual westward fall in sediment-level, at least during much of the Viséan period, from Scotland and Northumberland where it was generally above sea-level, to Ireland, where it was below.

Scottish Carboniferous stratigraphy is rendered particularly intricate by a heterogeneous tectonic foundation, marked lateral changes in facies and thickness, and vigorous and varied volcanic outbursts. There is also a paucity of zonal fossils, and correlation with Northumberland (similarly handicapped) and farther south is not simple, or wholly agreed upon. One result has been a confusing similarity of names for strata of different ages largely because they contain similar types of rocks. The main economic coal-bearing groups of Scotland are the Productive Coal Measures of Ammanian age and the Limestone Coal Group (Table 31). Until recently the latter formed the middle member in a tripartite grouping of the 'Carboniferous Limestone Series' of Scotland, and the seams were sometimes called 'Lower Carboniferous coals'. They are not

Dinantian, however, since the limestone-bearing strata of Scotland are late Viséan and early Namurian and the Limestone Coal Group is wholly of E age. The various 'Limestone Groups' therefore do not mean the same thing in northern England and Scotland. Most of the Dinantian of the Midland Valley is represented by the Calciferous Sandstone Measures—an exceedingly varied succession; the divisions shown in the table really only apply in the Oil Shale basin of the Lothians in the east and the age of the lowest beds is very uncertain.

Table 31. *Divisions of the Carboniferous rocks of Scotland*

Division		Subdivision	Zone
Upper Coal Measures, or 'Barren Measures'			
Middle and Lower Coal Measures, or 'Productive Measures'			
Passage Group, or Millstone Grit			E_2, R and ?G
Upper Limestone Group			E_2
Limestone Coal Group			E_1
Lower Limestone Group			P_2
Calciferous Sandstone Measures	locally divisible into	Upper Oil Shale Group Lower Oil Shale Group Cementstone Group	P_2 and below

The zones given are approximate and can only be applied where marine rocks are present.

This rather awkward grouping and nomenclature may become clearer if we recollect the conditions in Durham and Northumberland. On the Alston Block the Yoredale cyclic facies, with its marine limestones, continued up into Namurian strata, and the Millstone Grit was much thinner than in Yorkshire. In Northumberland this is a little accentuated: the Millstone Grit is a rather thin, poorly defined, arenaceous sequence below the Productive Coal Measures and some of the coals in the Yoredale facies below have become workable. The Scottish succession is a further step in the same direction. The Lower and Upper Limestone Groups, and intervening Limestone Coal Group, are all variations on a Yoredale theme, being a repetitive, and sometimes cyclic, sequence of marine and non-marine beds ranging from P_2 to E_2. In the middle group the coals are of considerable importance.

Calciferous Sandstone Measures

The base of this group cannot be determined palaeontologically. In some places there appears to be a transition, from dominantly red rocks to grey, or an alternation; certainly beds resembling Upper Old Red

Sandstone are known well above the accepted base of the Carboniferous and grey beds with impure limestones beneath. In other places there is a recognizable break, and even an overlap locally of Cementstones on to Lower Old Red Sandstone. It is the extent of this unconformity that is uncertain, and some authorities believe the Tournaisian to be missing altogether, both in the Midland Valley and around the Cheviot. A gap of this size would imply a general uplift, or at least non-deposition, over a long period; even so it would not be more exceptional than, for instance, the widespread absence of Middle Old Red Sandstone.

In the south-west of the Midland Valley the Calciferous Sandstone Measures are dominantly composed of cementstones. Typically they are not unlike those of Northumberland, shales and impure dolomitic limestones and minor, thin sandstones. The colour is normally grey, but may show red and green mottling; films of gypsum and salt pseudomorphs locally indicate rather saline conditions, and the sparse fauna includes *Lingula*, lamellibranchs and ostracods, together with fish scales. The group becomes thicker and more sandy to the north, and thinner, abruptly, over the Inchgotrick Fault and on to the Ayrshire 'shelf' to the south (Fig. 38). Here it does not exceed 1000 feet and is often less; along the line of the Lesmahagow inlier, for instance at New Cumnock, it is absent altogether. The Ayrshire shelf, backed by the Southern Upland Border Fault is thus a positive local massif which received very little Dinantian sediment, and has that much in common with the 'blocks' of other provinces. The sedimentary facies, however, is not very different from that in the Clyde basin, except where conglomerates and sandstones locally accompany erosion levels and overlap.

The west of the Midland Valley is famous for the thick, widespread volcanic rocks of this age, the Clyde Plateau Lavas; these are subaerial flows, particularly of olivine basalts. They may amount to as much as 4,000 feet near Glasgow, and in most places overlie cementstones, of varying thickness, but locally overlap these on to Old Red Sandstone; they did not extend on to the Ayrshire shelf. The irregular surface of the flows was very largely submerged before the end of the period, and an Upper Sedimentary Group deposited in the hollows. This differs from the lower cementstones in being more varied. It lacks the fresh water, dolomitic rocks but contains some coals and a few marine limestones; and the faunas include brachiopods and corals, *Dibunophyllum* among them. The beds represent the first local marine invasions from the west, before the more extensive submergence at the base of the Lower Limestone Group above.

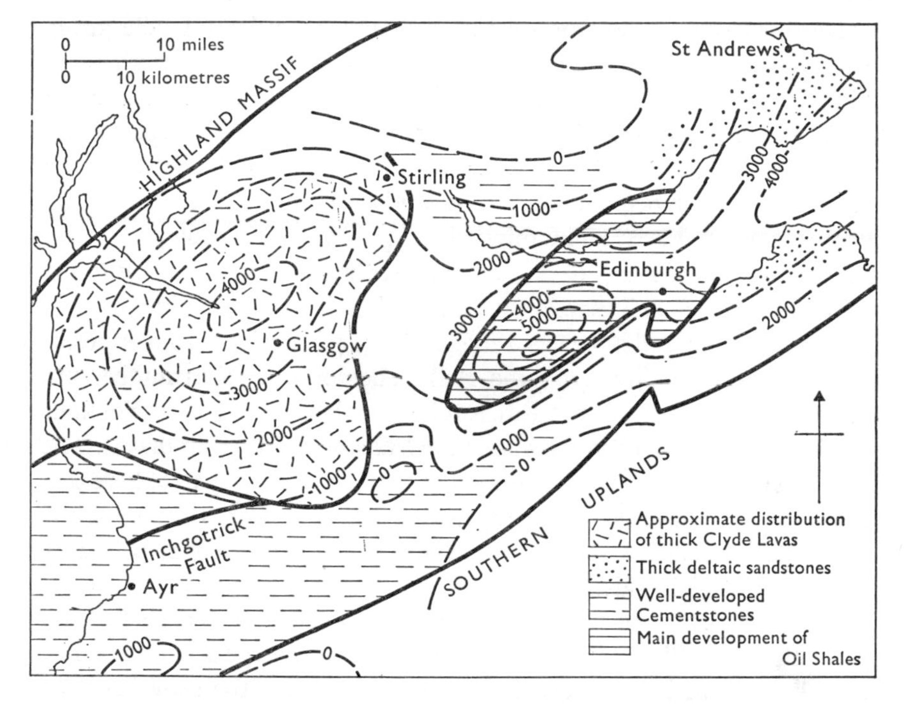

Fig. 38. Map showing relation of the principal facies of the Midland Valley of Scotland to the isopachs of the Calciferous Sandstone Measures (the thickness of the Clyde Plateau lavas near Glasgow is uncertain and isopachs here are conjectural). (Redrawn from George, 1958, p. 208.)

Farther north and east typical cementstones are not plentiful (though sometimes found near the base) and two other sedimentary facies take their place—the Oil Shales of West and Midlothian, on the south side of the Forth, and more sandy beds in East Lothian and also in Fife. The typical Oil Shale Group occupies a deep sedimentary basin west of Edinburgh, where up to 6,000 feet of fine-grained sediments accumulated. Much of them consists of shales, thin sandstones and occasional limestones; the true oil shales are thin seams of laminated bituminous muds, products of decaying vegetation in stagnant pools. Most of the impure limestones in the Oil Shales are freshwater, but there are a few marine beds, either limestones or shell beds; they disappear westwards, presumably against the barrier of the Clyde Plateau Lavas. One of the most useful is the Pumpherston Shell Bed in the lower part of the group, which contains rather fragmentary goniatites broadly indicative of the *Beyrichoceras* zone, or somewhere in the middle Viséan.

North-eastwards the true oil shales disappear and the whole group becomes thinner over the anticline of Burntisland, on the north shore of the Firth of Forth. It expands again in Fife where sandstones come in. These are locally thick, coarse-grained, current-bedded and exhibit many deltaic features, such as channelling. Plant debris and shells occur in the finer beds. Zonal fossils are lacking but the incursions are probably upper Viséan. The mineral and pebble content suggests that a source area, rich in Old Red Sandstone sediments and lavas, lay to the north or north-east. There is thus a linked series of deltaic invasions in eastern Scotland and northern England (Fig. 40), derived from approximately the same direction—the Fell Sandstone of Northumberland (? early to mid-Viséan), the Fife sandstones (late Viséan), the Yoredale sandstones (Viséan–Namurian) and the Millstone Grit (Namurian). The last, and greatest, belongs to the next chapter, but we may note here that there seems to have been a general, though not regular, advance of the deltaic facies southwards, till that of the Millstone Grit was halted by the Mercian Highlands.

East of the ridge of the Pentland Hills and the deep syncline of the Midlothian Coalfield, Calciferous Sandstone rocks are found again in East Lothian at North Berwick and Dunbar. Here the sediments are subordinate to the lavas; they include spreads of pyroclastic rocks and are cut by a large number of volcanic vents. The Garleton Hills are formed of another great group of lavas, basaltic and trachytic; they are probably continuous with similar flows in Fifeshire (e.g. at Burntisland), under the Firth of Forth, and as such a Forth Volcanic Group is on a scale comparable with the Clyde Plateau Lavas. Indeed borings in the central part of the Midland Valley suggest that there may be a continuous lateral passage from the Clyde Plateau Lavas eastward through intermediate groups to those of East Lothian, though this does not mean that there is strict contemporaneity of the various volcanic groups. Still farther south-east along the coast, where Old Red Sandstone and Carboniferous rocks lie with conspicuous unconformity on the Silurian of the Southern Uplands, marine limestones are found in north Berwickshire at Cove Harbour. They are apparently low in the Calciferous Sandstone Measures, here unconformable on Upper Old Red Sandstone, and contain well-preserved B_2 goniatites. Unfortunately these outcrops are small and isolated, and it is not possible to correlate them for certain with the marine bands of the Midland Valley itself or with the very few that may be of about this age in Northumberland. The existence of the limestones, however, and of the much more extensive Carboniferous sequence at Berwick on Tweed, 15 miles to the south-east, suggest that this was one of the lower gaps in the Southern Uplands barrier between the major troughs of the Midland Valley and Northumberland.

By way of summary, it may be seen that during Lower Carboniferous times the Scottish Midland Valley accumulated a variety of shallow water sediments and lavas in areas of diverse tectonic foundation, resulting in great differences of subsidence. The general process, nevertheless, was slowly towards a levelling up, or an increasing uniformity of sedimentary surface, for the succeeding major invasion of the Hurlet Limestone spread over nearly the whole region, covering oil shales, sandstones and cementstones alike, and interrupted only by occasional ridges of the Clyde lavas. Goniatites from or near the Hurlet level give a basal P_2 age, so that the succeeding Lower Limestone Group is actually the highest Viséan division in Scotland. Since, however, it initiates a new phase of sedimentary history, it is more convenient to postpone it to the next chapter, to be considered with the Namurian beds above.

The relationship of the Lower Carboniferous area of deposition to that of the Midland Valley trough fault is not easy to determine. Where the cementstones approach the westerly end of the Southern Uplands they certainly become thinner, but the conspicuous reduction is at the Inchgotrick Fault. In the outliers (Sanquhar, Thornhill and a few minor areas) on the surface of the upland, the Carboniferous rocks are almost all of later age, but a few small patches of Calciferous Sandstone have been found to indicate that the Southern Upland Fault was not a complete boundary to deposition. Little detritus seems to have been received from this direction, so that although the barrier was present it may have been only a low one.

On the northern margin the present outcrops approach the Highland Border Fault also only in the west, but on this side it is clearer that the fracture had little effect. There is a good development of cementstones, capped with a remnant of lavas, on the highland side of the fault between Loch Lomond and the Firth of Clyde, so that here at least there was an embayment in the local uplands.

On Arran any direct relationship is even farther to seek. Far from the Calciferous Sandstone Measures being thinner on the upthrow side of the boundary fault, they are thicker here (over 1,000 feet) where they lie on Dalradian rocks, and show both thinning and overlap of the upper over the lower members on the south-east side, where they lie on Old Red Sandstone. Moreover, the sediments are more sandy than those of the mainland, are plant-bearing and lack the marine limestones.

A local swell or island might be deduced, but not directly related to the fault or fault margin.

In the Machrihanish Coalfield, on the Mull of Kintyre, where Carboniferous rocks again overlap Old Red Sandstone on to Dalradian rocks, there are lavas of Calciferous Sandstone age.

Ireland: northern seas and shores

The general picture in north-central Ireland is of limestone-depositing seas transgressing northwards and abutting on the southern, irregular edge of the Dalradian highlands of northern Mayo and Donegal. Neither Southern Upland nor Highland Boundary Fault appears to have had much influence on this process and little detritus was derived from the Longford-Down Massif. On the other hand, drainage from the Dalradian highlands was vigorous and threw out great clastic wedges along the north-western shores of the Dinantian sea.

It seems probable that the Longford–Down Massif was more thoroughly covered by Lower Carboniferous sediments than was its eastern partner, the Southern Uplands of Scotland, which is another aspect of the general 'topographic rise' to the north-east during this period. Patches of limestone are found upon it, of which that in the Kingscourt Outlier on the borders of counties Louth and Cavan is the most important. Nearly 2,000 feet of varied limestones are found here, ranging from C upwards and the lower beds show some remarkable resemblances to the Clitheroe succession in the Craven Lowlands. Farther to the north-east much of Strangford Lough may be floored by Carboniferous Limestone (a through-way across the upland) for there is a small but significant patch of D_2 on the west side. The Dinantian submergence was thus probably later in the east, which fits in with the generally more extensive seas on the west.

In the north-central plain as a whole, as far as the Ox Mountains and Fintona outcrop of Old Red Sandstone, limestones or limestones and shales are the dominant facies. It is a matter of debate whether much, or any, Tournaisian is present, but it may form part of the lowest beds, filling in a somewhat irregular floor. At present the outcrops are broken by the Curlew Anticline; this appears to have had little contemporary effect on sedimentation, but that of the Ox Mountains formed a positive ridge or axis, over which the various beds become thinner, to expand again on the northern side. It also marks a facies-change, for there is an increase in terrigenous detritus to the north. Moreover, although the Highland Border Fault cannot be detected for certain between the

Fintona outcrops and Clew Bay, its line must follow approximately that of the Ox Mountains.

The northern facies is best exemplified in the down-warped basins on the 'highland' side of this line, especially in the south-to-north section provided by the synclines of Sligo and Donegal. The mixed facies of the former (C_2, S_1 and D_1 limestones, with intervening sandstones and shales) passes northwards into a more sandy, thicker sequence though in the northernmost outcrops only the lower Viséan beds remain. A similar, littoral or deltaic, facies is found to the west, across Donegal Bay, in northern Mayo, though there it is less complete; to the east it is well developed in the Omagh Syncline where the 8,000 feet of Lower Carboniferous rocks include great wedges of sandy beds.

This belt of detritus, derived from the Dalradian highlands, shows the direct influence of the Caledonian uplands as a source of sediment more clearly than anywhere else in the British Isles. The resulting rocks include current-bedded and slumped sandstones, boulder beds, and siltstones; the organic remains include drifted wood and there are occasional calcareous beds with shells. The sandstones are poorly sorted, and have mineral assemblages that can be matched in the Dalradian rocks and intrusions to the north. They usually pass southwards, or offshore, into shales and then into limestones, but sometimes the transition is remarkably rapid, from deltaic sandstones into massive bioclastic limestones in a few miles. This passage is worth noting because it does not seem to be accompanied by any physical barrier, showing that such barriers need not necessarily be invoked between clear water and thick detrital sediments of other ages and places.

There is no evidence that the Dinantian seas ever submerged these north-westerly mountains, but farther to the north-east similar detritus formed sandstones (apparently unfossiliferous) which now remain as irregular outcrops on the north-west side of the Antrim Basalts, for instance in Londonderry. There are also a few outliers south of Lough Foyle.

The small Ballycastle Coalfield, on the north Antrim coast, is only 25 miles from the Machrihanish Coalfield of Kintyre and the Carboniferous rocks of the two are rather similar. The 'Calciferous Sandstone' at Ballycastle, however, is more varied; in addition to lavas and tuffs, the 1,700 feet of measures include sandstones and conglomerates, shales and thin coals. There is thus a transition, only decipherable now in isolated outcrops, from a full, Scottish type of sequence in central Ayrshire, through Arran, Kintyre and Ballycastle, to Omagh, and finally to the

L. I. H. E.
THE MARKLAND LIBRARY
STAND PARK RD., LIVERPOOL, L16 9JD

fully marine limestone facies of the north-central Irish Plain. The most thorough facies-change, and the most difficult to reconstruct, lies underneath the Antrim Basalts, and not in the seaway that happens at present to separate Scotland and Ireland.

THE CULM: LOWER AND UPPER CARBONIFEROUS

A fairly precise line can be drawn in Ireland between the Dinantian limestones of the centre and south and the Culm facies, or 'Carboniferous Slate' in earlier terminology, of the extreme south and south-west. The lack of a comparable line of change in England is not much filled in by the tiny inlier of limestone (? D zone) of Cannington Park, in the north Somerset plain, as its boundaries are obscure and it may be a thrust mass.

DEVON AND NORTH CORNWALL

The Culm occupies the central Devon synclinorium, in which the detailed structures are extremely complex with much sharp folding and thrusting. Consequently it is very difficult to compile a full sequence of strata, and many areas are badly exposed and little known. Enough goniatites have been found to suggest that there is probably a complete succession from the lowest Tournaisian to G_2, or lowest Ammanian; non-marine shells extend a little farther up and reach the *communis* zone.

The term 'Culm' is an old one, referring to the soft, powdery carbon-rich layers that are not true coals but take their place at a few levels; microscopically they are shown to be detrital and consist of drifted plant debris. They occur chiefly in the upper beds, but have given the name to the whole of the Carboniferous succession in this province, where the traditional English terms are not applicable. There is also a common division into Lower and Upper Culm, which agrees approximately with that between Lower and Upper Carboniferous.

The broad structure of the region results in the Lower Culm cropping out in two fairly narrow strips on the northern and southern flanks of the synclinorium. The northern one extends eastwards from Barnstaple to Brampton and the New Red Sandstone unconformity, and on the south similarly from Boscastle on the Cornish coast to Launceston and round the northern edge of the Dartmoor Granite (Fig. 32). In both there is conformity with the Famennian below, and the Pilton Beds of the north (shales, siltstones and thin lenticular limestones) lie athwart the Devonian-Carboniferous junction. In the south this name is not used, but there is a gradual transition from green slates with Famennian fossils to black slates with silty bands and lenses.

Over 1,500 feet of black shales with bands of quartzite form the older part of the Dinantian north-west of Dartmoor and above them lies a thinner, more calcareous series of dark limestones and shales; goniatites of B, P_1 and P_2 age have been found near Launceston. Both series contain pyroclastic rocks. Near Barnstaple, the Pilton Beds (thickness uncertain) are overlain by a thin series of cherts, shales and siltstones (P_1–R_1) which do not exceed 500 feet. It seems probable that on this northern flank all the Dinantian and much of the Namurian is represented by relatively fine-grained sediments laid down in quiet conditions. Somewhere to the north there must have been a transition into the limestones of south Pembrokeshire and the Gower Peninsula, but how and where is unknown. Those limestones, however, were probably thicker than the geosynclinal deposits, at least in part of the Carboniferous succession—an interesting departure from the normal thickness contrast between shelf carbonate and clastic geosynclinal sequences. Vulcanicity was locally vigorous in the Lower Culm period, and there are spilitic lavas and tuffs in Cornwall, including pillow lavas; they are also known around Dartmoor but not in the northern outcrop.

The Upper Culm occupies all the centre of the Devon synclinorium, and has a wider extent than the lower group, but as yet less is known about it. On the south the quiet sedimentation of late Viséan times was brought to an end by an early Namurian incursion of greywackes, from which H and R goniatites have been collected near Exeter. Greywackes and shales also terminated the quiet conditions in the north, but at a later period, since the R_1 shales are succeeded by the greywackes of the Instow Beds (1,500 feet) which continue from R_2 to G_2. Above these there is an abrupt change to a facies with sandstone, shales and beds of culm, more nearly resembling normal Coal Measures but lacking the rooted vegetation and the peat. It perhaps represents a temporary southward extension from the true Coal Measures of South Wales. No comparable facies is found on the southern side and the influx may not have reached so far. The highest beds known, still in the Lower Westphalian, show a gradual return to marine conditions, with greywackes.

SOUTHERN IRELAND

The 'Carboniferous Slate' occupies the synclines in the Hercynian folded belt and being generally less resistant than the intervening Old Red Sandstone these outcrops form the long inlets such as Bantry Bay and Kenmare River (Fig. 31). The group has a twofold division into the Coomhoola Beds below and the Ringabella Beds above, with a total maximum thickness of possibly 8,000 feet in the south. The former resemble the Pilton Beds of North Devon in that their fauna, although sparse (chiefly brachiopods and lamellibranchs), includes Famennian and Tournaisian species. At their maximum (4,000 feet), however, they are much thicker and coarser than their English equivalents and include massive current-bedded sandstones, possibly derived from the south-west, together with plant debris. They are also much thicker than the contemporary Lower Limestone Shales to the north, so that here there is the normal expansion as the strata pass from the shelf facies into the subsiding geosyncline. It appears that the recognized boundary at the top of the 'Old Red Sandstone' may be diachronic and that grey beds of Carboniferous type were being laid down in the south contemporaneously with red and brown sandstones farther north. Elucidation of this junction needs more palaeontological evidence.

The Ringabella Beds follow, in places at least, after a gap, with an erosion surface and a basal conglomerate. They are dominantly argillaceous rocks, often cleaved, with thin calcareous beds that are locally rich in fossils—brachiopods, simple corals and rare goniatites. The ages range from Z to C_1 and possibly C_2; if Viséan beds are present there is little of them. The upper surface is nearly everywhere one of present erosion but there are a few areas of later Carboniferous shales, lying on the folded beds beneath and containing, in different areas, P and E zone goniatites. These suggest that some at least of the folding and cleavage in this belt is due to a late Viséan phase of the Hercynian movements.

The palaeogeographic picture reflected in the Culm facies is largely a confession of ignorance. There is no obvious northern landmass, as a source of all this detritus, whether in Ireland or north of Devon, and the existence of clearer seas of the limestone facies in South Wales and central Ireland is an added reason for considering no source to be available in this direction. There are certain similarities in lithology, and still more in fauna, with the German Dinantian rocks, and in Brittany there is some evidence of local land in the facies, unconformities and overlap in the rather fragmentary Dinantian outcrops. The Culm partakes of this Middle European clastic belt, much deformed by Hercynian movements, and such evidence as there is points to a landmass to the south or south-west of Britain as a source of Culm detritus.

Farther north there was a broad zone of shelf seas, in which clear, relatively shallow waters lapped around the remaining Caledonian uplands, some being finally submerged and others remaining above water-level. These residual masses were much more important in Wales, England and Scotland than in Ireland where the seas were uninterrupted over much of the country. The influence of the northern continent is seen, directly or indirectly, in north-west Ireland, eastern Scotland and northern England, but only in the first two were the rivers draining the residual Caledonian uplands vigorous enough to produce great masses of detritus. Elsewhere basins silted up, subsided, and silted up again, so that sometimes they were above sea-level and sometimes below. With certain conspicuous exceptions, such as the Fell Sandstone or those in Fife and Donegal, fine sediments were the rule rather than coarse. In Scotland the 'basin and block (or ridge)' environment was complicated by large outpourings of lava.

The Lower Carboniferous period in Britain, and in much of north-western Europe, is a fine example of the interplay of various factors operating in a *dominantly* marine environment, where a complex, inherited tectonic foundation was reacting to new strains. The variety in facies that resulted, with its accompanying diversity of faunas and floras and its correlation problems, was accentuated because the sea water was shallow enough for small changes in level to have far-reaching results.

REFERENCES

Lower Carboniferous stratigraphy and palaeogeography are dealt with comprehensively by George, 1958*b*, and relatively few further references are given.

Southern England. George, 1962*a*, *b*; Goldring, 1962; Green and Welch, 1965; House and Selwood, 1966; Prentice, 1960*a*, 1962; Welch and Trotter, 1961.

Midlands, Northern England and Scotland. Earp *et al.* 1961; Eden *et al.* 1964; Francis, 1965*a*, *b*; George, 1958*a*, 1960*a*, 1963*a*; Goodlet, 1957; Johnson, 1959; Moore, 1958, 1959; Rayner, 1953.

Ireland. Charlesworth, 1963, ch. 9; George, 1960*b*; George and Oswald, 1957; Smyth, 1950; Turner, 1952; Wilson and Robbie, 1966.

General. George, 1958*b*.

British Regional Geology. Midland Valley of Scotland, Northern England, Pennines, North Wales, South Wales, Bristol and Gloucester, South-West England.

7

UPPER CARBONIFEROUS

NAMURIAN, AMMANIAN, MORGANIAN, STEPHANIAN

In those parts of Britain where the traditional divisions, Carboniferous Limestone, Millstone Grit and Coal Measures, are applicable, the second is broadly equivalent to the stage Namurian and the third to Ammanian and Morganian (approximately Westphalian of the continent); the full table of stages and zones is given on p. 411.

NAMURIAN IN THE NORTH: MILLSTONE GRIT

South and Central Pennines

This part of the Pennine outcrop lies in Yorkshire, Lancashire, Derbyshire and north Staffordshire; it forms the southern part of the range, as far north as the Craven Faults, together with a westward extension into central Lancashire where the moorlands rise to the north of the Lancashire Coalfield. The Pennine ridge is rarely a simple anticline, though an asymmetrical fold can be found on the Yorkshire–Lancashire border between Halifax and Rochdale.

It is also the type area of the Millstone Grit, where millstones were once fashioned out of the coarser sandstones. The rocks range from shales to coarse, pebbly sandstones or gravel conglomerates. The sandstones, standing out as thick 'gritstone edges' or scarps, are much the most conspicuous, but being the products of relatively rapid deposition they probably represent a lesser proportion of Namurian time than their thickness at first sight suggests. An imperfect cyclic sequence has been observed in parts of the Millstone Grit, comprising in upward succession, marine shales, silty mudstones, siltstones, fine sandstones, coarse sandstones, coal. Fossils are virtually absent in the sandy and silty beds; the black shales contain faunal bands, locally rich in marine forms (chiefly goniatites and lamellibranchs) and there are occasional impure, silty limestones with an abundant shelly fauna.

The whole is the product of a delta complex, but this includes a number of environments. In particular the coarse pebbly, feldspathic sandstones, with conspicuous current-bedding and seams of well-

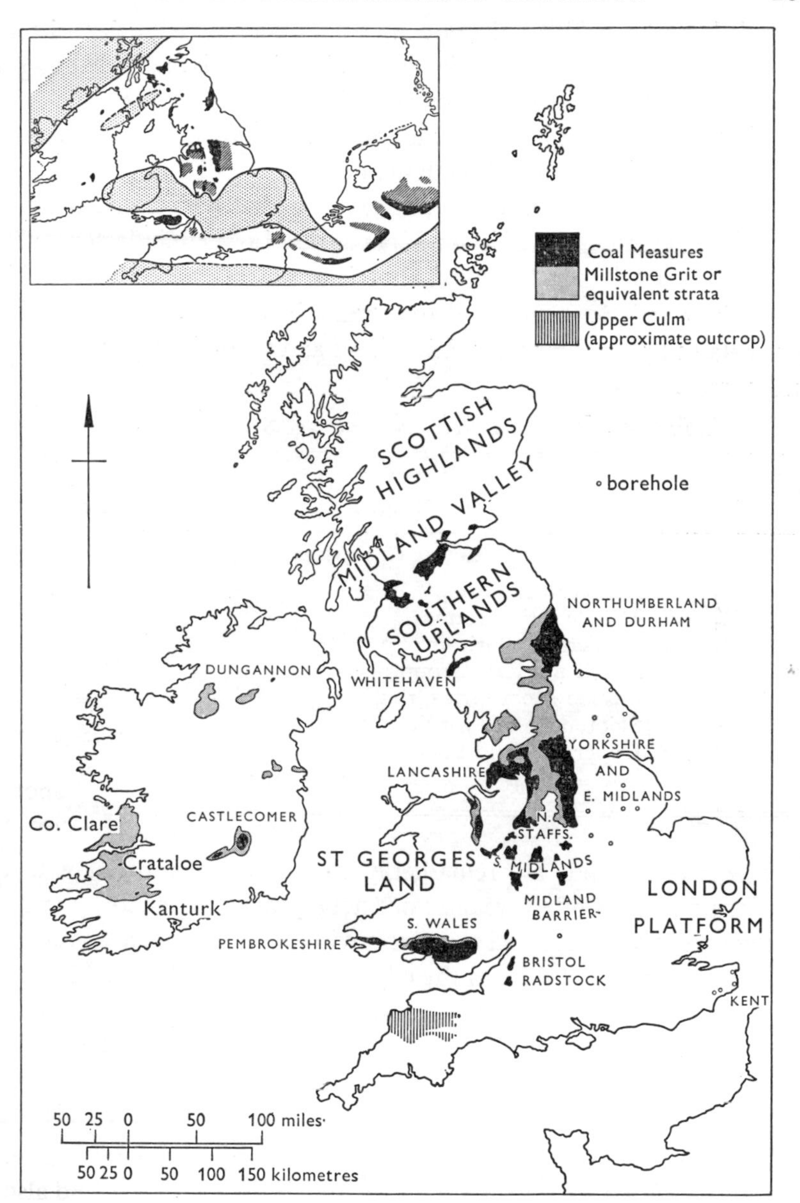

Fig. 39. Principal outcrops of Upper Carboniferous rocks: Namurian, Coal Measures and the Upper Culm. Boreholes show that the Upper Carboniferous strata extend eastwards from the Pennines, and also form the underground coalfield of east Kent. Inset: exposed and concealed coalfields (Lower Westphalian) of Britain, France, Belgium, Holland and North Germany, in their supposed palaeogeographic setting. (Redrawn from Wills, 1951, pl. ix.)

rounded quartz pebbles, do not resemble modern delta deposits but rather those of a fluviatile environment. They can be regarded as productions of a temporary emergence of the delta surface, on which river debris was spilled out in broad valleys, cutting into and channelling the previous deposits, angular fragments of which can be found as part of the gravel content. Such coarse, current-bedded feldspathic rocks are conspicuous among the Skipton Moor Grits, Lower Kinderscout Grit, and Rough Rock (see Table 32), but it would be misleading to think them characteristic of the sandstone units in general. In the south of the delta complex submarine delta-front and turbidite sediments were widespread.

Table 32. *Millstone Grit (Namurian) of West Yorkshire*

Zones	Groups	Average thickness (feet)
	(G_2—Lower Coal Measures)	
G_1	Rough Rock Group	225
R_2	Middle Grit Group	525
R_1	Kinderscout Grit Group	800
H	Middleton Grit Group	225
E_2	Silsden Moor Grit Group	1,000
E_1	Skipton Moor Grit Group Upper Bowland Shales	2,500
	(P_1–P_2—Lower Bowland Shales)	

The marine bands are remarkably widespread, compared with the sandstones, which often wedge out laterally to be replaced by another sandy unit. The same zonal sequence, with essentially same species, can be traced from Derbyshire to the edge of the Askrigg Block. The bands represent periods of unusually widespread submergence when quiet deposition of off-shore muds prevailed for a time over the delta complex as a whole. A turbidite facies has been recognized in the Mam Tor Sandstones and associated silty shales that underlie the Kinderscout Grits in north Derbyshire; they are marine sediments, very poor in fossils, which were laid down on the slopes of the advancing delta front.

The total thickness of the Namurian is greatest in the north of this Pennine outcrop, at approximately the northern boundaries of the Lancashire and Yorkshire coalfields; the type succession between Bradford and Skipton comprises over 6,000 feet of deltaic and fluviatile beds. Away to the west at Preston there is a conspicuous thinning, and also to the east, where the Millstone Grit passes under the Permian scarp north-east of Leeds. Along the Pennine range to the south, the distribution of the various grit groups show a comparable diminution as the sandy influx failed against the Midland ridge or Mercian Highlands. It will be seen from Table 32 that in the north half or more of the total Namurian succession is made up of the Skipton Moor Grit, so that in E_1 times there must have been a marked local downwarping to accommodate over 2,000 feet of beds, much of them coarse and sandy. But in south Yorkshire, and still more in Derbyshire, rocks of this age are much finer and

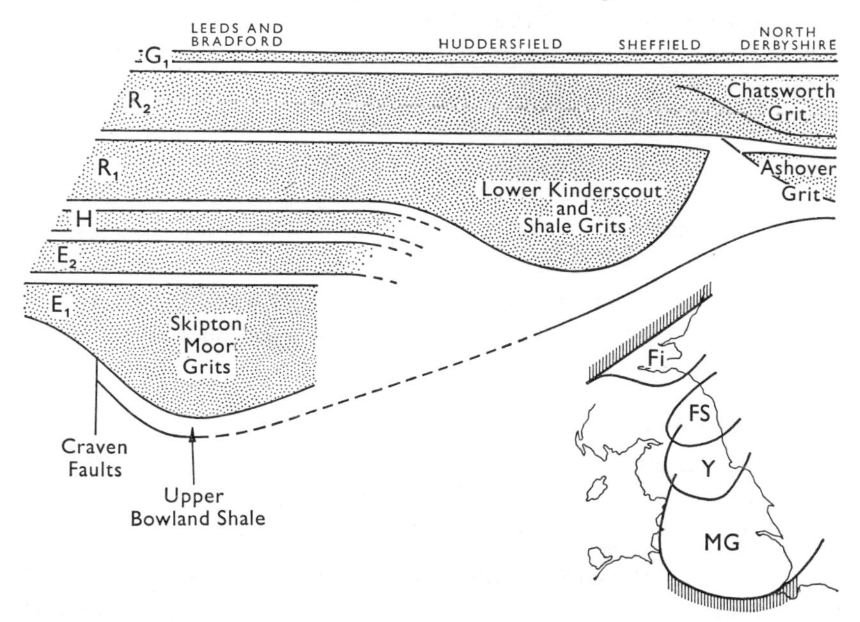

Fig. 40. Diagrammatic cross-section through the Millstone Grit. Three successive groups in particular (E_1, R_1 and R_2) contribute thick belts of coarse sediment in a southward progression. Not to scale. (After Ramsbottom, 1965, pl. 4.) Inset: Viséan and Namurian deltaic incursions; north to south, Fife sandstones, Fell Sandstone, Yoredale sandstones, Millstone Grit.

thinner and the individual sandstone units lose their identity; in north Derbyshire, for instance, the Edale Shales cover zones from E_1 to R_1. The southward thinning of the lower Namurian is a major factor in the reduction of the whole, which diminishes to nothing in Leicestershire and south Staffordshire.

This reduction is accompanied by another lateral change. The thickest zone, or grit group, as it is measured at successive places in this north–south traverse of the Pennines (Fig. 40), rises in the stratigraphical sequence. In the north, as already shown, it is E_1; south of Huddersfield it is R_1 (the Kinderscout Group) which has swelled to over 1,500 feet, to diminish again southwards; in Derbyshire the R_2 Middle Grit Group is the thickest. The highest unit, the Rough Rock Group, is always relatively thin, but is also the most persistent over the southern Pennines as a whole and maintains its identity farthest into the Midlands. From this pattern one may deduce that the belt of greatest subsidence gradually moved southward as the deltaic and fluviatile deposits were carried across the subsiding mass of earlier sediments.

The source of this great volume of detritus is considered to be north to north-easterly on various grounds. Although there is wide variation in the direction of foreset bedding, the dips are predominantly to the south and south-west; the general distribution and thickness just described preclude a southerly origin and there are many pointers to a northern landmass in Carboniferous stratigraphy as a whole. The mineral assemblages of certain feldspathic sandstones, notably the Rough Rock, in which the feldspars (orthoclase and microcline) are often fresh and unweathered, are compatible with a source area rich in granitic and gneissose rocks. This would fit a Pre-Cambrian terrain such as the Scottish Highlands, or Caledonian chain of which these were a part.

Northern Pennines, Northumberland, West Cumberland

There is no abrupt change at the Craven Faults but over a 20-mile belt in this region the Millstone Grit is reduced to less than a third. From the Askrigg Block northwards, into Durham and Northumberland, a rather ill-defined, thin, Millstone Grit can be recognized, but as mentioned in the last chapter it succeeds a Yoredale facies (the Upper Limestone Group) which in these northern counties persists into the E_2 zone or locally as high as R_1. This northern Millstone Grit is not typical in thickness or lithology, but is rather a variable series of sandstones, silts, shales and poor coals, which chiefly differs from the Limestone Groups below in the lack of marine limestones, and from the Coal Measures above in the lack of productive seams. As little as 300 feet of this intervening facies has been recorded in Northumberland.

In west Cumberland, at Whitehaven, Namurian beds are also unimportant, but for rather different reasons. The clastic components of the Yoredale facies failed to reach so far west, no cyclic sequence was established and the equivalent rocks are dominantly calcareous. Above them there is very little that can satisfactorily be called Millstone Grit; succeeding the top (or First) Limestone, which is the equivalent of the

Great Limestone of the Alston Block, there is a thin clastic sequence, in which the thickest sandstone is only 80 feet. From these beds only E_2 and G_1 goniatites are known and whether sedimentation was really continuous in between is uncertain. It thus seems that the major downwarping and the true delta were largely confined to the southern Pennines.

NAMURIAN IN SCOTLAND

The stratigraphical divisions fit the facies-titles, Carboniferous Limestone, Millstone Grit and Coal Measures, even less well in Scotland than they do in northern England. The following groups are established for beds of Namurian age and those slightly below and above:

Passage Group (previously Millstone Grit)	(Contains fossils of E_2 zone and Lower Coal Measures)
Upper Limestone Group	E_2
Limestone Coal Group	E_1
Lower Limestone Group	P_2

Lower and Upper Limestone Groups, Limestone Coal Group

These resemble a Yoredale facies in that they consist of cyclic sequences containing a marine stratum (which is usually a fossiliferous limestone, but may be a mudstone with shells), shales, sandstones and coals; but even more than in the Yoredale strata there is much local variation, local additions and local omissions. The basal limestone of the Lower Limestone Group is the Hurlet, which marks a widespread period of marine invasion. The Lower and Upper Limestone Groups resemble each other in that marine limestones are fairly abundant—or to put it another way the periodic subsidence that is a fundamental factor in this Yoredale type of facies was enough to bring the region frequently below sea-level. In the lower group this transgression often took place before the silting up had proceeded far enough for peat growth to become established and coal seams are consequently rare. Conversely, in the Limestone Coal Group the marine rocks (which are almost all shales) are not common, and the upper parts of the cyclic unit—the shales, sandstones and coals—are well developed and there are many workable seams.

These generalizations apply where the groups are thick and typical but thickness itself is another significant factor in variation. It was shown that the Calciferous Sandstone Measures were thin, and locally absent altogether, in certain areas—especially on the south Ayrshire block, where the Inchgotrick Fault coincided with the thickness change and acted as a kind of hinge. This type of variation, between areas of differing amounts of subsidence, whose margins approximately coincide with existing faults, is much easier to demonstrate in the three 'Limestone Groups'; certain beds, especially the limestones and the coals, can be correlated securely and used as a basis for estimating thicknesses. A traverse from north to south across Ayrshire, where the effects are well seen, shows abrupt changes in thickness over the lines of three major faults—of Dusk Water (near Ardrossan), Inchgotrick (Fig. 38), and Kerse Water (between the Inchgotrick and Southern Upland Fault). The greatest thickness is in the north-west and the least between the Inchgotrick and Kerse Water faults; south of the latter, where the Calciferous Sandstone Measures locally diminish to nothing, there is in fact one of the thickest sequences of the Upper Limestone Group. It thus appears that although the Lower Carboniferous pattern of subsidence is followed in part it is not altogether.

Towards the east and north-east there is a general but irregular thickening in the three groups, into basins of maximum sedimentation in Midlothian and central Fife, the latter containing 3,800 feet. An analysis of the Lower Limestone Group shows that there is also a lithological correspondence and that the thinner sequences tend to be more shaly, while sandstones swell the thickest. The belts separating the basin areas are probably tectonic in origin but are not so clearly related to fault-lines as in Ayrshire; in the Lower Limestone Group they are partly related to regions of thick lava flows; thus the main mass of the Clyde Plateau Lavas separates Ayrshire from the Central Coalfield, and the lavas of Burntisland (which range up into the Lower Limestone Group) lie between the latter and the Fife–Midlothian basin.

Passage Group or Millstone Grit

The former term has recently been preferred in Scotland because correlation with the Millstone Grit of the Pennines is rather uncertain, and the whole group is relatively thin. Goniatites of the H, R and G_1 zones are rare or not known in Scotland and part at least of the group ranges up into the Ammanian; lamellibranchs of the *lenisulcata* zone have been found 300 feet below the top in Ayrshire. In the centre and north the Passage Group amounts to 1,100 feet, but may locally be entirely absent, or, as in parts of Ayrshire, represented by volcanic rocks.

The sediments comprise a variable series of sandstones (in the Edinburgh region known as the Roslyn Sandstone) which are often coarse, current-bedded or pebbly, with subordinate shales, fireclays and occasional coals; marine bands, shales and impure limestones bearing an E_2 fauna are known in the lower part. Irregularity of sedimentation and subsidence is strikingly demonstrated in the tiny Westfield basin, one and a quarter square miles, which lies east of the Central Coalfield; here the seams vary in an astonishing manner over a number of north-east axes, and at their thickest there is 200 feet of coal in a total of 500 feet of measures, a thickness of peat accumulation that may well be unique in Britain.

NAMURIAN IN THE SOUTH AND WEST

Even outside the regions of massive deltaic sequences, the seas of the Carboniferous Limestone gradually silted up, to become ultimately regions of lowland swamps in which the Coal Measures were formed, and here also there is usually an intervening detrital phase which is commonly called Millstone Grit. The beds are never so thick as in the delta complex of the Pennines, but in the changing geographical conditions they played something of the same part. Goniatites are similarly the most important fossils, and, in fact, these outlying areas provide us with some of the best Namurian zonal successions. The black shales of Co. Leitrim (E_1–E_2) and Co. Clare (E_1–R_1) furnish such examples.

Wales and the South-West Province

In North Wales the Namurian follows conformably on the uppermost Dinantian, in which detritus had already invaded the earlier limestone seas. The only major outcrop extends southwards from Flintshire to Oswestry and the most fossiliferous sequence is found in the north. Apart from a sandstone at the top it is largely argillaceous—the 600 feet of the Hollywell Shales containing goniatites of all the major zones. Farther south a thinner development of calcareous sandstones is probably complete also; the facies is unusual in the Namurian and some beds contain brachiopods in addition to the more common goniatites and lamellibranchs. The North Wales outcrops represent a marine area of no very great importance in Namurian times, receiving only modest amounts of detritus.

On the south side of St George's Land the situation is more confusing and the limits of the division 'Millstone Grit' are particularly unsatisfactory. The most extensive outcrop lies north and east of the South Wales Coalfield, and here there are three traditional units:

Farewell Rock Middle Shales	G zone
Basal Grits	R, H and ?E zones

Goniatites from P_2 to E_2 are known only from borehole evidence, in the 'Upper Limestone Shales' beneath.

The three divisions are about 800 feet thick in the west but less than half that in the east. Most of the problems of correlation arise from the rarity of goniatites, the lower zones being poorly represented and that of *Eumorphoceras* scarcely at all. The lithological boundaries are almost certainly diachronous, and locally some or all of the Farewell Rock has been found to lie above the *Gastrioceras subcrenatum* band and thus strictly belongs to the Lower Coal Measures.

One important aspect of this 'Millstone Grit' is the way in which the base steps down, with an increasing unconformity, on to lower and lower members of the Carboniferous Limestone from west to east—a result of late Dinantian or 'pre-Millstone Grit' uplift; for a short stretch on the east crop of the coalfield the Basal Grits lie on a thin remnant of Lower Limestone Shales. The axis of the Usk Anticline is only a few miles away and early, mid-Carboniferous, uplift along this line may be held primarily responsible for the overstep.

There are comparable variations in thickness and lithology, for only 200–300 feet of Namurian beds are present in the north-east compared with 2,000 feet in Gower. Moreover, the conglomerates and coarse sandstones are more apparent in the north, and the whole series in the south-west (Pembrokeshire and Gower) consists of shales and thin sandstones which are not easily divisible. The influence of terrigenous detritus from St George's Land is clear in all this, but the relation between grade-size and stratal thickness is not that of the more normal Namurian 'basins', which tend to swell out with a greater proportion of coarser rocks. Here the coarser beds are inshore, and the thicker, offshore, sequence is the finer.

On the eastern side of the Usk uplift, in the Forest of Dean, there are no Namurian or Ammanian beds, and it is uncertain whether any were deposited. In the Bristol and Radstock coalfields arenaceous beds do exist between the top of the limestones and the overlying productive Coal Measures (some 600 feet at Bristol); but, as noted in the last chapter, the former are not uppermost Dinantian, and in the absence of goniatites (with the sole exception of a G_1 species of *Gastrioceras* in the complex thrust zone between the Radstock Coalfield and the Mendips) it is not possible to be certain how much of these sandy beds is truly Namurian. In any case it is no great thickness.

5. The Cliffs of Moher, County Clare. 600 feet of Namurian sandstones, siltstones and shales face the Atlantic on the west of Ireland.

Ireland

All Upper Carboniferous outcrops in Ireland are small and fragmentary, compared with the vast stretches of Lower Carboniferous, and (presumably) compared with their former extent. The Namurian beds can be divided into small residual patches lying on the limestone below, the marginal rims of certain coalfields, and the only one of any size, an outcrop in the west on either side of the Shannon Estuary. The facies tends to be fine, with much shale, some fine sandstones and siltstones, but little coarse sandstone. No one area shows a complete succession, but most of the stages can be recognized in one outcrop or another.

Namurian strata make up much of the west and south of Co. Clare, from the impressive cliff sections facing the Atlantic to the Shannon, and the same basin continues south of that river into parts of counties Limerick and Kerry. The beds lie unconformably on B_2 limestones and the lower part of the sequence (E_1–R_1) consists of the Clare Shales; above these lie a thicker, more varied, group (R_1 and R_2) of slumped sandstones, siltstones and shales. No higher beds are known, and though thin seams of anthracite led the early surveyors to attribute much of this outcrop to the Coal Measures almost all is in fact Namurian. The region formed a gently warped basin, the axis of which coincides approximately with the Shannon Estuary; here there are 4,500 feet of sediments of which the Clare Shales occupy about 1,700 feet; as the beds are traced northwards they show both thinning and overlap of the lowest zones, so that in north-west Clare the basal shales are of *Homoceras* age, and the total is 50 feet. A similar comparable overlap is found to the south. The Namurian thus occupies a shallow depression in the limestones which were gently folded along an east–west axis, the movement probably being late Viséan as the P zone has not been found.

Farther east, only 4 miles away from the Leinster Massif, 'Millstone Grit' is again present, surrounding the synclinal Leinster (or Castlecomer) Coalfield. A comparable overlap is seen here, for at the north end the D_2 limestones are overlain by E_2 shales, but at the south the basal shales are R_2. All the beds are thin, perhaps about 100 feet of strata all told, so that though the area probably represents an eastward continuation of the Shannon basin, it was near the Leinster margin, and subsided relatively little. Somewhere to the south there must have been a further area of Namurian sedimentation, possibly a westward extension of the Culm trough of Devon, but the only indication is a small isolated patch of E_2 shales near Cork.

In central and northern Ireland, south of the Donegal uplands, the remaining Namurian fragments are largely of the lower beds. There are a number of small outliers north of Dublin, only one (the Kingscourt outcrop) including beds as high as R_2; shales and siltstones are again dominant, and there is a slight overlap of the Balbriggan Massif. One of the fullest successions of E_1 and E_2 zones is found on Slieve Anierin—one of a number of outliers which rise up from the limestones on the borders of counties Leitrim and Cavan. The sequence is again shaly and contains some thin coal seams.

The Dungannon Coalfield lies west of the Antrim Basalts, in Co. Tyrone (Fig. 53); though small it contains the fullest Upper Carboniferous sequence in the north of Ireland, though much of it is known only from boreholes. The 2,000 feet of Millstone Grit are underlain by another 600 of marine mudstones, with goniatites of late P_2 or early E_1 age. The Millstone Grit is more varied than that in the south, and includes mudstones and sandstones, marine shales and coals, some of which are workable. E_2 fossils are known and also the *Gastrioceras subcrenatum* band at the base of the Coal Measures, but whether H and R beds are present is uncertain, and they cannot be thick. It will be recollected that these zones are elusive in parts of northern England and Scotland also.

The coal-bearing strata of the Ballycastle Coalfield, on the north Antrim coast, probably correspond to the Limestone Coal Group of Scotland, and consequently find a place in this section; a sandy sequence contains ten seams, ranging up to 4 feet in thickness.

THE COAL MEASURES

The Ammanian and Morganian (or Westphalian) strata in Britain contain the vast majority of the productive coal seams, and, in their bituminous coals, the country's greatest single natural resource. The base of both Coal Measures and Westphalian (the latter term being based on the German succession) is drawn at the *Gastrioceras subcrenatum* marine band. In most of the British coalfields this is also about, or just below, the level of the lowest workable seams—apart, of course, from the Namurian coals of Scotland already mentioned. The upper limit of the Coal Measures is much less uniform, chiefly on account of the varying effects of the Hercynian movements and of subsequent erosion. Stephanian beds are rare but are probably present in the southwest coalfields, the Midlands and possibly Ayrshire. Moreover, there is commonly a falling off in the thickness of the seams in the uppermost beds, which are often called 'Barren Measures'. The change is not everywhere at the same level, and a certain zone may contain productive seams in one coalfield and lack them in another. Some of these barren measures are grey, with the normal rock-types but the coals thin or absent; others may have been red deposits in the first place, but much of the 'red barren measures' seems to have been reddened by waters percolating downwards and introducing iron oxides or oxidizing the iron compounds already present.

The normal productive measures include a relatively small number of rock-types which are repeated over and over again in the full succession of any one major field—shales and mudstones, siltstones, sandstones, the coal itself and the marine bands. All these can be found in the Namurian rocks below, and to some extent the change from Millstone Grit to Coal Measures is a change in the proportion and emphasis of the different sedimentary types.

The marine bands are usually black carbonaceous shales, or more rarely a mixed rock of calcium, magnesium and ferrous carbonates; fossiliferous limestones, of 'Yoredale' type, are rare but sometimes found in Scotland. Goniatites and lamellibranchs are typical, but *Lingula* bands are common, which though probably fully marine may represent even shallower, intertidal conditions; occasionally there are other faunal bands of rather uncertain environment, but used locally in correlation, containing the small crustacean '*Estheria*', ostracods or foraminifera.

The non-marine mudstones and shales form the greatest proportion of the productive measures; they contain plants and lamellibranchs and their main variant is clay ironstone, in concretions or impersistent bands. The coarser rocks, which usually lack fossils though drifted wood is sometimes found, include siltstones and fine to medium sandstones. The former are often laminated, with small scale current-bedding or ripple drift; the latter may be flaggy or current-bedded and are usually lenticular in their distribution.

Most of the productive seams are bituminous, though anthracites are known from South Wales and cannels from several sources. We are not here concerned with the very complex subject of the microscopic structure and composition of coals, but can summarize their stratigraphical significance by saying that the coal seam represents a swamp vegetation, which later became peat, and that in the alteration to coal there was a great reduction in thickness. Thus a five-foot seam was once 25 or more feet of a lowland peat swamp, which subsided gently and continuously.

CHRIST'S COLLEGE
LIBRARY

Too rapid subsidence and the peat growth would be overwhelmed by detrital sediment and coal formation cease.

Conspicuous variations in the thickness or unity of a seam can usually be traced to local differences of subsidence or contemporary erosion. A splitting seam is a common occurrence, in which a single coal in one area splits within a few miles into two, or sometimes several, thinner coals, the total being separated by many feet of measures, as illustrated in Fig. 44 (p. 224). In such a case the split seam represents a greater amount of relative subsidence, the peat growth being interrupted by transported sediment. Wash-outs are local interruptions in a coal and its replacement by a sandstone. In this case the peat has been removed by the erosion of a stream flowing through a swamp forest, and its channel is filled with fluviatile sands. Some wash-outs can be traced in detail in underground workings and shown to take a wavering course through the seam—the same wavering course that was taken by the original stream. The more massive sandstones are built up similarly by the rivers of the delta swamps, which shifted their course from time to time.

In spite of all these local variations—splitting coals or lenticular sandstones—a country-wide view of the Coal Measures facies emphasizes its uniformity. Many seams can be identified over hundreds of square miles and individual marine bands, with their characteristic species, over even greater areas. Those that define the stratigraphical divisions in Fig. 41 can be found, under different names, in nearly all the British coalfields and most of those of northern France, Belgium, Holland and Germany.

This great areal spread of what are, relatively, very thin beds, is the clearest testimony to the extent and uniformity of the later Carboniferous swamp-lands. The long process of attrition of the Caledonian uplands and infilling of the basins between them resulted in very wide, swampy, forest and coastal flats, through which wandered streams, rarely capable of carrying anything coarser than silt or fine sand. That these flats were almost at sea-level is shown by the marine bands; during Ammanian times only a small subsidence was enough to bring in the sea over very many miles. Later the marine bands failed in north-western Europe as a whole; but the silting-up process continued and for a time so did the formation of peat, but subsidence was no longer able to bring the surface down to sea-level.

During the Hercynian orogeny the Coal Measures, in common with earlier rocks, were folded, uplifted and later eroded. The existing coalfields are the remnants that have escaped that erosion, and most of them have a synclinal, or part-synclinal form, though often a complex one.

They were also buried beneath the ensuing Permian or later strata, which have since been gradually removed. In some areas the removal is complete, and the coalfield is exposed, but elsewhere a partial or complete cover remains, and it is then a partly or completely concealed coalfield.

SOUTH WALES, FOREST OF DEAN, BRISTOL AND RADSTOCK

As in earlier Carboniferous times there was a single sedimentary basin in the south-west, bounded on the north by the relics of St George's Land and the Mercian Highlands. No southern limit is definable; the Culm measures of Devon include Westphalian beds, of both deltaic and greywacke facies but no productive measures, so that the forest swamps terminated seawards somewhere in that direction.

Structure

The South Wales Coalfield is divided by Carmarthen Bay into two unequal portions, the main basin and that of Pembrokeshire (Fig. 19). The highly incompetent Coal Measures of the latter are overfolded and overthrust as a result of pressure from the south. Most of the rocks exposed are Ammanian and, with many fault gaps, can generally be correlated with those of the main basin.

The main South Wales Coalfield is an east–west syncline with low dips on the north (the north crop) and higher ones, up to 45°, on the south. Superimposed on this structure there is a complicated system of folds and faults or fault belts. The most important of the last is the Neath (or Vale of Neath) disturbance, which cuts the coalfield in two, extending north-eastwards from Swansea Bay, and is one of several that follow an inherited, Caledonoid trend. The easterly portion of the field is dominated by north-east to east–west strikes; the dominant fractures are at right angles with a north-north-west trend. In the westerly portion the Tawe Valley disturbance runs parallel to that in the Vale of Neath, and there are other north-easterly structures; the principal faults are north–south or north-north-west. As a result of this structural complexity the Upper Coal Measures occupy a large but irregular outcrop, with partly faulted boundaries, in the centre of the coalfield and to the south-west and south-east.

The main basin of the Forest of Dean is gentle north–south syncline, with subsidiary bounding structures that diverge a little to north-north-east and north-west. The whole is in line with a belt of influence and north–south structures related to the Malvern Hills. Both South Wales and the Forest of Dean exemplify the 'open' type of coalfield in which the measures, usually in a dominantly synclinal structure, crop out at the surface.

The Bristol and Radstock coalfields (Fig. 23) are partly covered by Trias and Lias (Lower Jurassic). The northern syncline (called variously the Bristol, Gloucestershire or Coalpit Heath basin) is a fairly simple north–south structure, bounded by ridges of limestone which converge to the north; the measures are largely exposed and include an oval outcrop of Upper Coal Measures. Just north of Bristol the Kingswood Anticline, asymmetric and with steep dips on the northern side, brings up Lower Coal Measures; south of this the Radstock Coalfield has more extensive Mesozoic cover and a much more complex structure. It is dominated by great east–west fractures and the southern limb is overturned and overthrust, as the block of the Mendips has been pushed northwards against the less competent Coal Measures.

Succession

In these south-western coalfields there is commonly a conspicuous three-fold lithological grouping that was for many years used as a formal stratigraphical division:

Upper Coal Series
Pennant Sandstone (or Pennant Sandstone Series)
Lower Coal Series

As a classification, however, this had its shortcomings. The Lower and Upper Series did not correspond with the Lower and Upper Coal Measures in other coalfields (themselves not wholly consistent) and the Pennant sandstones lack clearly defined boundaries; they are rather a group of measures in which thick, persistent sandstones are unusually prevalent, but their onset varies in age from one area to another.

In the revised classification the beds beneath the Pennant sandstones, where these are well developed, agree fairly well with the Lower and Middle Coal Measures, and the sandstones and all the beds above belong to the Upper Coal Measures. The majority of the workable seams are in the Lower and Middle Coal Measures; they reach 8 feet in the Radstock Coalfield and considerably more in the east part of South Wales, where thick coals result from the amalgamation of thinner seams to the west. The *lenisulcata* zone, at the base, is often sandy, and locally includes part of the Farewell Rock of Wales and 'Millstone Grit' of Bristol.

Marine bands are plentiful in South Wales, and their distribution and

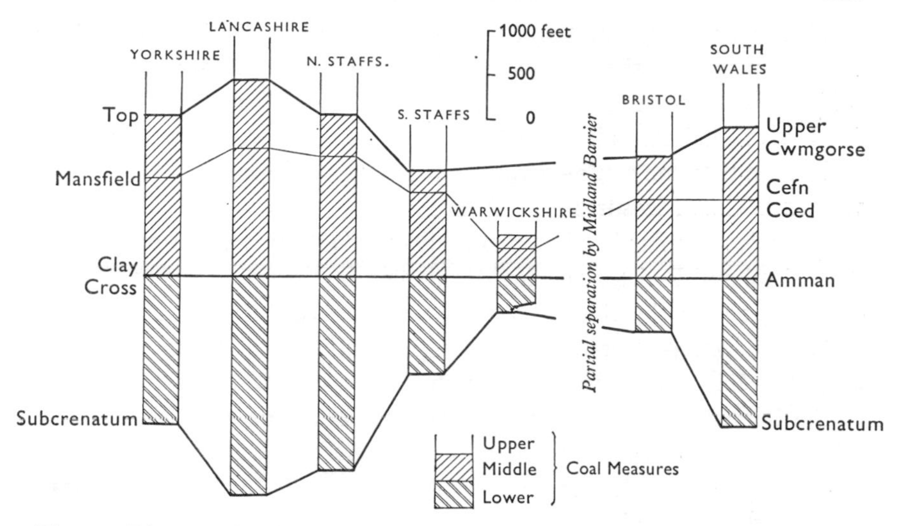

Fig. 41. Diagram showing the thickness of Lower and Middle Coal Measures in selected coalfields and their correlation by means of the major marine bands; the names of these are given for the Yorkshire and South Wales coalfields only. (Adapted from Stubblefield and Trotter, 1957, p. 1.)

fauna give some support to the postulated direction of more open sea to the south-west. In the *lenisulcata* zone there may be as many as five marine levels in 100 feet of measures, and the *Gastrioceras subcrenatum* band at the base itself reaches 50 feet, whereas it is only a few feet in the Pennines. The fauna also is unusually varied, particularly in the south and west; a 6-inch shale in the Cefn Coed band in the south of the coalfield yielded some eighty marine forms including such rarities (in the Coal Measures) as crinoid ossicles and a zaphrentid coral.

The Lower and Middle Coal Measures of South Wales are thickest in Gower and at Swansea and thinnest in the north and east, in Monmouthshire; there is a comparable northerly thinning in the Bristol Coalfield. The differences recall those of earlier Carboniferous times when subsidence was also greater in the south and the Usk Anticline exercised a positive effect to the east.

The Pennant sandstones of Wales are typically coarse, feldspathic rocks, often massive and current-bedded. They contain conglomeratic bands with pebbles of quartz and also of ironstone, shale and coal, which indicate contemporary erosion of the beds below. Even where typical and thickest (4,000 feet), near Swansea, these sandy measures still contain thin coal seams. From here eastward there is a gradual change: the lower beds become less sandy and more normal measures, the total is thinner, and in the east and south-east of the coalfield several minor breaks appear and there is a marked attenuation, affecting Lower and Middle Coal Measures and the Pennant group. Thus near Pontypool 600 feet of sandstones lie on the same thickness of productive measures and in the extreme south-east, near the edge of the sedimentary basin, very little Coal Measures are left at all.

There is a similar but less pronounced effect on western margins of the Bristol and Radstock coalfields, along a line from east of the Forest of Dean, across the Severn, and down the east bank to Avonmouth and then Weston-super-Mare. Although the information is more meagre along this line, there appears to be erosion and unconformity at the base of the Pennant group, which lies on more than one earlier formation. The sandstones resemble those in South Wales; they contain only a few thin seams and reach a maximum of 2,500 feet in the Radstock basin.

The Pennant sandstones have no precise upper limit, but workable seams enter again in the *tenuis* zone and the sandstone beds gradually become less important. The most complete sequences in these uppermost measures come from the Swansea region and the Radstock basin; in the former the upper surface is one of erosion and in the latter there is the Mesozoic cover, so it is not known how long the Coal Measure swamp conditions lasted. The floras, however, are characteristic of the I zone (*prolifera* in the lamellibranch sequence) and they are probably equivalent to the lowest Stephanian on the continent.

The succession in the Forest of Dean is both like and unlike that just described. The Lower and Middle Coal Measures are entirely lacking; the Upper Coal Measures (of *tenuis* and *phillipsii* zones only) consist of a lower sandy group, equivalent broadly to the Pennant sandstones, and an upper, the Supra Pennant Group, of productive measures. The lower lies with conspicuous unconformity on everything from Lower Old Red Sandstone in the east to Drybrook Sandstone (S_2) in the west. This transgressive base of the Upper Coal Measures is accordingly an even more striking effect than that seen in South Wales. Whether earlier Coal Measures were never deposited, or whether they were laid down and later entirely removed, is not certain.

THE MIDLANDS AND NORTH: THE PENNINE COALFIELDS

The several coalfields in the Midlands and north of England differ greatly in size, structure and economic importance, but are linked together by being the remnants of a single area of swamp flats that stretched north from St George's Land and the Mercian Highlands to the Southern Uplands of Scotland. Some of the principal marine bands extend into Durham and Northumberland and a few into Scotland, so that the residual uplands must have been low and incomplete. The

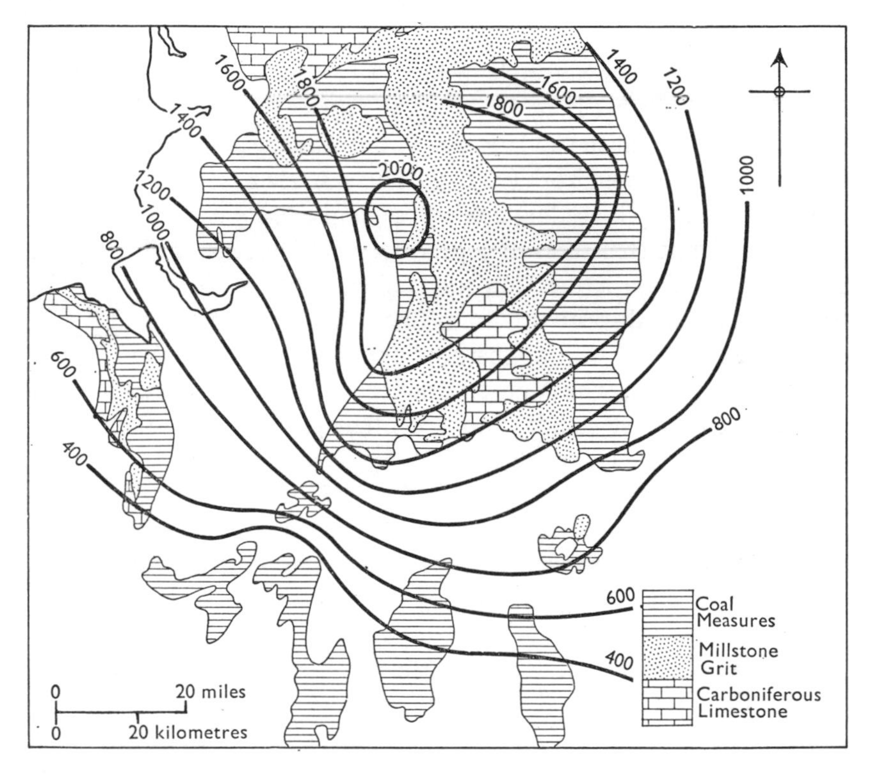

Fig. 42. Pennine coalfields and neighbouring areas. The isopachs of the *modiolaris* and lower *similis-pulchra* zones show the area of greatest thickness to be near Manchester. (After Trueman, 1947, p. lxxv).

centre of the Lake District may have been one of them but the Pennines were not; the Lancashire and Yorkshire coalfields are linked not only by marine bands but by individual seams that were once continuous from side to side. The three Pennine coalfields (North Staffordshire, Lancashire, and the Yorkshire and East Midlands or 'East Pennine') lie near the centre of the basin, where the measures are thickest; the East Pennine has the greatest output of any field in the country and is also the largest.

Structure

There is a general tendency for the beds to dip away from the Pennines, but whereas on the east the effects are fairly straightforward, on the west more complex influences enter and in places divert or suppress this tendency.

From Yorkshire to Nottinghamshire the exposed measures form one side of a gentle, rather irregular basin, the eastern part of which is concealed under a Permian and later cover, but which extends at least as far as Lincoln and beyond York (p. 229). The easterly to south-easterly dip is low (only locally exceeding 20 degrees, and usually considerably less) and so moderate structures may produce local reversals of dip; conspicuous among these is the Don monocline, which with accompanying faults extends north-eastwards from Sheffield to Doncaster, and several concealed anticlines with a Charnoid trend known from the oil-fields of Nottinghamshire. The faults tend to form a lattice pattern with north-east and north-west trends, and are often grouped into fault belts; throws rarely exceed 500 feet.

The eastern part of the Lancashire Coalfield is chiefly governed by a series of large faults trending north or north-north-west; nine of them at some point throw as much as 1,000 feet, nearly all to the east, but the effect is compensated by the strong westerly or south-westerly dip. These fractures are largely or entirely post-Triassic, and their effect can be seen on even a small-scale map in the jagged margin between the outcrops of Coal Measures and the New Red Sandstone to the south.

There are high westerly dips along the Lancashire side of the Pennine Anticline, but farther north east-north-east folds predominate, the two most important being the Burnley Syncline and the Rossendale Anticline, which bring up a broad area of Millstone Grit between the Burnley and East Manchester coalfields. The generally steeper dip west of the Pennines means that the Coal Measures are carried deep below the Trias of the Cheshire plain in a relatively short distance and less of the concealed coalfield is workable on this side than on the east of the Pennines.

In Cheshire, south of Macclesfield, the thin strip of Coal Measures bounding the Pennines is interrupted by Triassic overlap, but the measures reappear in the North Staffordshire Coalfield. The main, Potteries, coalfield is triangular, with overlapping Trias on the south and west; the dominant structure, a syncline trending north-north-east and plunging southwards, has gentle dips on the east and steep ones on the west.

Lithology and succession, productive measures

The Lower and Middle Coal Measures in the three Pennine coalfields are all similar and in many ways the East Pennine Coalfield is nearest to a standard succession for the country as a whole. The Lower Coal Measures lie conformably on the Millstone Grit, and though the *lenisulcata* zone contains fewer economic seams than do the higher zones, the other rock types are normal and there are several marine bands. In the *communis*, *modiolaris* and lowest part of the lower *similis-pulchra* zones marine bands are sparse, the Clay Cross or Sutton Manor, at the base of the Middle Coal Measures, being an exception. There is a return of marine invasions higher up, including the widespread Mansfield (Dukinfield, Gin Mine) band, and the Middle Coal Measures terminate with the Top (Prestwich Top, Lady) marine band.

Variations in lithology and thickness have not the same dramatic effects as in the more unstable South-West Province. Ten feet is a fairly massive seam, such as the Barnsley or Top Hard of Yorkshire. Sandstones are usually fine-grained and lenticular and not nearly so important as in the south-west. The productive measures are thickest in Lancashire (about 4,500 feet), slightly less in Yorkshire, and fall to about 2,500 feet in Nottinghamshire (Fig. 41). The centre of the subsiding pile of deltaic swamp sediments in northern England thus probably lay in east Lancashire.

Sedimentary cycles have been described in detail from the Middle Coal Measures of Nottinghamshire and Derbyshire, and are especially well developed in the upper part of those measures, where nine cyclothems are found in some 700 feet of strata. The ideal cyclothem runs as follows:

coal
seat-earth
sandstone
siltstone or silty mudstone
pale-grey mudstone with shells
dark-grey to black shales with marine fossils
(coal)

The sequence represents the inundation of broad coastal flats by the sea, the gradual silting up to above sea-level by first fine-grained and then coarser sediments, and the final redevelopment of swamps on the stabilized land surface. This is, of course, only a theoretical example and in the East Pennine Coalfield, as elsewhere, there are many divergences from it—usually in the absence of either the coal (though the seat-earth with rootlets may be present) or of the marine episode. Where the marine beds are well developed, they sometimes make up quite a considerable proportion of the total cycle, though they may be intercalated with non-marine levels. The thickest is the Mansfield, in which marine fossils (including a variety of molluscs and chitinous brachiopods) are found through 30 feet of rock or roughly a third of the total cyclothem. In the Yorkshire section of the coalfield the marine bands are not so abundant nor the cyclothems so clear; conversely, the thick lenticular sandstones are more common.

Upper Coal Measures

On the east side of the Pennines these beds are largely hidden under a Permian cover; the *phillipsii* zone is present and probably the *tenuis* also. The lower, grey, measures amount to some 1,000 feet in the north but much less in the south; they are normal in lithology except that marine bands are absent and the coals few and thin. The red measures are entirely concealed. They come in irregularly, and sometimes unconformably, above the grey and consist chiefly of mudstones, red-brown to yellow, with very few fossils or coals but local, coarse sandstones. Some are grey measures that have been reddened, but some may possibly be originally reddish.

On the Lancashire side both *phillipsii* and *tenuis* zones are known and together reach over 2,000 feet at Manchester. The lower zone consists of grey measures and a number of thin coals; the upper is more sandy. Both contain freshwater limestones, sometimes dolomitic with ostracods, lamellibranchs and *Spirorbis*. Reddened beds are locally thick, up to 1,000 feet, but since the coloration may affect any level from the Millstone Grit upwards there is little doubt here of its secondary origin.

Upper Coal Measures are particularly well developed in North Staffordshire; the whole has a maximum thickness of 2,500 feet and forms four lithological divisions:

Keele Group, brown and red sandstones and mudstones.
Newcastle Group, grey sandstone and shales with occasional limestones.
Etruria Marl, red and purple marls.
Black Band Group, grey measures with thin coals and ironstones.

There are broad alternations between grey measures with plants, lamellibranchs and thin coals, and those which introduce red muds and sandstones. The Etruria Marl is largely unfossiliferous, but *Anthraconauta phillipsii* and a few plants have occasionally been found in the ironstone bands. The Newcastle Group belongs to the *tenuis* zone; the Keele Group contains *Spirorbis* limestones and ostracods, and is probably in the *prolifera* zone.

THE SOUTH AND WEST MIDLANDS AND NORTH WALES

These coalfields comprise several which lie in a broad arc along the northern margin of the Midland Barrier and St George's Land, from Leicestershire to Flintshire; they are separated, and some still partly covered, by the New Red Sandstone of the Midlands and the Cheshire basin. They are especially characterized by several structural and stratigraphical features which arise from the rather unstable foundations on this southern margin of the northern province, and in certain Midland coalfields the Upper Coal Measures are unusually extensive and there are important links with early Permian rocks above.

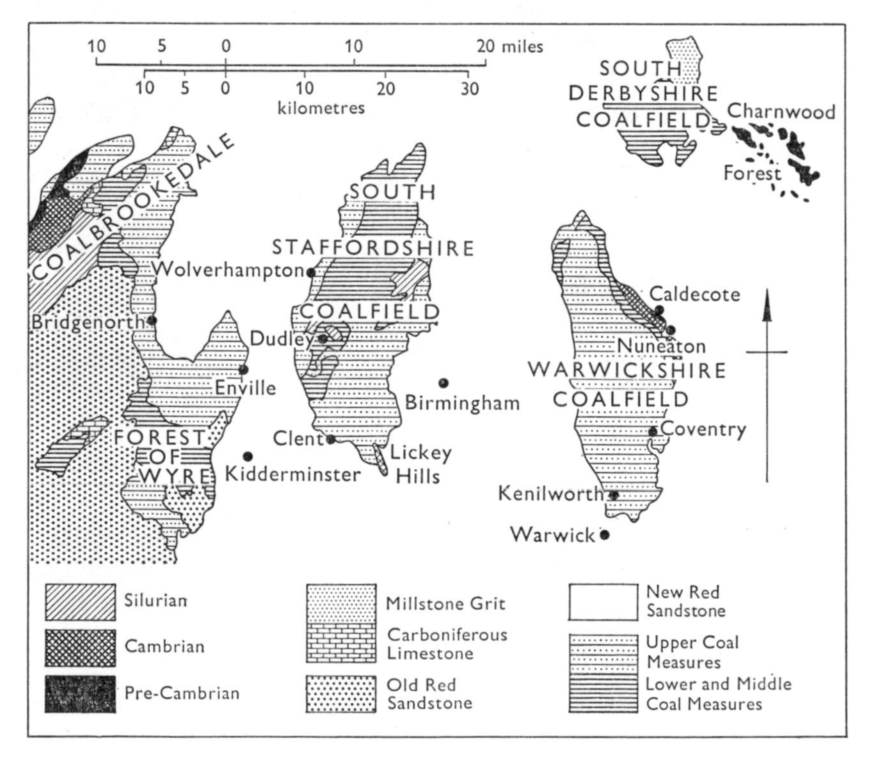

Fig. 43. The Midland coalfields and neighbouring areas.

Structure

In very general terms the dominant strike swings round with the arc of the outcrops, from north-westerly on the east, through northerly in the centre and south, to north-easterly (Caledonoid), as might be expected, on the borders of Wales. In the Midlands there is a tendency for the coalfields to protrude through the Trias as some form of rather complex horst, on which the Coal Measures lie with their own folding and faulting.

A major anticline runs north-north-west through Ashby de la Zouch and forms the central structure of the joint coalfield of Leicestershire and South Derbyshire, the two sections forming parallel synclines on the east and west respectively. This is the Charnoid direction, which also governs the major faults. Unconformable patches of Trias remain even in the centre of this coalfield.

The Warwickshire Coalfield is one of the simpler, and conforms fairly well to the structure of 'a syncline on a horst'. The synclinal axis trends north–south and the Cambrian of Nuneaton forms a strip inside the north-eastern boundary fault. The South Staffordshire Coalfield also has important north–south structures, especially in the south—an anticline flanked by the two troughs of Birmingham and Wolverhampton—but much of the coalfield is decidedly complex. The Coal Measures rest on a Lower Palaeozoic basement, chiefly Silurian, which is brought up in sharp, local anticlines. The visible boundary is chiefly one of Triassic overlap, except where faulted on the north-east, but there are sub-Triassic faults elsewhere.

The relatively unimportant coalfields of the Forest of Wyre and Coalbrookdale, on the eastern edge of the Welsh Borders contain Caledonoid and Malvernoid (north–south) structures; the former govern the Shrewsbury Coalfield, where, cut by large north-easterly faults, the beds dip north-eastwards under the Trias of the Cheshire plain. An easterly dip under the Trias is also fundamental in North Wales, but here the coalfield is divided into the northern Flintshire and southern Denbigh fields by the north-easterly Llanelidan Fault, and strike faulting, throwing down to the west, compensates for the dip to some extent.

Lower and Middle Coal Measures

The succession in the Midlands is broadly similar to that in the North Staffordshire Coalfield; the lowest measures are conformable on the Millstone Grit, and red Upper Coal Measures are conspicuous. The productive measures, however, are always thinner than in the Pennine group, because subsidence was less near the Midland Barrier.

Minor variation on this pattern can be seen where local structures in the basement affected the steady process of accumulation and subsidence. There is, for instance, a marked easterly thinning from the South Derbyshire to the Leicestershire Coalfield; this is in the general direction of the Pre-Cambrian of Charnwood Forest, although no close connexion with a pre-existing ridge can be found. In Warwickshire the lowest recognizable zone is *communis*, lying on Cambrian rocks; above this Lower and Middle Coal Measures show a pronounced southerly thinning together with amalgamation of seams; the Thick Coal, of over 20 feet, is formed in the centre of the coalfield by the coalescence of a number of seams that are separate in the north, and itself becomes reduced farther south.

The irregular foundations are particularly apparent in south Staffordshire where underground ridges of Silurian rocks coincide with areas of thin, or even absent, seams, and the measures are banked up against them. The famous Thick Coal of south Staffordshire, which straddles the junction of the *modiolaris* and lower *similis-pulchra* zones, is the equivalent of 200 feet of strata to the north containing three named seams; as in Warwickshire the Thick Coal itself becomes thinner to the south. The top and bottom of the productive measures exhibit this diminution most strongly, with the lowest coals failing southwards (though sometimes the seat-earths remain) and the replacement of some of the highest seams by the Etruria Marl (Fig. 44).

Irregularities in deposition, and their relation to structure, can be seen in Flintshire. The productive measures are thinner, or even show contemporary erosion in areas that were later to become post-Carboniferous anticlines. A comparable tendency for thicker measures to coincide with later synclines is also known from the Yorkshire Coalfield. In the Shrewsbury Coalfield no Lower or Middle Coal Measures are found.

Upper Coal Measures

The lithological divisions in the south Midlands follow those in north Staffordshire with some local changes in nomenclature; not all the divisions (Table 33) can be found in each coalfield and some are reduced or missing through actual unconformity.

One conspicuous variation is in the amount of grey productive measures present above the uppermost marine band. For instance, in the north of the South Staffordshire Coalfield there is as much as 250 feet, but to the south the base of the Etruria Marl falls, so that these grey measures are progressively reduced: in a few places the marls rest on Middle Coal Measures. A comparable variation affects the base of the Etruria Marl in Warwickshire. Since the junction appears to be everywhere conformable, it is probably a diachronous one, and red muds were deposited near the Mercian Highlands while peat swamps persisted farther north (Fig. 44).

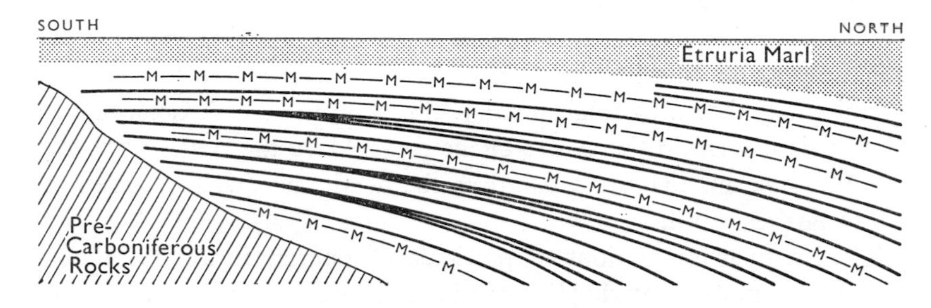

Fig. 44. Diagram illustrating the northward splitting of coal seams in the South Staffordshire Coalfield and their correlation by marine bands; the uppermost is equivalent to the Mansfield of Yorkshire (Fig. 41). The Etruria Marl facies appeared earlier in the south while peat swamps (producing the uppermost two seams) still persisted in the north. (From Edmunds and Oakley, 1948, p. 38.)

Table 33. *Upper Coal Measures of South Staffordshire*

(Enville Breccia, ? Carboniferous or Permian)
Keele Group, red and purple sandstones and clays
Halesowen Beds, grey beds with thin coals
Etruria Marl, red and purple
Black Band Group (productive measures)
——— Bay or Lady marine band ———
(Middle Coal Measures)

Another sign of widespread instability is the unconformity which develops locally at the base of the Halesowen Beds. In some parts of the Warwickshire Coalfield this group rests on 300 feet of Etruria Marl, while as little as 13 miles away, in the north-east of the field, the base has transgressed right across the marls on to the productive measures. A similar effect is found in parts of south Staffordshire and even more spectacularly in Coalbrookdale. Here below the Keele Group are the Coalport Beds, red and grey marls and sandstones, thin coals and *Spirorbis* limestones. They lie with marked unconformity on the Productive (Middle) Coal Measures, which were folded along Caledonoid axes and eroded before the Coalport Beds were laid down. As a result there is strong local variation in the number and value of the seams present; along the narrow outcrop linking the Coalbrookdale and Forest of Wyre coalfields erosion has been complete and only Upper Coal Measures remain. Although this is a normal unconformity its effect in cutting out seams has led to the name the 'Symon Fault'.

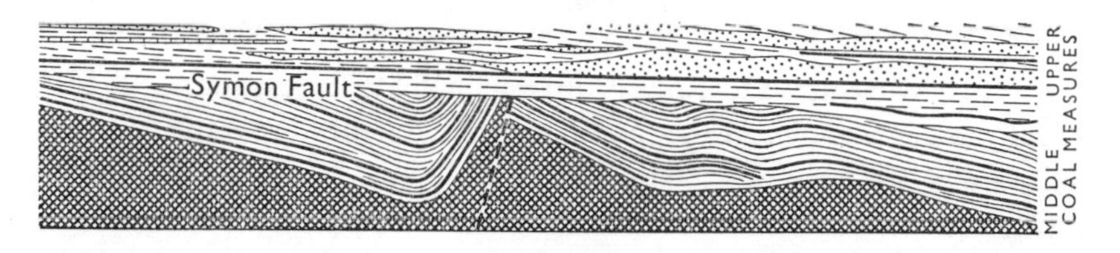

Fig. 45. The 'Symon Fault' of the Coalbrookdale Coalfield. The Middle Coal Measures, which overlie Silurian rocks, were strongly folded and faulted before the Upper Coal Measures were laid down (only the lowest beds of the latter are shown). (Adapted from Edmunds and Oakley, 1948, p. 47.)

At Shrewsbury the Upper Coal Measures rest on Lower Palaeozoic and Pre-Cambrian strata, and include representatives of the Keele and Newcastle Groups and Etruria Marl. Only the middle group is well known, and contains seams up to 3 feet. In North Wales there is a full succession of Coal Measures; they overlie the Millstone Grit conformably and are thickest in Denbighshire. The Upper Coal Measures are transgressive at the base in the north and south of the outcrops, in Flintshire and at Oswestry. They can be broadly correlated with north Staffordshire as follows:

Erbistock Beds	Keele Group
Coed-y-alt Beds	Newcastle Group
Ruabon Marl	Etruria Marl

It is possible that the Black Band Group is represented in the uppermost productive measures of Denbighshire, but marine bands being absent exact correlation is doubtful.

If we enlarge our view for a moment and examine the Coal Measures sequences nearest to St George's Land, whether on north or south, we find that the tendency for a transgression or unconformity at or near the base of the Upper Coal Measures is remarkably widespread. It may appear as a sub-Pennant unconformity, the Symon Fault, the break below the Halesowen Beds, or the transgressive base of the Shrewsbury Coal Measures. It is one of the best known Hercynian phases, whose later manifestations can usually only be called 'post-Carboniferous'.

COALFIELDS OF THE NORTH-EAST AND NORTH-WEST

The two major coalfields in this region, those of Whitehaven and Durham and Northumberland, have a good deal in common; the upper and lower measures especially are poorly represented.

The structures in Durham and Northumberland are fairly simple; an irregular syncline trends north-north-west and the eastern limb is largely under the sea or, in the south, under a Permian cover. The exposed Coal Measures accordingly dip regionally to the east or north-east, usually at a low angle. At the mouth of the Tyne they cross the synclinal axis, and a westerly dip is also found under the Permian. The major faults run east-north-east and include the Ninety Fathom Fault (or 'Dyke') north of Newcastle and the Butterknowle Fault in South Durham; it is probable that these lines are related to the north and south boundaries of the Alston Block of the Pennines to the west.

The lowest productive measures belong to the *communis* zone, neither the *lenisulcata* zone nor the *G. subcrenatum* band being known—a paucity in line with that in some of the Namurian zones below. The 2,000 feet of the productive measures continue up into the *phillipsii* zone, above which follows unconformable Magnesian Limestone. Most of the workable seams are confined to the middle of the three local groups, the upper and lower being more arenaceous.

Upper Group, mainly sandstone	*phillipsii* upper part of *similis-pulchra*
Middle or Productive Group	lower part of *similis-pulchra* *modiolaris* upper part of *communis*
Lower or Ganister Group	lower part of *communis*

The lower measures are poor in marine bands, and most of those that have been found contain *Lingula* and not goniatites; the Harvey Marine Band, however, is the Clay Cross of Yorkshire and so defines the base of the Middle Coal Measures, and the Ryhope is the Mansfield of Yorkshire. Few of the problems that affect the Midlands or south-west are found here, but the seams, which are nine feet thick at their thickest, are liable to abrupt changes in thickness or in quality; in east Durham, for instance, several seams become thinner where the associated sandstones thicken.

The exposed Whitehaven Coalfield forms a strongly faulted strip of ground on the coast north of St Bees Head, and is part of the Carboniferous rim to the Lake District. Like that rim as a whole, the rocks tend to dip away from mountains, here to the north-west. Undersea workings are extensive and in these the dip flattens out. The main faults of the coalfield are also north-westerly and are post-Triassic, but some on the south-east margin and north of Maryport moved in pre-Triassic times and have a Caledonoid trend.

Conditions characteristic of productive measures began here in the late Namurian and continued almost to the end of the Middle Coal Measures. The *Gastrioceras subcrenatum* band has its most northerly occurrence in Britain, and the Solway band marks the base of the Middle Coal Measures; the highest band is probably equivalent to one just below, not actually at, the top of the Middle Coal Measures. The rocks are normal in lithology but rather thin, being some 1,200 feet, and the Upper Coal Measures are largely arenaceous and usually reddened. The coloration ceases downwards irregularly and sometimes affects the top of the Middle Coal Measures. This group is usually known as the Whitehaven Sandstone, but its two principal criteria as a formal stratigraphical unit are doubtful; the red colour is probably secondary, and the small unconformity, once thought to occur at the base, is not now recognized. The Whitehaven Sandstone, up to 1,000 feet thick, is probably best considered as a thick, arenaceous type of Upper Coal Measures.

The Ingleton Coalfield is small in extent and economic value, but helps to fill in the long gap between the Pennine and northern coalfields. It lies on the south, or downthrow side of the Craven Faults (Fig. 24); in the centre of the syncline there is a patch of red breccias that have been called Permian. The essential feature of this little outcrop is the unconformable separation of the grey, productive measures below from the red Upper Coal Measures above. The former are thin, 1,200 feet in the *lenisulcata*, *communis* and *modiolaris* zones, and the latter rather thicker, and include *phillipsii* and *tenuis* zones. They are dominantly sandy and sometimes coarse-grained, but also contain more usual rock-types, such as shales with plants, rootlet beds and rare lamellibranchs.

The Ingleton Coalfield fits in fairly neatly between Lancashire and Cumberland and in spite of local characters emphasizes the essential unity of the Coal Measures facies from the Midlands to southern Scotland.

CARBONIFEROUS ROCKS IN EASTERN AND SOUTHERN ENGLAND

It is convenient at this point to review both the Lower and Upper Carboniferous rocks that are known underground in the south and east of England, and as in earlier chapters the description can be related to the London Platform and the ridge of Pre-Cambrian basement rocks from Leicestershire to Norfolk.

Around or outside the margins of the London Platform Dinantian rocks have been recorded in a few deep borings in Sussex, Surrey and Kent and also near Cambridge and Northampton; they do not, as far as is known, make up much of the centre of the platform. In all these places the facies is calcareous and resembles that of the South-West Province; no Culm has been encountered. Millstone Grit, or arenaceous developments of that age, have not been recognized and the only productive underground coalfield is that in east Kent. The lowest Coal Measures here rest on Carboniferous Limestone over most of the field, but overlap

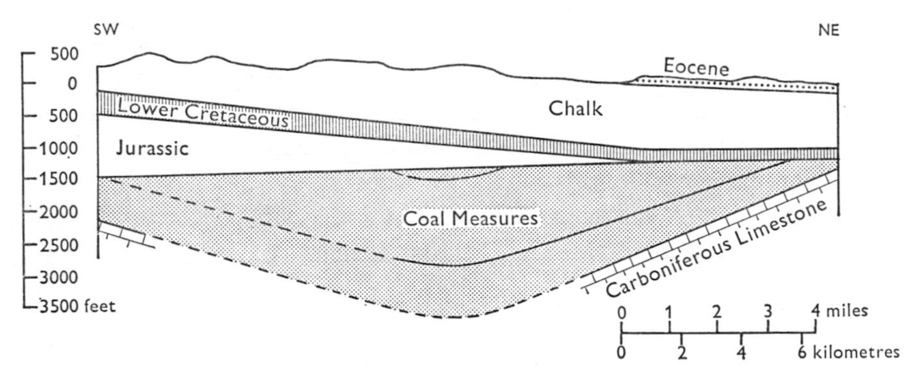

Fig. 46. Section through the Kent Coalfield showing the synclinal structure of the Carboniferous rocks and overstep in the Mesozoic cover. (After Trueman, 1954, p. 156.)

to the north on to the Silurian edge of the London Platform. The structure conforms well to the theoretical buried syncline, the axis of which plunges south-eastwards towards the Straits of Dover. In the south-west section the cover is of Jurassic rocks and in the north-east of Lower Cretaceous. The 2,800 feet of Coal Measures are clearly of south-western type and resemble the Radstock succession in the thick Upper Coal Measures, with its Pennant-like sandstones in the *tenuis* zone.

There is no other wholly concealed coalfield quite comparable, but a thick basin of Upper Coal Measures probably underlies Mesozoic rocks on the borders of Gloucestershire and Oxfordshire—encountered for instance in borings at Stow-in-the-Wold and near Burford. 4,000 feet of measures are known, again with 'Pennant Sandstones' and rich in seams. Farther north-east on the edge of the London Platform Lower Carboniferous rocks are directly overlain by Mesozoic.

To the north of the Mercian Highlands and the Charnian ridge there lies the great Carboniferous basin of north-eastern England, whose components appear from under their Permian cover along the eastern edge of the Pennines. Apart from oil and coal borings, in Nottinghamshire and nearby, the records are too widely or irregularly spaced to show detailed changes in structure or facies, but it is fairly clear that many of the Pennine features continue eastwards; Carboniferous rocks certainly extend to the Yorkshire coast, and probably to that of Lincolnshire as well except possibly in the extreme south (Fig. 47).

In the north, from about the latitude of York to Tees-side, the rocks under the Permian and Mesozoic cover seem to have been gently folded in a manner comparable to that in the Pennine outcrops. Lower Carboniferous shales and limestones are present in an anticline from Middlesbrough to the mouth of the Tees; farther south later rocks come in, up to Middle Coal Measures, in a syncline plunging north-eastwards whose axis possibly crosses the coast some miles south of Whitby. Inland there is a wide spread of Millstone Grit. The Dinantian beds are largely of basin facies, with much shale and relatively thin limestones, and it seems likely that the edge of the Askrigg Block is not far east of the present outcrop.

Farther south there are two components to the subsurface Carboniferous stratigraphy—the concealed Coal Measures themselves and an earlier pattern of blocks and basins (the latter here called troughs or gulfs) that has much in common with the structures seen in the exposed areas of Carboniferous Limestone and Millstone Grit. These two components are partly related, however, and where the earlier groups expand into the gulfs there is commonly a thickening in the Coal Measures as well.

The three gulfs shown in Fig. 47 have been discovered through oil borings in the east Midlands; they are chiefly defined by an expansion of the Millstone Grit, which is 2,000 feet or more in each, compared with much thinner sequences on the intervening shelves. The Gainsborough Trough is a shallower and narrower, south-easterly continuation of the Craven Lowland Trough. Similarly the shorter Edale Gulf extends south-eastwards from the region of basin facies north of the Derbyshire Massif (p. 178). The Widmerpool Gulf is the most extensive and illustrates the thickness changes particularly well. From 300 feet on the northern flank (resting on D limestones), the Millstone Grit expands to over 2,600 feet in the gulf where it is underlain by 1,500 feet of P shales. Few of the boreholes have penetrated Lower Carboniferous rocks completely, but at Eakring, north-west of Newark, 3,000 feet of

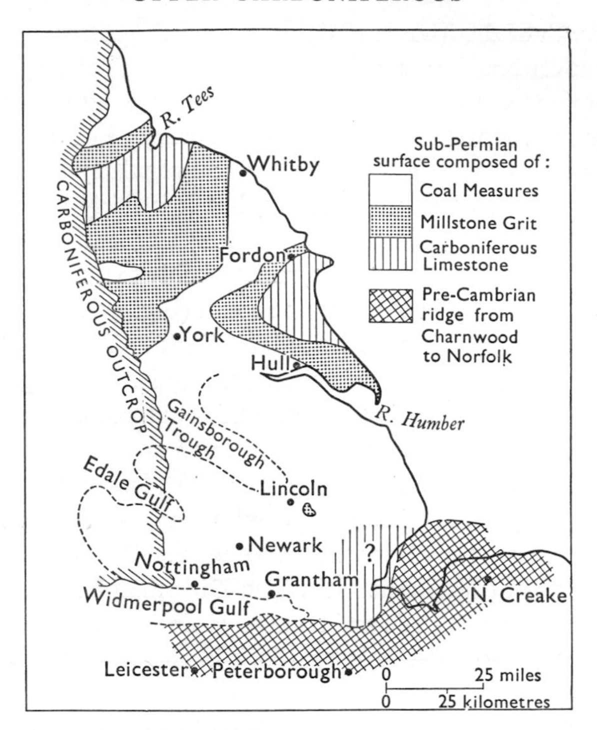

Fig. 47. Generalized interpretation of the rocks underlying Permian and Trias from Tees-side to Norfolk, together with three subsurface 'troughs' in the east Midlands. (Adapted from Kent, 1966, pp. 335 and 343.)

limestones (C_1–D_2) with sandstones and conglomerates are underlain by 1,700 feet of red conglomerates; the latter may be Old Red Sandstone or a basal Carboniferous facies.

All the Carboniferous divisions become much thinner eastwards on to a south Lincolnshire 'shelf'. Pre-Permian erosion was also active here and the Coal Measures are reduced for this reason as well; in fact on the crest of a ridge south-east of Lincoln the Permian beds rest locally on Millstone Grit.

An unexpected feature of this underground exploration is the discovery of much basic igneous rock, at several horizons. In the Lower Carboniferous these include dolerites similar to those of Derbyshire, but there are thicker layers, both intrusive and extrusive, in the Millstone Grit and Coal Measures. They are most prevalent in the east–west fracture-belt that bounds the Widmerpool Gulf on the north side, and here over 400 feet of altered basalts have been found replacing part of the Millstone Grit and nearly all the Lower Coal Measures. At one place on the south side of the gulf there are 330 feet of igneous rocks in 565 feet of Coal Measures.

All the components of this northern Carboniferous basin disappear southwards (probably by overlap) against the Pre-Cambrian ridge from Charnwood to Norfolk, in the same way that the Lower Carboniferous and Namurian beds are seen to do in the outcrops to the west.

COAL MEASURES IN SCOTLAND

The Scottish coalfields comprise several separate, largely synclinal areas in the Midland Valley and a few small patches outside the faulted rift. The principal regions are:

The Ayrshire Coalfield, an irregular and much faulted outcrop from Saltcoats in the north to Dalmellington on the Southern Upland border.

The Douglas Coalfield lies to the east of the Ayrshire and is a small outlier in central Lanarkshire.

The Central Coalfield, the largest in Scotland, extends east from Glasgow and north to Falkirk; it is geologically continuous with the Clackmannan basin north of the Forth, whose northern boundary is the Ochil Fault.

The West Fife and East Fife Coalfields, the latter being the larger and reaching the Firth of Forth.

The Midlothian Coalfield, a deep north-north-east syncline south-east of Edinburgh and the Pentland Hills.

The East Fife and Midlothian coalfields extend under the sea and the Coal Measures may be continuous under the Firth of Forth. There is no cover to these Scottish coalfields except a relatively small area of red sandstones, probably Permian, at Mauchline in Ayrshire.

Productive Measures

Owing to the separation of these several fields, which are also much faulted, the Scottish coal seams have been given an exceptionally large number of local names, but some generalizations can be made on the stratigraphical succession in the Midland Valley as a whole. The coal-bearing measures, sometimes called the 'Upper Productive Measures' in contrast to those of the Limestone Coal Group, range from the *lenisulcata* to the lower *similis-pulchra* zones. Lamellibranchs and plants are fairly common, but marine faunas less so, and the bands contain *Lingula* more often than goniatites. *Gastrioceras subcrenatum* is among the absentees, so that, lacking a palaeontological datum line, the base of

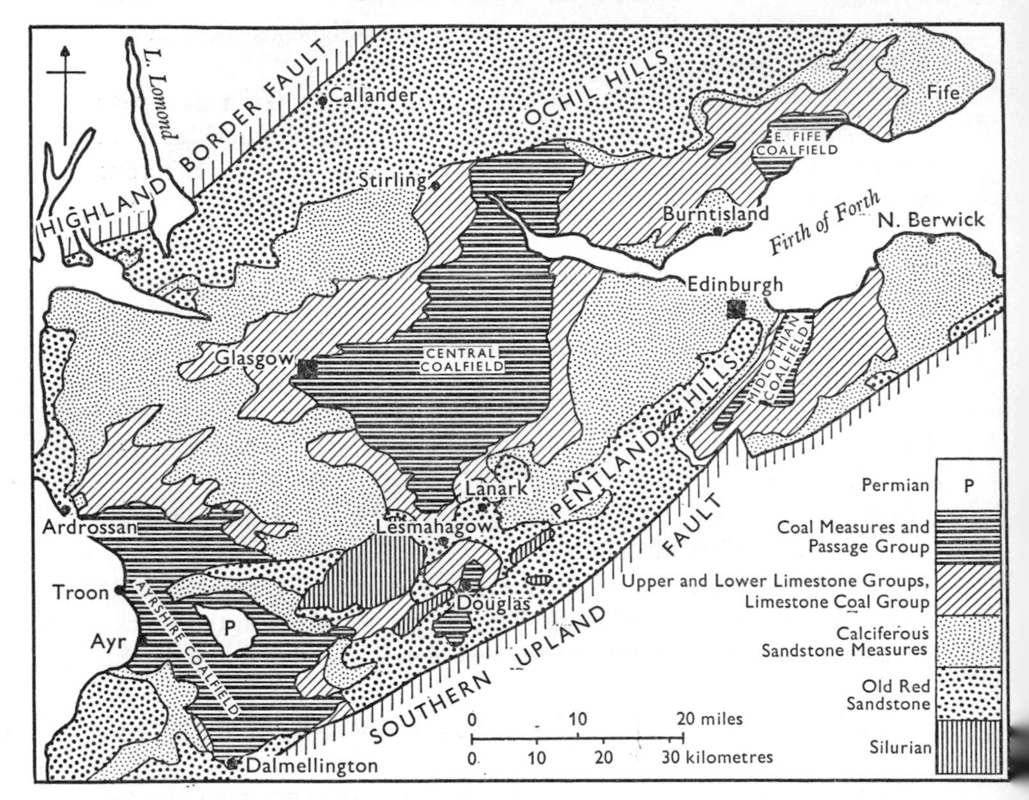

Fig. 48. Geological sketch map of the Midland Valley of Scotland. Igneous rocks are not differentiated; they are important, however, among Old Red Sandstone and Lower Carboniferous outcrops and in particular occupy most of the Calciferous Sandstone Measures west, north and south of Glasgow.

the Coal Measures in Scotland has to be taken arbitrarily in each coalfield.

In the middle of the *modiolaris* zone, the Queenslie marine band, with *Anthracoceras vanderbeckei*, is widespread; as the equivalent of the Clay Cross, Amman, etc., it marks the base of the Middle Coal Measures. The upper limit of the latter is more difficult to define. The most important Scottish marine incursion is the last—Skipseys marine band—which has a varied fauna of *Lingula* and other brachiopods, nautiloids, lamellibranchs and goniatites, including *A. hindi*. It is not, however, equivalent to the Top marine band of Yorkshire, but to the Mansfield, at the junction of upper and lower *similis-pulchra* zones. This is the level of a general lithological change in Scotland, and consequently the base of the locally defined Upper Coal Measures is drawn at a slightly lower horizon than in England and Wales.

There is a general tendency for marine strata to be more abundant in the west than in the east; for example, the lower *similis-pulchra* zone contains four marine bands in Ayrshire, but only two in Fife and Midlothian. Regional changes in thickness are also apparent. In the Midland Valley as a whole the thickest successions occur in the north and south, and a thinner belt in the centre. In north Ayrshire there is locally as little as 500 feet of Productive Measures, without apparently an internal unconformity, whereas 1,500 feet are recorded at Dalmellington in the south, and 1,700 in the Douglas Coalfield. In the opposite direction, the southern part of the Central Coalfield contains about 900 feet of these measures, but some 1,200 feet farther north. In Midlothian and East Fife there are 1,300 and 1,500 feet respectively.

Upper Coal Measures

Above Skipseys marine band the coals are few and thin, and the shells and plants much less plentiful. The beds show regional changes in thickness comparable to those in the Productive Measures below, but do not reflect differences in subsidence so closely because the upper surface is always one of erosion and only an incomplete and variable amount remains. About 1,000 feet are known in Fife, 900 in the Douglas Coalfield and slightly more on the west side of the Central Coalfield. The upper measures are thickest, 1,500 feet, in the southern part of the Ayrshire basin, where they are overlain by 500 feet of olivine basalts, the Mauchline lavas. Till recently these were thought to be Permian, as were the Mauchline Sandstones above. However plants found in shales intercalated in the lava flows include species that are probably Stephanian and so the Mauchline Lavas should perhaps be formally included in the highest Carboniferous. The Upper Coal Measures have also yielded species of the *phillipsii* and *tenuis* zones.

The term 'Barren Red Measures' has commonly been used for these Upper Coal Measures, but there is a considerable body of evidence to suggest that much, and perhaps all, of this colour is secondary and that the rocks are merely reddened, in spite of their great thickness. The base of the red zone is often uneven; grey beds with poor coals may occur above Skipseys marine band, as in the Douglas Coalfield, or red beds well below that horizon.

These variations have been studied particularly in Ayrshire, where in addition to reddening (essentially the oxidation of ferrous iron compounds with hematite as an end-product), there are other secondary changes which have been attributed to percolation of waters beneath a late Carboniferous or early Permian land surface. Of these the most spectacular is the replacement of coals by carbonate rocks, usually dolomite. How far such processes are generally responsible for all Carboniferous 'Barren Red Measures' is uncertain. The Scottish red beds do not appear to differ significantly from the productive measures below in being much more arenaceous, though some of the sandstones are coarser. They have not therefore the primary lithological distinction of, for instance, the Whitehaven Sandstone.

It is not easy to draw a regional geographical picture of the Midland Valley during the deposition of the Coal Measures, partly because, being chiefly in the centre of synclines, the youngest Carboniferous beds are less extensive than the older divisions. The slight increase in marine strata to the west is in line with a similar direction of facies-change in earlier periods, and the extra downwarping to the north might perhaps reflect a greater supply of detritus in that direction; on the other hand there is little to suggest that the Southern Uplands were a similar source. A little more information can be gained from the Coal Measures outside the Midland Valley.

Outlying coalfields

The principal outcrop on the Southern Uplands is the Sanquhar basin, where nearly all the Carboniferous rocks preserved are some kind of Coal Measures. The Productive Measures overlap the small amount of Millstone Grit and limestones (probably the Limestone Coal Group) on to steeply folded Ordovician rocks; they are in turn overlain by Barren Red Measures. There was thus progressive overlap on the edge of this small basin in Namurian and Westphalian times. Essentially the same relationship is present farther south at Thornhill which is the other Carboniferous outlier preserved in this north-westerly trending depression across the Southern Uplands. But here the limestones are thicker and probably include some Viséan strata, and the Coal Measures are overlain fairly extensively by New Red Sandstone.

A significant westward extension is indicated by the Productive Measures on Arran, where beds of *communis*, *modiolaris* and *similis-pulchra* zones are known, lying on lavas that resemble those of Namurian age in Ayrshire. The Coal Measures lack coals, and show signs of overlap among themselves, so that they may be local marginal deposits.

A somewhat similar state of affairs is found in the Machrihanish

Coalfield on the Mull of Kintyre, where over 500 feet of Productive Measures, possibly with a small amount of barren red beds above, again rest on lavas. Here, however, there are seams over 4 feet thick, though less productive than those in the Limestone Coal Group below. Two small outcrops are found much farther to the north-west. Three hundred feet of Lower Westphalian beds are known from Morvern, north of the Sound of Mull; they are chiefly sandstones, but include plant-bearing shales and a thin coal seam. At the Bridge of Awe, Argyll, there are a few feet of measures probably of Upper Carboniferous age.

Isolated and restricted though these outliers are, they are at least evidence of sedimentation over a wide area west of the Midland Valley, and show that in places peat swamps persisted also. They are the last, northernmost signs of the process of erosion, sedimentation and subsidence that finally invaded even outlying portions of the Caledonian highlands.

COAL MEASURES IN IRELAND

Ammanian beds only are known in Ireland except possibly in the small Kanturk Coalfield of Co. Kerry. The largest coalfield is that of Leinster, or Castlecomer, of about 150 square miles, close to the western margin of the Leinster Massif; the smaller Slieveardagh Coalfield 20 miles to the south-west has a very similar succession and belongs to the same province. Both are synclinal and, as in all the coalfields in southern Ireland, the upper surface is one of erosion and there is no cover of later rocks. Over 1,000 feet of Coal Measures are known in the Leinster Coalfield; their correlation is clearest at the base, the *Gastrioceras subcrenatum* band being present and the *lenisulcata* zone. A conspicuous bed is the 200-foot Clay Gall Sandstone, named after its rounded mudstone pebbles; its base is erosive and sometimes transgressive. Anthracites and bituminous coals are known from the Leinster field, in seams up to 2 feet in thickness.

The large Namurian outcrop of County Limerick and northern Kerry is overlain by two small outliers of Coal Measures; a very small one at Crataloe in the north-centre, and a slightly larger and much more important one at Kanturk (Fig. 31), on the Blackwater river in the south. At Crataloe the succession does not amount to more than 300 feet, most of which is taken up with the Clay Gall Sandstone and beds below. The Kanturk Coalfield, however, though only 12 square miles contains over 3,000 feet of Coal Measures, steeply dipping and much faulted as it lies in the belt of strong Hercynian folding and thrusting. There are many seams in a varied lithological sequence and marine bands occur in the upper part; it is not well known as yet, however, but may provide interesting information in the future.

The Kingscourt Outlier (Fig. 27) includes a small amount of Coal Measures in Co. Monaghan. It is all within the *lenisulcata* zone, the *G. subcrenatum* band is present and there are faunal associations with the Pennine coalfields.

At Dungannon, in Co. Tyrone, the Coal Measures of Coalisland follow conformably on the Namurian beds; about 900 feet are known, in the *lenisulcata* and *communis* zones. There are twelve seams, with a maximum thickness of 9 feet. This is a considerably richer sequence than is found elsewhere, but the area is small, faulted, and lies under a cover of Trias and drift. The general position of the Tyrone coalfield, in relation to the Longford–Down Massif, Southern Uplands and Irish continuation of the Midland Valley of Scotland, links it structurally with the last; the stratigraphical succession, however, is not Scottish in type, but nearer that of the Pennines, in so far as the *G. subcrenatum* band is present and the *lenisulcata* zone is well developed. This is another minor instance of the way in which structural and stratigraphical affinities do not always go hand in hand across the Irish Sea.

These few paragraphs on the Irish Coalfields form a somewhat melancholy postscript to a chapter describing the extensive, and still valuable, Coal Measures of England, Wales and Scotland. Stratigraphically we are grateful for the glimpses we get of a wide westward extension of the forest swamps and coastal flats of the Coal Measures. Economically they are just enough to show what Ireland has lost through the erosion that stripped those measures all too thoroughly.

REFERENCES

The stratigraphy of the British Coalfields is covered by Trueman, 1954, and is described in many memoirs of the Geological Survey.

Millstone Grit, etc. Earp *et al.* 1961; Hodson, 1954, 1957, 1959; Hodson and Lewarne, 1961; Ramsbottom, 1965; Ramsbottom *et al.* 1962; Stephens *et al.* 1953.

Coal Measures. Edwards, 1951; Edwards and Stubblefield, 1948; Ford, 1954; Fowler and Robbie, 1961; Nevill, 1961; Stephens *et al.* 1953; Stubblefield and Trotter, 1957; Woodland and Evans, 1964.

General. Charlesworth, 1963, ch. 9; Francis, 1965*a*, *b*; George, 1962*a*, *b*, 1963*a*; Trotter, 1960; Wills, 1956.

British Regional Geology. Midland Valley of Scotland, Northern England, Pennines, North Wales, South Wales, Central England, Bristol and Gloucester, Wealden District.

8

NEW RED SANDSTONE OR PERMIAN AND TRIASSIC SYSTEMS

The various names that have been given in Britain to the formations that overlie the Coal Measures and underlie the Jurassic System reflect a real difficulty in the identification and correlation of beds that are often non-marine and largely unfossiliferous. Being the second big group of red continental beds they were early called New Red Sandstone in contradistinction to Old Red Sandstone.

There have been various attempts to divide these, and similar rocks elsewhere, into the more conventional Permian and Triassic systems, but none has been entirely satisfactory. The term Permian was coined by Murchison in 1844 for strata near the city of Perm in the Ural mountains; although these strata form only part of the modern Permian, the system can be fairly well defined in Russia, and in the south-central part of the United States, marine rocks occurring in both countries. North Germany lacks the open-sea, marine facies of Russia, but there is a widespread division of the Permian into the red Rotliegendes below and a saline facies, the Zechstein, above.

Table 34. *Permian and Triassic Classification in Central Europe*

TRIAS	RHAETIC
	KEUPER
	MUSCHELKALK
	BUNTER
PERMIAN	ZECHSTEIN
	ROTLIEGENDES

The name Trias is derived from a threefold sequence in southern Germany: the middle member of this, the Muschelkalk, is marine, though without ammonites, which are found in the eastern Alps. Germany is thus an important area for the classification of the New Red Sandstone, and is the source of all the major component names except the Rhaetic. This, the highest division, is derived from the eastern Swiss Alps and is often taken to be the base of the Jurassic System; the reasons for its inclusion in the Trias are considered on page 269. Recently microfloras have helped in the correlation of continental Triassic formations.

In Britain only the Rhaetic and Zechstein have been generally recognized and equated with their type areas. The former is found as the long north–south outcrop east of the Pennines; beds of Rotliegendes age are probably present in the Midlands and north-western England, but are incomplete and their age is rather a matter of deduction. The Rhaetic is normally conformable with Lower Jurassic above and Keuper below, and in north-eastern England the 'Bunter Sandstone' also follows the Magnesian Limestone without a major gap. Consequently somewhere in the British Trias there should be beds equivalent to the four German divisions, though possibly with breaks. A marine incursion of Muschelkalk age left its traces in the Midlands, but not in its typical German facies, so that we have normally recognized only beds of Bunter-like or Keuper-like facies. Of these two names the Keuper correlates with its German counterpart at least at the top, being overlain by the Rhaetic, but the lower 'Keuper' beds, and in some outcrops the 'Bunter' also, are much less satisfactory units. The confusion is aggravated when it is found that 'Bunter Sandstones' of the east Midlands pass northwards into the Magnesian Limestone (Zechstein) of north Yorkshire and Durham. It follows that the names used in Britain are largely those of widespread rock-types, or facies, and only here and there of strata well-defined by time-lines. Where even the Zechstein sea did not penetrate and there is only a variety of red continental rocks, any division between 'Permian' and 'Trias' is very largely arbitrary.

This is a short explanation for the title of this chapter; in recent years there has been an increasing practice of treating all these beds, continental or marine, together and referring to Keuper Marl, Bunter Sandstone or Magnesian Limestone as lithological units or groups. The term New Red Sandstone may be revived for the whole, or fossiliferous strata referred to Permian or Trias.

After the principal Hercynian movements in Britain, in late Carboniferous or early Permian times, the region was one of varied topographic relief. The irregularities and basins became gradually filled up and the later deposits overlap the earlier; to the east and south-east there is the additional Jurassic and Cretaceous cover. Subsurface exploration is therefore a great help in describing and interpreting the Permo-Triassic rocks, particularly in the evaporite-bearing strata of the Midlands and north of England. In the south-east it is less vital. Triassic rocks do occur underground to the west and north of the London Platform (for instance

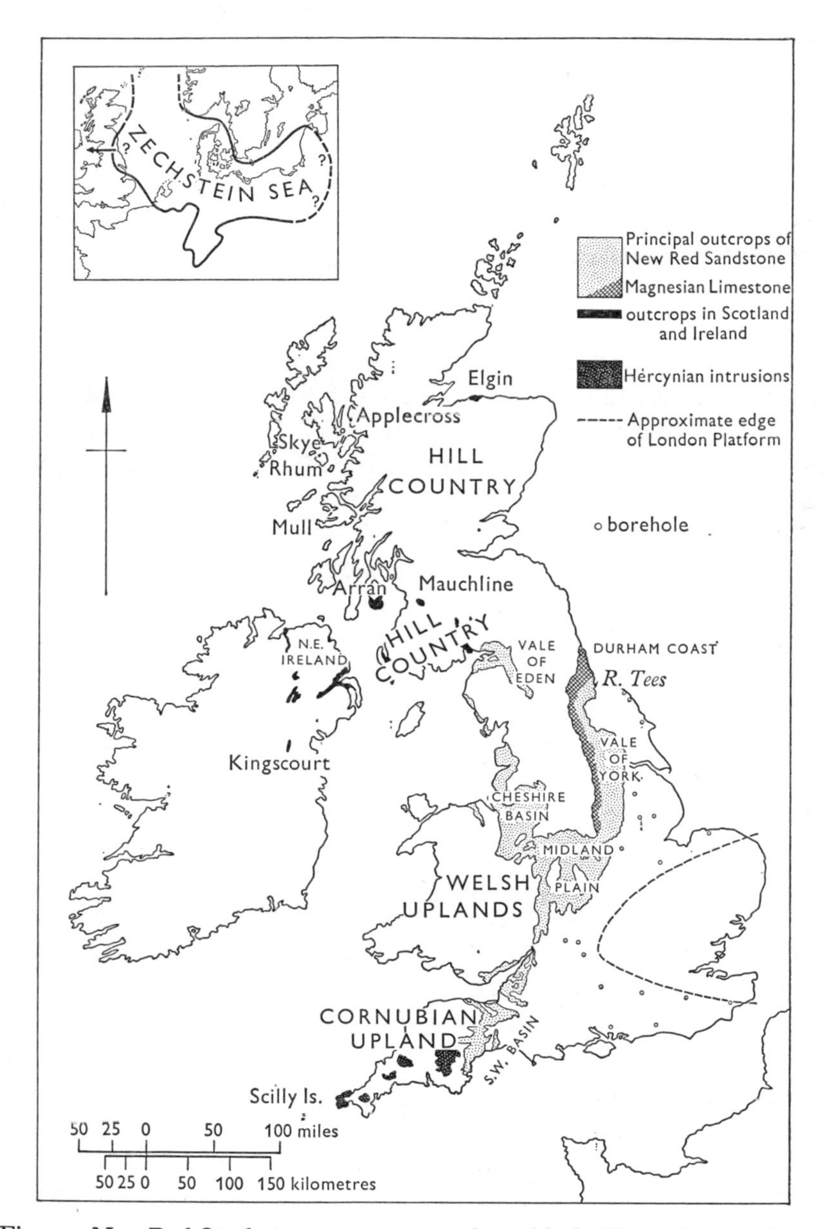

Fig. 49. New Red Sandstone outcrops, together with the Hercynian granites of Devon and Cornwall. Boreholes reaching New Red Sandstone (including Rhaetic) in eastern and southern England are shown and the approximate outline of the London Platform.

at Burford, Oxfordshire and in north Norfolk) but they are less extensive than the later Mesozoic formations and in the Weald are thin or absent. This is because the submergence of the low, but remarkably persistent feature, was a very gradual process. Each major rock group becomes thinner as the platform is approached and is overlapped by the beds above, as they extended a little farther over the shelving edge. The highest portion, from north London to East Anglia, was not covered till mid-Cretaceous times.

THE ZECHSTEIN SEA

This sea occupied a large area of northern Germany, where Zechstein outcrops are found on the edges of the Hercynian massifs, stretched across Holland and into north-eastern England. The Magnesian Limestone and associated rocks can be correlated by lithology, and occasionally by fossils, with the German succession and so there is little doubt of their age as Upper Permian. Moreover, one of the firmest correlations is at the base—the equivalence both in the general succession and in fish faunas of the English Marl Slate with the German Kuperschiefer, so that Rotliegendes in this part of England must be represented by the large unconformity beneath the Magnesian Limestone. The visible discordance is usually small, but along the 100-mile outcrop, basal Permian beds rest on anything from Viséan limestones to Upper Coal Measures and the same diversity is found underground.

The English Zechstein deposits may conveniently be described in three regions:

(1) Surface outcrops in Co. Durham, on the north-western side of the West Hartlepool Fault, with some boreholes from the Hartlepools area.

(2) Beds underground known in numerous boreholes in and around the saltfields of Tees-side, and extending to north-east and east Yorkshire; also in the anhydrite mine at Billingham (Tees-side).

(3) The Yorkshire and Nottinghamshire outcrops, boreholes between York and Hull and in north and south Lincolnshire.

There was a marked lateral variation in the Permian deposits as they were originally laid down in these various areas—for instance in the relative proportions of the detrital components (chiefly sand and marl) to those that were chemically precipitated. There were also differences among the second group in the type and thickness of the evaporites. To these original differences have been added a variety of secondary effects. Evaporites, being the precipitates of the more concentrated brines of a saline sea, are highly soluble: consequently in a humid climate they are not found in surface exposures, where both the composition and the structure of the rocks may be altered by their removal. Then there are also the changes in the evaporites themselves from the early stages of their accumulation onwards—diagenetic changes involving the replacement of one mineral by another. It is this last factor in particular that makes the interpretation of these rocks, which can only be examined underground or in borehole cores, such a complex problem of geology, mineralogy and chemistry. The causes of the diagenetic changes are many, but probably important among them is the downward migration of brines at a late stage in the evaporation (the more saline brines being the heavier) into the 'crystal mush' already precipitated, and the upward migration of connate solutions still present in the spaces between the crystals as the weight of successive layers of strata compresses those beneath.

In a body of normal sea water evaporating under arid conditions, the less soluble salts are precipitated first and the more soluble later. This produces the primary mineral phases of a marine evaporite cycle:

Calcite —	gypsum —	gypsum +	halite —	halite
$CaCO_3$	$CaSO_4.2H_2O$	$CaSO_4.2H_2O$	$NaCl$	$NaCl$

A land-locked basin evaporated to dryness will, however, produce only a small thickness of evaporites—for instance, 79 feet for a basin the depth of the Mediterranean. The major evaporite sequences, such as the Zechstein, may be ten times this figure, so replenishment from oceanic waters is an accepted mechanism. Hence arises the concept of the 'barred basin' into which new supplies of sea water flow from time to time over a lip or bar. This replenishment of the Zechstein sea appears to have resulted from relatively abrupt subsidences, giving periodic access of less saline waters. In such conditions the salts being deposited will return temporarily to an earlier, less soluble, mineral phase, and the cycle of evaporation (or partial cycle if the influx is less) begins again. In certain instances the late, highly soluble, phases were protected by a covering of marl before the reversal took place.

In north-east England three such periods of replenishment at least can be detected, forming three cycles of evaporation. Towards the deeper parts of the basin, that is in north-east and east Yorkshire, there may be additions or complications to this pattern. In its simplest form each cycle begins with a carbonate zone of limestone, often partly or wholly dolomitized, succeeded by a sulphate zone, chiefly anhydrite ($CaSO_4$) but also gypsum, and that by a halite zone or rock salt. Sequences approaching this are found towards the edge of the basin. The later products of evaporation, the potash salts (especially sylvine, KCl, and the magnesium chlorides) are only found in quantity farther east. In particular the Lower Evaporite Bed at Whitby and Fordon contains important concentrations of polyhalite ($Ca_2MgK_2(SO_4)_4 . 2H_2O$).

A general vertical zoning within the evaporites is thus characteristic, but there is a comparable lateral zoning as well. The potassium and magnesium chlorides are only found in north-east and east Yorkshire; halite disappears northwards, in Tees-side and south-east Durham, and also westwards in south Yorkshire; still farther south-west in Yorkshire the calcium sulphates and finally carbonates die out in the direction of the presumed shoreline (p. 247). It is probable that at a certain time, when halite of the Middle Evaporite Bed was being deposited in the deeper regions, anhydrite was precipitated in an intermediate belt, and carbonates near the shore. The various regional successions can now be described in relation to this rather complicated mineral distribution.

County Durham

North of the West Hartlepool Fault the surface area of Magnesian Limestone is supplemented by boreholes in the south-east and a few miles offshore. In detail this succession is peculiar to Co. Durham, but the Upper and Lower Magnesian Limestones are approximately equivalent to divisions of the same names in Yorkshire.

Table 35. *Permian succession in Co. Durham*

Lithological formations		Maximum thickness (feet)
Upper Permian Marls		*c.* 550
Upper Magnesian Limestone	Hartlepool and Roker dolomites	430
	Concretionary Limestone	65
Middle Magnesian Limestone	Dolomite and anhydrite	520
	Reef dolomites, etc.	470
Lower Magnesian Limestone		225
Marl Slate		14
Basal Yellow Sands with breccias		104
(Carboniferous rocks)		

The Basal Sands, locally mixed with breccias, lie unconformably on Carboniferous rocks and sometimes exhibit strong aeolian bedding; they are the shore dunes and sands of the encroaching Zechstein sea and also occur intermittently along the length of the outcrop into Yorkshire and Nottinghamshire. The detrital grains and angular fragments of the breccias are derived from the local Carboniferous sandstones and limestones that formed a gently shelving shore.

The Marl Slate is a fissile, silty, dolomitic rock, sometimes bituminous, with thin impersistent laminae; locally it contains plants, or more often plant debris, as do many of the carbonate rocks above, and is well known for its fish fauna, mostly Palaeoniscids. The Marl Slate represents the first marine influx, in very quiet conditions, before the Zechstein sea became abnormally saline. Above lie the three divisions of the Magnesian Limestone, amounting at a maximum to over a thousand feet.

The Lower Magnesian Limestone exemplifies the central rock-type—a well-bedded dolomite, or dolomitic limestone, in which fossils are locally quite common, especially casts of brachiopods. A small amount of collapse breccias points to the original inclusion of some evaporites, now removed by solution. The Middle Magnesian Limestone is more varied, and includes the well-known Permian reef, which extends as a sinuous belt about a mile wide, slightly inland from the present coastline, from Sunderland to Hartlepool. The reef-building organisms were chiefly algae, with local spreads of polyzoa; hollows in the reef were occupied by brachiopods, gastropods and crinoids. As in modern reefs the fauna was divided into those inhabiting the outer and inner parts and there was an upward impoverishment of the fauna as a whole.

There are some signs of a seaward, or eastward, growth of the reef-belt with time, the upper beds being built out on the seaward slope over the reef talus, or steeply inclined beds with breccias that accumulated on the reef-front. On the shore, or westward, side the 'back-reef' deposits consisted of bedded dolomites, often oolitic, with a restricted fauna of small lamellibranchs. The offshore facies, in front of the eastward face of the reef, has been found in offshore borings to include substantial beds, up to 500 feet, of anhydrite.

The Upper Magnesian Limestone, which overlies these diverse facies, is a return to bedded dolomites, with a thick development of oolites towards the top. The curious concretionary beds, found only in this upper division in Co. Durham, probably represent dolomites which have undergone calcification (the reverse of the more usual dolomitization of limestone), the source of the secondary calcium being possibly the leaching of gypsum and anhydrite in the underlying reef complex.

Tees-side, Whitby and East Yorkshire (Fordon)

On the south-eastern, or downthrow, side of the West Hartlepool Fault, the Magnesian Limestone is overlain by red marl, succeeded by a sandstone usually called Bunter, and then by Keuper Marl. Farther away there follows an increasing cover of Jurassic rocks of the north-east Yorkshire moors. All the information in this region comes from underground sources, where extensive beds of evaporites are present. The succession at Billingham, just north of the Tees, exemplifies the cyclic sequence at the edge of the evaporating basin; it is relatively thin, and the potassium salts that are formed only in the later stages are absent. The evaporite horizons of the three cycles, sometimes known as the Lower, Middle and Upper Evaporites (and not to be confused with the Lower, Middle and Upper Magnesian Limestone of Durham) are here represented by anhydrite with a little gypsum. The most reliable datum plane is the top of the Upper Magnesian Limestone; that division is generally correlated with the Upper Magnesian Limestone of Durham, and there is some slight faunal support for this, especially in the problematic fossil (?algal) *Tubulites permianus*.

Table 36. *Permian succession at Billingham, Co. Durham*

Lithological formations	Evaporite cycles
('Bunter Sandstone') Upper Permian Marl	
Upper Anhydrite Marl	Third Cycle
Main Anhydrite, with a little halite Upper Magnesian Limestone	Second Cycle
Lower Anhydrite, and dolomite Lower Magnesian Limestone	First Cycle
Marl Slate and Yellow Sands	

Around the estuary of the Tees substantial beds of halite come in above the anhydrite of the first and second cycles—that is, they form part of the Lower and Middle Evaporite Beds, and these thicken towards the south-east. In the Whitby (or Eskdale) boreholes there is a marked thickening of all three evaporite beds, more halite, and potassium chlorides appear as late products in the upper two; the maximum

Permian thickness is over 3,000 feet. Here and at Fordon, south of Scarborough, the Lower Evaporite Bed is greatly expanded with a thick sequence of polyhalite at the top. Fordon was probably farthest from the Zechstein shore and this lowest cycle is complicated by a subdivision into three partial cycles, as a temporary freshening of the brines produced a return from the chloride to the calcium sulphate facies.

The transition from Tees-side to east Yorkshire shows clearly how the evaporite sequences become both thicker and more complex away from the contemporary shore, with the latest, most soluble salts appearing at the top of the cycles. It is possible that the very thin Top Anhydrite at Whitby is the beginning of a fourth cycle which did not attain completion.

South Yorkshire, Nottinghamshire and Lincolnshire

A traverse southwards from mid-Yorkshire shows another shoreward approach, but in a rather different fashion. First, there is the disappearance of all but small amounts of evaporites; second, in the extreme south, even the carbonates (dolomites and limestones) disappear also, to be replaced by a sandy facies; and third, there is a marked thinning of all beds traditionally ascribed to the Permian. In the Hayton boring, between York and Hull, the Permian beds are estimated at about 1,000 feet; there are still substantial amounts of anhydrite in and above the Lower Magnesian Limestone, and above the upper division.

Table 37. *Permian succession in south Yorkshire*

Lithological formations	Approximate thickness (feet)
Upper Permian Marl, with evaporites	200+
Upper Magnesian Limestone	140
Middle Permian Marl, with evaporites	200
Lower Magnesian Limestone	700+
Lower Permian Marl	100
Basal Sands and Breccias	80
(Carboniferous rocks)	

In the surface outcrops of south Yorkshire, and boreholes between Doncaster and Hull, anhydrite is relatively unimportant and the succession is thinner. The Middle Permian Marls correspond to the Lower

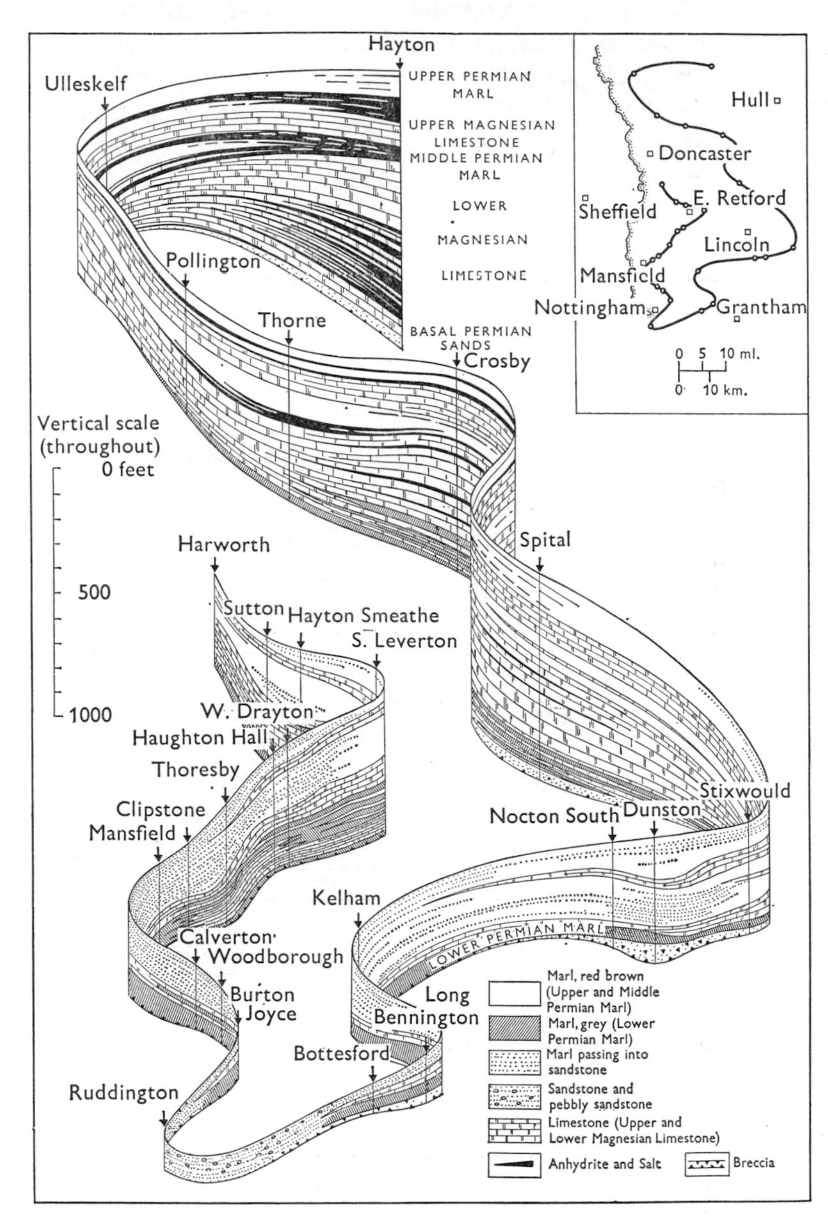

Fig. 50. Ribbon diagram of the Permian rocks of south Yorkshire, Lincolnshire and Nottinghamshire, to show the facies-change and the lateral passage into thin sandy beds in the south. (From Edwards, 1951, p. 97.)

Evaporite bed farther north, but they lose their evaporite content south of a line between Sheffield and Lincoln; the Upper Magnesian Limestone is tapering out at this latitude also, passing into sandstones. The Lower Magnesian Limestone is the most extensive and is found as a very thin bed as far south as Nottingham, but has disappeared in borings six miles beyond. Finally all that remains is a sandy pebbly bed of Bunter Sandstone type.

It is this facies change in particular, from well over 2,000 feet of Zechstein carbonates, evaporites and marls in north and east Yorkshire, to 'Bunter Sandstone' in Nottinghamshire, with many signs of lateral passage and none of major unconformity, that makes the division between Permian and Trias so impossible to define in Britain. The same lateral transition seems to lead inescapably to the conclusion that part of the 'Bunter' of the east Midlands, at least, is of Zechstein age since the Magnesian Limestone and its evaporite cycles can be correlated fairly satisfactorily with the German succession.

PERMIAN RED BEDS OF THE MIDLANDS

Within and around the south Midland coalfields there is a variety of red beds, often including conglomerates and breccias, that are overlain by the widely transgressive Bunter Pebble Bed and, in places, themselves overlie the Keele Group or its equivalents. Since the Keele Group is late Carboniferous, a Permian age is probable for most of these red beds. For reasons which appear later they include, under the title of Dune Sandstone, rocks that in earlier classifications were called the Bunter Lower Mottled Sandstone and hence ascribed to the Trias.

The late phases of the Hercynian earth movements uplifted the Mercian Highlands to the south and probably rejuvenated the inherited structures of the coalfields as well—Charnoid, Caledonoid and others less definable. The irregular uplands so produced were subject to erosion under arid or semi-arid conditions; between them, and spreading out on the lower ground to the north, there accumulated spreads of sands, pebbles and angular scree; occasional sheets of water evaporated to give thin impure limestones similar to the cornstones of the Old Red Sandstone.

Beds formed in such conditions naturally show great local variation, in thickness and in the composition and distribution of the breccias and conglomerates. Fossils are almost absent, for life must have been sparse and conditions of preservation exceedingly poor. These red beds have received many local names in and around the individual coalfields, and correlation between them is by no means clear. Nevertheless, the following grouping can be attempted:

Shropshire	South Staffordshire	Warwickshire
Bunter Pebble Bed	Bunter Pebble Bed	—
Dune Sandstone	—	—
Enville Breccia	Clent Breccia	Kenilworth Breccia
Enville Conglomerate	Calcareous Conglomerate Group	Conglomeratic Group

Of these the lowest, conglomeratic, group in each area may be Lower Permian or Upper Carboniferous.

The Conglomerate Groups

In the west these beds have also been called the Bowhills Group; both Bowhills and Enville are on the margins of the Forest of Wyre Coalfield and the group is generally characteristic of the west Midlands. It consists of red marls and sandstones with important bands of conglomerate, together with thin and impure limestones formed in brackish waters and occasionally containing the worm tube known as *Spirorbis*. The pebbles include a high proportion of Carboniferous rocks, not so much of the Coal Measures, which are conspicuous in the region today, but of limestones and cherts, which are hardly known. It has therefore been deduced that considerable outcrops of Carboniferous Limestone once existed in the south-west Midlands (possibly an eastward extension of the small outlier on the Clee Hills) which have now been removed. Lower Palaeozoic rocks, such as Cambrian quartzites, Llandovery sandstones and Wenlock Limestone are recognizable, and also a Uriconian type of volcanic rock. The distribution is often very irregular, as if local spurs and hills were undergoing erosion and the ill-sorted debris dumped at their foot and nearby. In general the conglomerate fans fade out to the north and give place to finer beds in north Staffordshire and the Shropshire plain.

Clent and Enville Breccias

A large number of breccias and associated sandstones are included in these groups but they may well not be strictly contemporaneous. The thick breccias are characterized in particular by a predominance of Pre-Cambrian volcanic rocks, the angular unsorted chips and boulders often being coated with hematite. Again we must suppose a more extensive source area of Uriconian and perhaps Charnian and Caldecote rocks, in Permian times, than is present in the slim outcrops of today. The Clent group reaches its maximum development in the Clent Hills, south of

Birmingham, where there is as much as 400 feet, consisting solely of breccia, very coarse with fragments more than a foot in length.

In Warwickshire the Kenilworth Breccias are probably an easterly attenuated representative, farther from the main source of detritus. It is from here that the few fossils have been found that help to date the Midland red rocks as a whole. Many years ago an amphibian, possibly Lower Permian, and a reptilian jaw were collected near Kenilworth, and species of the conifer *Lebachia* (also known as *Walchia*) have also been found. The last supports a probable Lower Permian age, but the genus has a fairly long time range.

Dune Sandstone and quartzite breccias

The Dune Sandstone comprises a group of red, aeolian sandstones in the west Midlands and elsewhere; in Worcestershire it is also known as the Bridgnorth Sandstone and more widely as the (Bunter) Lower Mottled Sandstone. On the west side of the boundary fault of the South Staffordshire Coalfield the Dune Sandstone amounts to about 1,000 feet, but on the east its place is taken by a series of thin, local, breccias—the Quartzite Breccia of Birmingham, and the Moira and Hopwas breccias farther east. In their dominant component, possibly a Cambrian quartzite, they differ conspicuously from the Clent Breccia. To what extent these breccias and the Dune Sandstone are strictly contemporaneous is not determinable, but both underlie the transgressive base of the Bunter Pebble Bed and, in places, overlie either the Clent or Enville breccias.

There are other aeolian sands in the west and north that have been correlated with the Dune Sandstone on a basis of similar lithology and mode of accumulation. These include the Lower Mottled Sandstone of the Cheshire plain and south-west Lancashire, the Collyhurst Sandstone of east Lancashire, and the New Red Sandstone of the Vale of Clwyd in North Wales—the last capped by a small remnant of Pebble Beds. Then farther afield more tentative correlatives include the aeolian sandstones in the Penrith Sandstone of the Vale of Eden and others in the basins of south-west Scotland. These are all mentioned later in this chapter, but it is worth noting that the thin, conformable cover of Magnesian Limestone to the southern part of the Penrith Sandstone is one of the few stratigraphical pointers to the age of the Dune Sandstone. The few fossils in Warwickshire suggest a possible Lower Permian, or Rotliegendes, age for the Clent group; it is thus consistent that the Dune Sandstone may be Zechstein, though perhaps early Zechstein. On such an

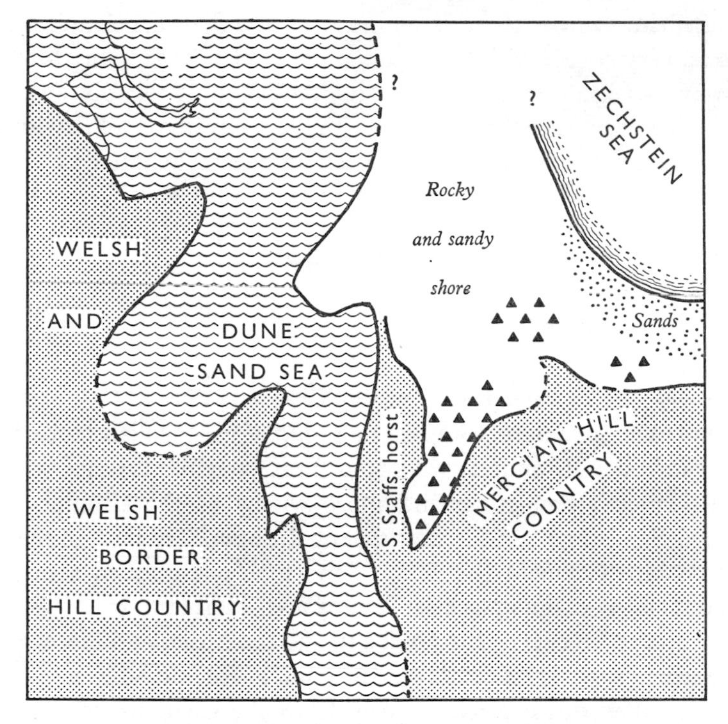

Fig. 51. Palaeogeographic reconstruction of the English Midlands during the accumulation of the Dune Sandstone. The 'sand sea' occupied the Cheshire plain and part of the west Midlands; to the east were spreads of scree, north of the Mercian Highlands, and (if the Dune Sandstone is rightly correlated with part of the Zechstein) to the north-east the Zechstein sea invaded a sandy and rocky shore. (Adapted from Wills, 1956, p. 104.)

interpretation the desert dunes extended widely over the slightly irregular coast lands that lay west and south-west of the saline seas (Fig. 51).

In some places, for instance in Warwickshire, the Permian red beds and breccias follow without obvious unconformity on those taken to be Upper Carboniferous, and it may be that in some of the hollows north of the Mercian Highlands sedimentation continued without major interruption. But in semi-arid conditions of varied relief and local basins, the accumulation of screes, gravels, sand and desert dust is likely to be intermittent. A succession of conglomerates and breccias does not necessarily imply persistent sedimentation nor a local unconformity more than an unimportant period of erosion.

NEW RED SANDSTONE IN NORTH-WEST ENGLAND AND IRELAND

The most important outcrops in north-west England are those that partly surround the Lake District, outside the more complete Carboniferous rim. New Red Sandstone occupies the relatively low-lying Vale of Eden, between the Lakeland hills and the Pennine scarp, and continues northwards into the Carlisle plain and the margins of the Solway Firth. On the west red sandstones stand out at St Bees Head, and from here southward there is a narrow coastal strip, overlying either Lower Palaeozoic rocks or Carboniferous Limestone, as in the Furness district of Lancashire. The red rocks here were once continuous with those of the Lancashire coastal plain and hence with the northern margins of the Cheshire basin; a westward extension is indicated by similar beds on the northern tip of the Isle of Man and in several places in north-east Ireland. Several of these scattered outcrops include a small thickness of Magnesian Limestone, the faunas of which on the whole most resemble the Lower or Middle division of the Durham succession.

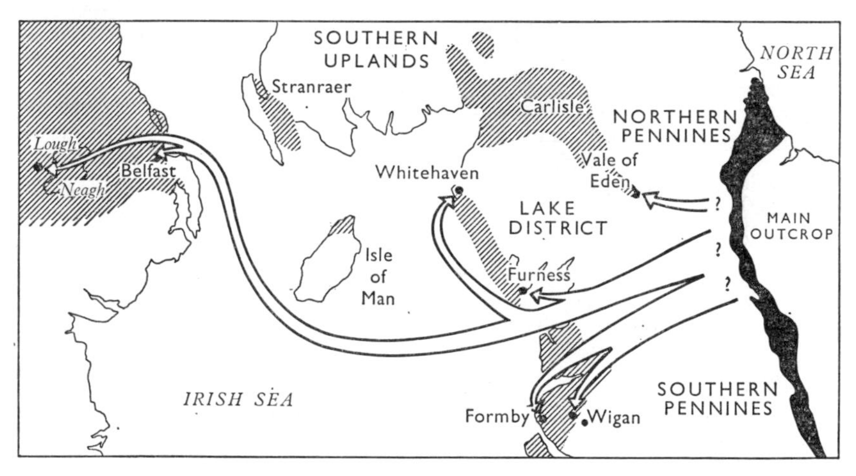

Fig. 52. Sketch map of northern England, the Irish Sea and north-east Ireland to show the possible routes of the north-westerly invasions of the Zechstein sea. New Red Sandstone basins and outcrop areas shaded; Magnesian Limestone outcrops black.

The route by which the Zechstein sea reached the north-west is problematical. There is little positive evidence of a ridge in the south or mid-Pennines in Zechstein times, and, in view of the shoreward transition southwards into a Bunter facies in Nottinghamshire, the very thin remnants of Magnesian Limestone in Lancashire, and total absence from north Edenside to the Carlisle plain, a westward transgression in north-west Yorkshire or south Durham is at least a reasonable hypothesis (Fig. 52). From whatever source this arm of the Zechstein sea probably lapped round the south side of the Lake District, did not invade the Isle of Man, but stretched north-westward to Belfast.

North-West England

The mainland outcrops in the north-west include all or most of the following formations, which are given with their traditional grouping. This classification in Table 38 is inevitably lithological, but, since there is a Lower Jurassic outlier on the Carlisle plain, the identification of the Keuper Marl is at least reasonably sound; the Magnesian Limestone and beds below are similarly Zechstein.

Table 38. *Generalized succession of New Red Sandstone in north-west England*

Traditional grouping	Lithological formations	Maximum thickness or range (feet)
KEUPER	Keuper Marl (or Stanwix Shale of Carlisle)	900
BUNTER	Kirklinton Sandstone, fine-grained red sandstones St Bees Sandstone, red, with shale bands	800–3,200 +
PERMIAN	St Bees Shale, red and grey, locally with gypsum and anhydrite	600
	Magnesian Limestone	60
	Grey marls or silty dolomites, locally with gypsum (also Hilton Plant Beds of south Edenside)	200
	Penrith Sandstone (of Edenside) Brockrams, local and variable breccias	2,000

The difficulties of applying the terms Permian, Bunter, Keuper are mentioned in the text.

The lowest beds are considerably more variable than the upper ones, which is not very surprising since they were the first deposits to accumulate on the margins of these Hercynian hills; whatever the topography in the Pennine region the Lake District formed an important local upland. Some basal breccias, if only a few feet, are generally present, their thickness varying according to the hollows in the pre-Permian land surface. Typical 'brockram' refers to a coarse breccia in which the angular fragments of Carboniferous Limestone are embedded in a red sandy matrix, but other rock-types, both Carboniferous and Lower Palaeozoic, are also found. In southern Edenside there are at least two great wedges of brockram; the upper cuts into the Penrith Sandstone and has a more varied clastic content than the lower. It also contains very rare fragments of quartz dolerite similar to the Whin Sill suite. This Hercynian intrusion has a isotopic date of 281 million years and cuts Coal Measures in Durham; it is thus late Carboniferous. The sill, or a related dyke, was therefore within reach of erosion by the time the brockrams were formed, but there is no evidence of a major Pennine ridge at this time.

The 2,000 feet of brockram in southern Edenside are replaced northwards by the Penrith Sandstone and locally by the Hilton Plant Bed. The latter consists of dolomitic sands and silts whose flora includes some species common to the Marl Slate and Kupferschiefer. Like the 50 feet of Magnesian Limestone above it does not extend into north Edenside or the Carlisle plain where red marls and shales with gypsum ('St Bees Shale') lie directly on the Penrith Sandstone. The lower part of this sandstone is one of the conspicuous aeolian red rocks comparable with the Dune Sandstone of the Midlands. The grains are characteristically very well rounded and sorted, coloured red by a surface film of hematite and cemented with secondary quartz; over 1,000 feet are recorded in Edenside. Far less is known of the overlying red sandstones and marls and exposures in the Carlisle plain are few. On the Scottish side, across the Solway in Dumfriesshire, the Annan Sandstone is probably to be correlated with the St Bees Sandstone.

In west Cumberland, near Whitehaven and St Bees Head, there is a very thin representative, 2 or 3 feet, of Magnesian Limestone, but only at and near the present coast. Inland, that is, towards the Permian as well as the present hills, it is replaced by brockram, which extends higher and higher till it also replaces all the St Bees Shale and possibly some of the St Bees Sandstone as well—striking evidence of scree accumulating in wedges at the foot of the contemporary highlands. Southwards the Magnesian Limestone tends to pass into massive anhydrite and the brockrams are reduced or absent. The St Bees Shale thickens in the same direction and so do its evaporite members. Some 80 feet of massive anhydrite has been encountered underground south of St Bees Head.

The outcrop of the St Bees Sandstone continues down the coast into south Cumberland and across the Duddon estuary into the Furness district of Lancashire. Here it is underlain by a variable series, grouped together as the Kirksanton Beds, which include brockram, and shales with gypsum which resemble the St Bees Shales but which are interbedded with the brockram; certain of the boreholes contain up to 60 feet of Magnesian Limestone or a comparable thickness of evaporites (gypsum and anhydrite); as at St Bees, the latter die out in the landward direction, here to the north. These changes indicate a gradual transition away from the contemporary shore, which evidently lay to the east or north-east and although both carbonate and sulphate phases are much less in thickness and extent (and much less well known) than those east of the Pennines, their relationships are similar.

West of the Cumberland Boundary Fault, Keuper Marl underlies drift on Walney Island, off the coast at Barrow. The 1,000 feet or so of red and grey marls contain beds of salt, comparable with those in the Isle of Man and near Fleetwood on the south side of Morcambe Bay.

West and South Lancashire, Permian Beds

Along the southern margins of the Lancashire Coalfield, from St Helens eastward to Manchester and south to Stockport, the following formations are recognized:

Keuper Marl
Keuper Sandstone
Bunter Sandstone
Manchester Marls
Collyhurst Sandstone

On the west of the coalfield there are only small outcrops in country much obscured by drift, but they are supplemented by a group of boreholes near Formby, between Southport and Liverpool. The Bunter and Keuper rocks are comparable with those of the Midlands (p. 258), but the Manchester Marls and Collyhurst Sandstone, which have good claim to be called Permian, are considered here.

The Collyhurst Sandstone lies with pronounced unconformity on various divisions of the Carboniferous from low Namurian (near Formby) to Upper Coal Measures. It is also remarkable for abrupt changes in thickness; in one of the Formby boreholes the sandstone is over 2,300 feet thick, and yet is absent altogether in another a few miles away. Often such changes coincide with faults, as if these basal Permian sands accumulated in arid landscape of Hercynian fault scarps. Lithologically the Collyhurst Sandstone is a soft, poorly cemented rock with well-rounded grains coloured red by iron oxide and with marked aeolian bedding; it has been broadly correlated with the Dune Sandstone of the west Midlands.

This stratigraphical position is supported by the fauna of certain carbonate rocks in the Manchester Marls above. The lower part of these include impure limestone bands and fossils similar to those of the Magnesian Limestone, including species of *Schizodus*, *Bakevellia*, *Pleurophorus* and *Liebea*. Thus in spite of their unusual lithology part of the marls at least were marine, perhaps forming in shallow water mud-flats. This view is strengthened by local, thin but important Magnesian Limestone. Yellow dolomites have been found in one small exposure on the western margin of the Wigan Coalfield and also in the Formby boreholes; presumably they represent the southern edge of the Zechstein embayment of Edenside and Furness. To the west the formation becomes more and more sandy, and in the Liverpool district the Manchester terminology can no longer be applied; all the red beds beneath Bunter Pebble Bed are traditionally called Lower Mottled Sandstone. As in the Midlands this equivalence depends to some extent on the reliability of the Pebble Bed as a stratigraphical horizon, and it becomes less satisfactory northwards. In the Stockport region it is a recognizable bed, but at Manchester and Wigan the Bunter Sandstone above the marls carries only scattered pebbles.

North-Eastern Ireland

New Red Sandstone in Ireland occurs principally in two provinces. Most of it lies in the faulted rift that is the continuation of the Scottish Midland Valley; at the same time these outcrops form part of the incomplete border of Mesozoic sediments around the southern edge of the Antrim Basalts. A much smaller area is included in the Kingscourt Outlier on the Longford–Down Massif. Nearly all the surface outcrops are of Trias (or presumed Trias) and recognizable Permian is only known from a few, and in boreholes. A characteristic sequence has been described from Grange, in Co. Tyrone, west of Lough Neagh.

Table 39. *Permo-Triassic succession in Co. Tyrone*

Lithological groups	Maximum thickness (feet)
Keuper Marl	1,050
Keuper Sandstone	100
Bunter Sandstone and Marl	1,690
Marl with gypsum ('Upper Permian Marl')	15
Magnesian Limestone	75
Basal Sands	25

The brockrams in the basal beds do not form great wedges like those around the Lake District, although at Belfast they reach 180 feet; but in boreholes and small separated exposures they overlie a great variety of rocks, from Ordovician to Carboniferous, another indication of the strong erosion that followed the Hercynian movements. The Magnesian Limestone from borings in Belfast and small outcrops on Belfast Lough

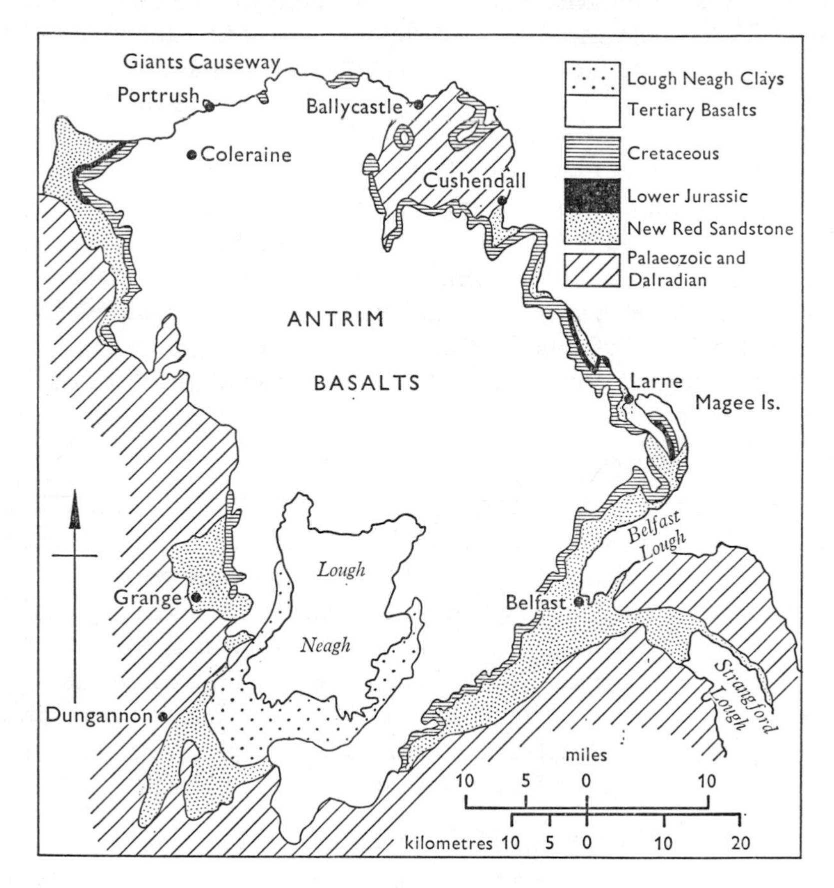

Fig. 53. Geological sketch map of north-east Ireland showing the major outcrops of Mesozoic and Tertiary rocks; Palaeozoic and Dalradian not differentiated.

maintains about the same thickness as in Co. Tyrone. A sparse fauna, chiefly lamellibranchs, gastropods and polyzoa, is fairly widespread. The junction of 'Upper Permian Marl' and 'Bunter Marl' is largely based on lithology: the former locally contains evaporites and the latter evidence of desiccation in mud cracks and mud-flake breccias; similar indications, with ripple marks and units of aeolian bedding, are found in the thick 'Bunter Sandstone' above.

The Bunter beds are more extensive than the underlying Permian and in many places overlap them on to a variety of earlier rocks. In these circumstances they are often underlain by thick conglomerates, derived from the local uplands—Dalradian rocks in the north-west, Old Red Sandstone at Cushendall on the north-east, and Carboniferous rocks in Counties Tyrone and Armagh on the west and south. Fossils are very rare in the Bunter and Keuper rocks, but the small crustacean *Euestheria* occurs at several horizons in the former, and Bunter Sandstone was apparently the source of a rich bed of piled-up fish remains in Tyrone (*Palaeoniscus catopterus*), which was found in the 1830s. The Keuper Marls contain salt and gypsum, the first in beds of workable thickness which have been exploited at Larne.

In the Kingscourt Outlier (Fig. 27, p. 124) there is a thin strip of New Red Sandstone along the western, faulted, margin consisting of about 400 feet of marls below ('Permian') and about 1,500 feet, chiefly of sandstones, above ('Trias'). The lower group contains massive beds of gypsum and anhydrite near the base, and grey shales with plant remains between these and the basal beds. The facies is Permian-like, but there is no absolute proof of age, since the Zechstein sea did not reach so far and the plants are not determinable.

Where recognizable Permian rocks are present the conditions in Ireland form a logical extension of those east of the Irish Sea. Structurally, however, the New Red Sandstone of Belfast and Tyrone is in the Midland Valley, and the small strip of Trias that connects Belfast and Strangford Loughs is analogous to the basins of the Southern Uplands at Stranraer and farther east. But the Zechstein sea never invaded Scotland and there is no Magnesian Limestone there; consequently, as in earlier periods, particularly the Lower Carboniferous, there is a divergence in facies and conditions between Scotland and northern Ireland, and a corresponding divergence between tectonics and sedimentation on either side of the Irish Sea.

It is very difficult to draw a broader picture of Ireland during Permo-Triassic times. The desert sands and marls and thin impersistent sheets of water presumably extended far beyond the present outcrops, and many of the deductions made on the Trias of the English Midlands might apply here. But the evidence has been removed, and sedimentary history in Ireland during Mesozoic and Tertiary times is almost entirely a blank, except in the north-east where limited information can be gained from the incomplete border of Mesozoic rocks which have been protected from erosion by the Antrim Basalts.

RED TRIAS IN THE MIDLANDS

The removal of the Lower Mottled Sandstone leaves four major continental formations in the Midlands to which the name Trias is customarily given, the Rhaetic being treated separately on p. 269:

Keuper Marl, with Waterstones facies locally at the base
Keuper Sandstones
Bunter Sandstone, or Upper Mottled Sandstone
Bunter Pebble Beds

Bunter Pebble Beds and Bunter Sandstone

These two divisions are best considered together, because essentially they consist of a group of sandy beds which in certain parts of the Midlands contain scattered or lenticular seams of pebbles (especially towards the base) or, more rarely, thick banks of shingle with little sandy matrix. In the west Midlands the distinction of Bunter Sandstone ('Upper Mottled Sandstone') overlying Pebble Beds (sandstones with pebble seams) is fairly clear; each amounts to about 400 feet, and there is an unconformable junction with the Dune Sandstone or Permian breccias below. To the north-east, however, the pebbles gradually become smaller and the seams fewer so that they are inconspicuous in a slightly pebbly Bunter Sandstone, and a similar dwindling takes place northwards into Lancashire. Conversely, in Leicestershire, all the Bunter is pebbly.

Compared with the earlier breccias the pebbles of the Bunter are not only much better rounded, especially the smaller ones, but are more varied in type. Quartzite and vein quartz make up a large proportion and there are lesser amounts of chert, limestone and volcanic rocks. In some areas local rocks predominate, particularly in the coarse basal conglomerates, and some quartzites can be matched in the Cambrian outcrops of Nuneaton or the Lickey Hills of south Staffordshire; the fossiliferous Llandovery sandstones and Carboniferous Limestone are probably local also. Nevertheless, many of the components have no known source, particularly much of the quartzite; equally enigmatic are rare pebbles whose nearest known source rocks are far to the south—including those with tourmaline and other minerals of Devon and Cornwall, Devonian sandstones, slates with the Devonian *Cyrtospirifer verneuili*, and large cobbles (up to eight inches) containing Ordovician brachiopods known in Normandy but not in the Welsh Borders. A southerly source for some of these gravels therefore is probable, but it need not necessarily have been as far away as France; it is at least possible that outcrops of, say, the Ordovician quartzites were available in southern England in areas now buried, or perhaps on the site of the English Channel.

The Pebble Beds as a whole formed thick spreads of sand and gravel on the country of plains and shallow lakes that lay to the north and west of the Mercian Highlands, and tailed off northwards into sands. The succeeding Bunter Sandstone is a fine red sandstone with bands of marl and is current-bedded, but not aeolian-bedded. From Shropshire it thins eastwards, and in the eastern Midlands it is not known, or not under that name. This may be due partly to a lateral facies-change or partly to absence—a break at the base of the overlying Keuper Sandstone. To the north-west, however, in the Wirral and Cheshire it continues to be a recognizable division, and as the pebbles of the beds below become rarer, the whole forms the Bunter Sandstone, 1,000 feet or so thick.

Keuper Sandstones, Waterstones and Keuper Marl

The Keuper rocks overlie the Bunter in the Midlands with an erosional break, or disconformity, which in places represents a marked gap in deposition. In general there is an upward decrease in grain size and a progression from fresher to more saline waters, although this is diversified by many local variations. Much of the Keuper Sandstones show fluviatile characters, being buff, yellow or red-brown rocks, commonly current-bedded. In the west Midlands the lower part is conglomeratic and to the east the group is spasmodically developed, thin, or locally absent. The name 'Waterstones' refers, a little obscurely, to the lustrous effect of micaceous bedding planes, resembling watered silk. It does not relate to the rock as an aquifer though in fact both Bunter and Keuper will yield water, albeit hard. In terms of environment the facies is probably transitional between the fluviatile sandstones below and the marls, with their thick saline intercalations, above. It is variably developed and the facies-boundaries may well be diachronic.

The Keuper sandstones are exceptional among the red Triassic rocks of the Midlands in that they locally contain a few fossils, of several different groups. Allowing for the unsuitability of a porous sandstone as a medium of fossil preservation, the sparse records are a reminder that while these plains with wandering water courses and fluviatile deposits were arid they were not a lifeless desert. The water-living forms, which include the lungfish *Ceratodus* and *Euestheria*, are a indicative that the waters were not highly saline. Reptiles and amphibia were washed in from the land, and left their footprints on the shores. Plants include the conifer *Voltzia*, and Cordaitean genus *Yuccites*; *Schizoneura* was a horsetail that probably lived at the edge of pools. Fossil scorpions are among the curiosities of Keuper life. There is also the significant discovery of *Lingula* in the Waterstones of Nottinghamshire. At the present day *Lingula* is fully marine, though living in very shallow, and sometimes intertidal, conditions. A comparison between Britain and the continent suggests that these shells reflect a temporary, Muschelkalk, invasion from the east, since sediments of this type have been found not far from the English coast during North Sea gas explorations.

There is a gradual lithological transition from Keuper Sandstone to Keuper Marl, and the latter, in addition to red and green marls, contains lenticular sandstones, beds of gypsum and salt and many other signs of evaporation such as mud cracks and salt pseudomorphs. Nearly 4,500 feet of Keuper marls have been proved in the deepest basin, that in Cheshire, and these include two thick salt-bearing divisions of about 600 and 1,300 feet.

Typical Keuper Marl is a reddish siltstone; the colour is due to minute granules of hematite, and the bulk of the rock is composed of fine angular grains of quartz and evaporite minerals, which may be dolomite or gypsum. The grey-green layers and patches are probably due to the later reduction of the iron oxide rather than to original differences of deposition. The uppermost bed is more uniformly green than the remainder, and is sometimes separated off as the 'Tea Green Marls', but even this colour distinction may be secondary.

The lenticular sandstones are known as 'skerries'; they are commonly greyish and in addition to fine-grained quartz contain small rhombs of dolomite. The more massive among them show current-bedding, horizons of slumping, and salt pseudomorphs. The Arden Sandstone of Warwickshire is a particularly persistent example, and very unusual in its debris of reptilian and amphibian bones, fish scales and spines, *Euestheria* and rare, poorly preserved shells. There is no clear indication of the environment of these forms but very shallow or intertidal conditions are a possibility.

The principal economic precipitates in the Keuper Marl are rock-salt and gypsum, especially the former. Over the Midland lowlands and their northerly extensions, beds of salt were formed in several places and at more than one period. In addition to the exceptionally thick beds of the Cheshire basin, deposits are found in Worcestershire, Staffordshire and at Fleetwood in Lancashire. In north-east England, near Whitby, the Keuper evaporites include salt, though on a much smaller scale than in the Zechstein below. Nottingham appears to have been outside the main belt of salt deposition, but gypsum is known in the Newark region in beds up to 60 feet thick, as well as in veins and patches in the upper marls.

The original environment of precipitation is rather uncertain. There is no clear succession of mineral phases that is so well demonstrated in the Zechstein marine evaporites. It is true that the carbonates are widespread in the form of dolomite crystals in the marls, and that both the sulphates and halite are present at one place or another, but not in their theoretical cyclic superposition. For example, the Cheshire salt beds are not underlain by comparable beds of anhydrite or gypsum. Nevertheless it is difficult to account for the widespread halite, thickest in Cheshire but found in other basins, except by persistent influx of marine waters.

The gradual extension of the Keuper sedimentary basins, and the accompanying reduction of the Uplands, is shown by the widespread overlap of the Keuper beds over those below. This is not very conspicuous around the Cheshire plain, except on the margin of the Pennines near Macclesfield, where the marls rest on Coal Measures and Millstone Grit. In the Midlands, however, Keuper Sandstone overlies the Cambrian at Nuneaton, the Coal Measures of Leicester and South Derbyshire, and Carboniferous Limestone in the small inliers north of this coalfield.

The most striking, and best known, overlap is the covering of Keuper Marl on the Pre-Cambrian rocks of Charnwood Forest and the neighbouring Mount Sorrel Granite, a small Caledonian intrusion. The marl at Charnwood is being removed by contemporary erosion, and there are good sections in the quarries, since the harder Charnian rocks are a valuable road metal. The marls can be seen banked up against steep slopes and filling in valleys between Pre-Cambrian spurs; since the infilling is relatively soft the present streams often come to lie above a Triassic desert buried valley, with the ridges of older rocks rising up on either side. In this partly resurrected topography there are glimpses of a Triassic landscape, when the marls and silts were gradually lapping round and burying the slates, lavas and granite boss.

The residual upstanding masses were subject to the rigorous, arid erosion in a desert climate. The Mount Sorrel Granite, for instance, is remarkably fresh immediately under the marl cover, with the feldspar bright and unweathered. There are signs of insolation (weathering under pronounced variations in temperature) and some of the emergent granite surfaces are grooved and polished by the wind. Although this last feature has been claimed as a Pleistocene effect (arid erosion can operate in a dry cold desert as well as a dry hot one), the total effects of the Triassic burial and erosion of the adjacent cliffs and buttes are well supported and form a striking exhumation. Such were the Keuper lagoons and very shallow saline waters, to be succeeded by the fuller Rhaetic marine invasion ushering new conditions and new faunas.

NEW RED SANDSTONE OF THE SOUTH-WEST

Along the lower Severn Valley and into the area of the Severn Estuary, around Bristol and into central Somerset, Keuper beds appear at the surface; most of them are Keuper Marl, but Keuper Sandstone, similar to that of the west Midlands, is known. As in the Midlands there is conspicuous overlap on to the surrounding uplands, here chiefly the residual hills of the Hercynian chain, formed of folded Carboniferous rocks. The marls contain lenticular sandstones and also evaporites towards the top, including both gypsum and rock-salt. Celestine (strontium sulphate) is a more unusual mineral; it is worked north of Bristol, and has also been found south of the Mendips. In all about 1,500 feet of Keuper Marl are known on the English side of the Bristol Channel, including 100 feet of 'Tea Green Marls' at the top.

On the Welsh side Keuper Marls hold a similar position in south Glamorgan where they overlie an irregular floor of Old Red Sandstone and Carboniferous rocks. There is evidence of a former westward extension in the few patches of Keuper conglomerates in the Gower Peninsula, and probably also in the 'gash breccias' of south Pembrokeshire. These are large swallow holes, or the remains of underground caverns, filled with angular blocks of limestone and dolomite, set in a red calcareous or mudstone matrix. The Triassic land surface has long since been denuded from these westerly limestone plateaux, but this unusual type of subterranean evidence remains.

The Dolomitic Conglomerate

Where Triassic and lower Jurassic rocks are banked up against the Carboniferous Limestone, in Glamorgan, the Mendips or in the outcrops near Bristol, there is a local marginal facies called the Dolomitic Conglomerate—though in fact boulders may be angular and formed of limestones as well as dolomite, set in a red sandy or silty matrix. This fringe of debris accumulated against the Carboniferous outcrop and exceptionally may be as much as 8 miles wide, as in places north of the Mendips. In their early stages the limestone cliffs and scarps stood up above the arid Keuper plains, and later formed the shore of the encroaching Rhaetic and early Jurassic seas.

In Glamorgan and the southern Mendips (for instance, north of Shepton Mallet) burrows and solution hollows have been found in the limestone and in them traces of small terrestrial animals. Their bones were washed into underground watercourses and are now embedded in the silts and muds that fill these old cavities. Some of the remains probably date from the time when the limestone hills were islands in the Rhaetic or Liassic sea, but others are earlier and include several small carnivorous dinosaurs and lizard-like reptiles. From these rather unusual sources we get more information about Triassic quadrupeds than from any other region in Britain except the sandstones of northern Scotland.

South Devon, Somerset and Dorset

The red rocks that underlie the Keuper Marls are best known from the sections on the south coast, from Torquay to Teignmouth and into the westernmost cliffs of Dorset. There are also useful sections in north Somerset from Porlock to Watchet and in the long tongue that follows the Crediton Valley in the synclinorium of Carboniferous rocks in central Devon.

Table 40. *New Red Sandstone of south-east Devon*

Lithological groups	Maximum thickness (feet)
Keuper Marls	(200+)
Red Sandstones	400
Pebble Beds	80
Red Marls	700
Red sandstones and breccias	450
Breccias and conglomerates, including volcanic rocks	950
Watcombe Clays (local)	0–200
(Devonian and Carboniferous rocks)	

Table 40 is a somewhat composite sequence from the south coast. Near Exeter there is also a series of volcanic rocks at or near the base (the Exeter Traps); and the Keuper Marls at the top are poorly exposed on the cliffs of the Channel coast, because they are much reduced by the erosion preceding the deposition of the Upper Cretaceous Greensand. Further north, in Somerset, they are more normal and 1,300 feet thick. The dark red Watcombe Clays are found chiefly in a few faulted outcrops around Teignmouth.

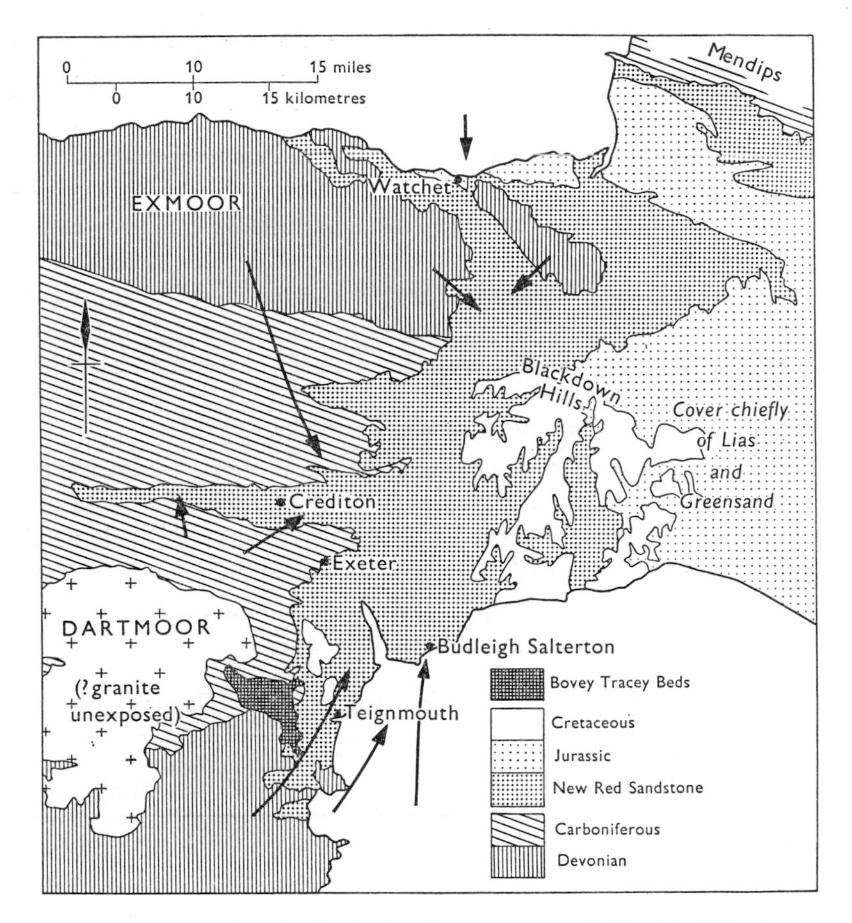

Fig. 54. South-West England, showing outcrops of New Red Sandstone and sources of detritus (from various published sources).

The remaining groups comprise rock-types already familiar in other outcrops of New Red Sandstone, and the dune sandstones near Exeter are similar to the other, Permian, aeolian rocks of the Midlands and north. Many of the pebbles in the lower breccias and conglomerates can be matched in the neighbourhood. A fourfold succession has been established in the Crediton Valley, each group characterized by a more extensive suite of pebbles than that below; thus the lowest contains fragments of the local Culm only, and the upper ones lavas, tourmalinized rocks and potash feldspars. Most of the detritus appears to have come from nearby Dartmoor upland and reflects the exposure, during erosion, of the tourmaline-bearing rocks. There is little indication of the contact aureole, however, and if the granite was exposed it was only in a small area, or possibly drainage from it did not flow in this direction. The Exeter volcanic group is a general term for several types of potash-rich lavas, mostly trachytic, but some basaltic. They now form scattered outcrops around the city at or near the base of the breccias, and in this area contribute largely to the pebbles.

The upper pebble beds (also called the Budleigh Salterton Pebble Beds) are much more like the Bunter Pebble Bed of the Midlands. Quartzites and sandstones are common, sometimes rounded and sometimes sub-angular, and layers rich in pebbles are separated by seams of sand in which they are sparse. Devonian rocks with *Homalonotus* and *Cyrtospirifer* have been found, tourmaline-rich rocks and also the quartzite with orthids resembling those of Normandy. The fossiliferous pebbles decrease northwards, to be replaced in north Somerset by sandstones and Carboniferous Limestone. It thus seems that rivers, possibly intermittent, entered this south-westerly basin from north, south and west.

In north-west Somerset, between the Quantock Hills and Exmoor, there are comparable breccias and conglomerates near the base of the New Red Sandstone with angular fragments of the nearby Devonian rocks and more rounded pebbles of Carboniferous Limestone. A buried landscape, now partly uncovered, can be detected here also; the local uplands and scarps were of Devonian sandstone and not of the more striking Carboniferous Limestone, nevertheless the marginal facies is similarly present where the red marls of the plains pass locally into sandstones and then breccias as the Devonian uplands are approached.

The age of the New Red Sandstone in south-west England is only certain in the Keuper beds, overlain by Rhaetic and Jurassic. The reptiles of Glamorgan and the Mendips have been ascribed to the Upper Trias but, occurring as they do in limestone swallow-holes and burrows, they cannot be correlated for certain with any one level in the bedded deposits. The Budleigh Salterton Pebble Beds have been equated with those of the Midlands, chiefly on the grounds of similar composition, and thus called Bunter. The Permian age sometimes ascribed to the breccias is based partly on their position, below the Pebble Beds and the intervening red sandstones, and partly on a comparison of the volcanic rocks with those of the Upper Rotliegendes of Germany. These are not very secure bases of comparison, and the term New Red Sandstone seems amply justified in this province.

BASINS AND OUTLIERS IN SCOTLAND

The South of Scotland

The Southern Uplands of Scotland are traversed by several elongate depressions or basins sculptured into the Ordovician and Silurian greywackes, mudstoncs and shales, and for the most part trending between north and north-west (Fig. 25, p. 117). Those of Berwick–Cockburnspath and Thornhill–Sanquhar have already been mentioned because they were through-ways across the uplands in Carboniferous times, and being infilled in some degree by sediments of that period are still marked by Carboniferous outliers. The Thornhill Coalfield retains an incomplete, unconformable cover, and farther south-east, down Nithsdale, there is a deep basin of New Red Sandstone centred on Dumfries. A little to the east the River Annan runs in the similar basin of Lochmaben and red sandstones continue up the valley as far as Moffat. The third of this group of cross basins has been eroded by the sea, and remains as a low neck of land at Stranraer, leading into Luce Bay on the south and Loch Ryan on the north. It has already been seen how the neck of land between Belfast and Strangford Lough lies in a very similar position on the Irish side. Geophysical evidence suggests that there may be further areas of New Red Sandstone under the Irish Sea, for instance east and west of the Isle of Man.

The rocks in these hill-girt basins are dominantly red sandstones, breccias and conglomerates; they contain no signs of life except for the reptilian footprints found near Annan. By a rather broad analogy with the Penrith Sandstone they are sometimes considered to be Permian, but the title is not very securely founded. The three largest basins are surprisingly deep for their area. That at Dumfries is less than 8 miles across, and at Stranraer less than 5, but over 3,000 feet of sandstones and breccias have been deduced, by gravity surveys, in the former and even more in the latter. It is probable that they were depressed somewhat as they filled and that the Stranraer and Lochmaben basins were defined by a contemporary fault on one side, while the other acted as a hinge. The sandstones and breccias reflect the vigorous erosion of the local rocks of the hills nearby.

Comparable red rocks at Thornhill and Sanquhar form a link between the basin of Dumfries and the red sandstone in Ayrshire to the north.

In the Thornhill outlier the Coal Measures are overlain unconformably by a varied cover of sedimentary and volcanic rocks which comprise olivine basalts below and red, dune-bedded sandstone above. At Sanquhar no sediments are known but there are small patches of the lavas.

The Mauchline basin in Ayrshire is a larger outcrop of the same sort, and contains the only post-Carboniferous sediments in the Midland Valley (Fig. 48). Here about 1,500 feet of red, dune-bedded sandstone lie on 500 feet of olivine basalts. The latter have recently been discovered to be late Carboniferous; by analogy those of Sanquhar and Thornhill may be well also, though there is no direct evidence on this point. The sediments are remarkably consistent in the lack of any pebbles or signs of water-sorting and in their great wedges of aeolian bedding and well-rounded grains. Of the several volcanic necks that penetrate the Carboniferous rocks, and were presumably connected with the Mauchline basalts below, none cut the Mauchline Sandstone, which has been another reason for separating it off as Permian.

Arran and the Inner Hebrides

It has been seen how, during Carboniferous times, there was some kind of extension of the Midland Valley area of deposition westwards and northwards for, although the evidence was very scanty, sediments were found overlying Dalradian schists. A lowland area in this direction, west of the Scottish Highlands and probably connected with the deposits in north-east Ireland, becomes clearer in later systems, and Mesozoic sediments of some kind are associated with most major outcrops of the Tertiary Volcanic Province. Like those of Antrim they have largely been preserved by a covering of the Tertiary lavas. Red Sandstones and marls are found at the base of some of these Mesozoic outcrops, and except for the lower group on Arran all have been called Trias.

The southern half of Arran is composed of the central, Tertiary, ring complex and a broad area of New Red Sandstone cut by numerous Tertiary dykes. The 3,000 feet of unfossiliferous red sandstones, breccias and marls are in places apparently conformable on the Carboniferous beds below. Locally the breccias contain many basalt fragments and the interbedded aeolian sands, also with basalt chips, resemble the Mauchline Sandstone. Other debris can be matched with the local Carboniferous and Highland Border rocks, together with schist, quartzite and quartz. Some pebbles are rounded, but some are angular, like dreikanter. The 'Trias' is finer in grain, and much of it is waterlaid and resembles Keuper Marl. The Keuper age of the upper beds is supported from the fragments in the central ring complex where they are associated with fossiliferous Rhaetic beds.

Table 41. *New Red Sandstone of Arran*

'TRIAS'	Red marls, red and white calcareous sandstones, and impure limestones	c. 1,000 feet
'PERMIAN'	Massive variegated calcareous and red sandstones; breccias and red, dune-bedded sandstone (? includes some Carboniferous beds)	c. 2,000 feet

The evidence from Arran points to the existence of a further Permo-Triassic basin in south-west Scotland, in which red beds accumulated over a long period, perhaps from late Carboniferous to late Triassic times.

The remaining outcrops in western Scotland can be dealt with more briefly. The rocks vary from breccias and conglomerates rich in local rocks, especially schist and vein quartz, to red calcareous sandstones and impure limestones ('cornstones'). A southern group includes the outcrops of south-eastern and western Mull, where the beds are capped by Rhaetic, Ardnamurchan, and the Loch Aline district of Morvern, north of the Sound of Mull. The last succession is the thickest, over 300 feet.

Farther north there is a small outcrop on the island of Raasay, near Skye, where sandstones and conglomerates contain pebbles of fossiliferous Durness Limestone. On the mainland coast, at Applecross there are conglomerates with schists, quartzites and Durness Limestone, together with sandstones and cornstones. In both areas the Trias underlies, or is very close to, Lower Lias.

Elgin

On the opposite side of Scotland, in the relatively low-lying country north of the Grampian Highlands and south of the Moray Firth, there is a poorly exposed outlier of Permo-Triassic sandstones. These rocks have no obvious connexion with those of any other province, and are highly unusual in containing a number of fossil reptiles.

At least two groups of beds are included in the 10-mile stretch of the Elgin sandstones, but the outcrops are so sparse in drift-covered country that their relations are not clear. The coarse red sandstones of Cutties Hillock, with well-rounded grains, have produced a late Permian fauna. The Lossiemouth Sandstones are pale buff to pink, with angular grains and in places a cement of barytes and their varied reptilian fauna shows them to be Upper Keuper in age.

Interesting as these occurrences are, as additions to our all too rare Permian and Triassic vertebrates, they are so exceptional and isolated from all other outcrops that they do not add much to the regional picture of the New Red Sandstone. Nevertheless, with the Keuper fossils of the Midlands, the reptilian skeletons in the limestone of the Mendips and Glamorgan, and the occurrence of footprints where no actual bones remain, they are a reminder that animals could survive at least in some places and at some times, among the inhospitable conditions of Permian and Triassic Britain. Further it may not be accidental that the most fossiliferous beds are related in some way to the marginal uplands, rather than to the plains and saline lakes and seas, and most of our sediments and evaporites come inevitably from the latter.

THE RHAETIC SERIES

The problems of the Rhaetic Series in Britain arise particularly from the contrast, or even clash, between the depositional environment of these beds and their formal stratigraphical grouping. They are the first stage in the invasion of the Keuper saline facies by the shelf seas that dominated sedimentation in north-western Europe for the remainder of the Mesozoic era. As such they have an obvious link with the Jurassic beds above. But conditions of deposition, however important, are not an international arbiter of stratigraphical grouping, which in Mesozoic systems is based upon the ammonite faunas of the marine facies. Ammonites are lacking in the British Rhaetic, but in the type area of the eastern Swiss Alps they are available, and in the Alps as a whole there are some Triassic ammonites. The consensus of opinion in this region is towards including the Rhaetic in the Trias, and this applies to some extent in Germany. Accordingly, although it may at first sight be a little illogical in the English setting, the international system is followed, and the Rhaetic is included in this chapter, as the uppermost series of the Trias.

The main English outcrop is comparable with that of the Jurassic above (Fig. 55), in that it extends across country from the south-west (the Bristol Channel) to the north-east; only in the former region, however, are the beds well exposed in cliff sections, being hidden in the north under alluvium at the mouth of the River Tees. In the absence of ammonites the divisions are largely lithological, with some assistance from the commoner lamellibranchs. As such, their definition, and separation from the Tea Green Marls below and lowest Jurassic (Lower Lias) above, is not always easy. An extreme opinion would in fact advocate the abolition of the Rhaetic Series and the division of its constituent formations between the Keuper and the Lias.

South-West England and South Wales

The type sections are on the Somerset coast of the Bristol Channel, near Watchet, and there are others farther up on either side of the Severn Estuary, and on the Glamorgan and Dorset coasts. The beds in Somerset that at one time or another have been called Rhaetic are shown in Table 42.

Table 42. Rhaetic Series, Somerset

Lithological formations	Maximum thickness (feet)
(Blue Lias—Lower Jurassic)	
Watchet Beds	8
White Lias or Langport Beds	25
Cotham Beds	19
Westbury Beds	47
Grey Marls or Sully Beds	14
(Tea Green Marls—Keuper)	

Of these five, however, the lowest and the upper two are often not accepted as Rhaetic. The Sully Beds of Glamorgan and Grey Marls of Somerset are a somewhat variable group of shales, siltstones and calcareous mudstones that are found between the unfossiliferous Keuper below and the distinctive Westbury Beds above. The latter lie non-sequentially on them, and were it not for their marine fauna (which includes the lamellibranch *Pteria contorta* typical of later Rhaetic horizons) these lowest groups would not form a separate entity nor are they found outside this small area around the Bristol Channel. It is thus not unreasonable to consider the Sully Beds, as is often done, to be the latest Keuper beds of the district, but which happened to be laid down in the first, south-westerly and restricted, marine invasion.

The succeeding Westbury invasion was much more extensive, and over the rest of the main outcrop it is these beds that form the recognizable base of the Rhaetic Series. Where Westbury Beds lie on the Keuper Marl north of Bristol (for instance at Aust Cliff, on the Severn) the basal conglomerate is also a bone bed, in which are rolled together the remains of lung fishes, probably washed from fresh waters, and marine reptiles. Bone beds are characteristic of the lower Westbury beds, though not always at the base.

The much thinner Cotham Beds are also widespread; they are a group of grey, silty and calcareous mudstones and their rather rare fossils include *Estheria*, marine pectens and *Pseudomonotis fallax*. North of the Mendips there is a freshwater intercalation with liverworts, suggestive of quiet shallow conditions behind a gently shelving shore. Cotham Marble is a local name for a calcite mudstone with irregular, nodular or arborescent structures, probably of algal origin.

The Westbury and Cotham beds make up the Rhaetic in its most restricted sense, but it is more usual to include the White Lias with them. This is an impure limestone facies found chiefly in Somerset and on the Devon–Dorset borders—a white or cream calcite mudstone with marly partings. Characteristically it shows many signs of shallow water or emergence, in contemporaneous conglomerates, mud-cracks and eroded or bored surfaces. Contemporaneous slump structures have been described from the Dorset coast. The White Lias dies out in north Gloucestershire, but reappears for a stretch in the Midlands. The remaining division, the Watchet Beds, are a local marly facies at the top of the White Lias on the north Somerset coast and of no regional significance.

Littoral facies of Rhaetic age are not very conspicuous east of the Severn, though they include part of the conglomerate spreads north of the Mendips and near Bristol. On the opposite side the main formations (Sully Beds to White Lias) are for the most part very similar, but littoral beds appear in west Glamorgan near Bridgend. Here the Cotham Beds are recognizable, but the black Westbury Shales beneath are largely replaced by two sandstones—20 and 35 feet thick, with 6 feet of shales between; this unusual Rhaetic facies may be derived from the erosion of the Millstone Grit and Coal Measures to the north. There are also various 'islands' of Carboniferous Limestone to the south; the characteristic limestone conglomerates are banked up against these, and at Cowbridge the nearby Rhaetic Beds are formed largely of oolitic and sandy limestones.

The Midlands, Lincolnshire and Yorkshire

Rhaetic beds are rarely exposed along most of the main outcrop, as they lie under drift, at the foot of the first low scarp of Jurassic limestone or ironstone that rises above the lower country of the Trias. Our knowledge, however, has been much increased by borings in the east Midlands and north-east Yorkshire, and from these there appear to be two fairly distinct Rhaetic formations present, which on fossils and lithology can be correlated with Somerset:

Upper Rhaetic, or Cotham Beds
Lower Rhaetic, or Westbury Beds

Both attain their greatest known thickness (just over 40 feet each) in Lincolnshire. Here the Lower Rhaetic lies with an abrupt junction and some signs of erosion on the Keuper Tea Green Marls; it consists of black pyritous shales with an abundant lamellibranch fauna, including *Pteria contorta* and *Protocardium rhaeticum*, and also fish bones. The Upper Rhaetic is typically grey shales with silts and sandstones and impure limestone nodules. Like the Cotham Beds of the south it is not very fossiliferous, but includes *Estheria minuta*.

Comparable sequences in the Yorkshire boreholes are also made up of these two distinctive beds, though not much more than half the total thickness. In both areas there is an occasional influx of Keuper-like marls into the Rhaetic deposits—red marls near the top in north Lincolnshire, and grey marls alternating with the basal Westbury shales in certain Yorkshire sections. Nevertheless, the general impression is one of remarkable uniformity, in beds so thin, and it is a testimony to the flatness to which erosion and deposition had finally reduced the Keuper plains.

North-Western England, Scotland and Ireland

At various places thin Rhaetic beds may be found in these regions, between red Triassic beds and the Lower Jurassic. There are two Liassic outliers in north-west England—the Prees outlier on the Cheshire plain, and another on the Carlisle plain. Both are largely drift covered and no Rhaetic beds have been found associated with the second. Fossils in black shales collected long ago from the first suggested that the Westbury Beds at least are present.

On Arran the Rhaetic shales and thin limestone are only known from the downfaulted and jumbled fragments in the central ring complex, but the lamellibranchs, including *Pteria contorta*, are a sufficient indication of age. That the seas also invaded the Keuper lowlands west of the Scottish Highlands is shown by the sections in western Mull. Here there are about 40 feet, mainly of dark calcareous sandstones with lamellibranchs; again they are probably the Westbury Beds, to which local erosion has added a sandy component. The Cotham Beds may be represented by grey-green shales and impure limestones.

On the other hand, no recognizable Rhaetic has been found associated with the upper red Trias or the Lias of Raasay or Applecross, and in the former there is apparent conformity between these two. This, together with the sandy influx in Mull, may mean that we are here at or near the margin of this northerly Rhaetic gulf. The southern part of the gulf has much more extensive outcrops near Belfast and at several places around the Antrim Basalt Plateau. Underground Rhaetic beds are known in the boreholes of Coalisland, Co. Tyrone and through the basalts of Lough Neagh. The sequence has many features in common with those in various parts of England. At Belfast there are 50-odd feet, with typical black shales containing *Pteria contorta* overlain by black and grey shales with *Protocardium rhaeticum*; there is a local influx of red marls at the top. A bone bed, like those in Gloucestershire, occurs near the base of the lower shales and similarly contains bones and teeth of fishes (including *Ceratodus*) and marine reptiles. The Irish outcrops do not add much that is new to the Rhaetic environment, as deduced from the English outcrops, but, in extending the geographical picture, underline the wide uniformity of this closing scene in the long process of continental erosion, deposition and reduction of relief that finally reduced the abrupt Hercynian mountains and ridges to relatively gentle uplands bordering the new seas.

REFERENCES

Arkell, 1933, ch. 4; Bott, 1964; Charlesworth, 1963, chs. 11 and 12; Craig, 1965*b*; Dunham, 1960; Dunham and Rose, 1949; Elliott, 1961; Edwards, 1951; Francis, 1959; Hallam, 1960*b*; Hollingworth, 1942; Hutchins, 1963; Rose and Kent, 1955; Sherlock, 1947; Smith, 1958; Stevenson and Mitchell, 1955; Stewart, 1954, 1963; Taylor *et. al.* 1963; Tyrrell, 1928; Wills, 1948, 1956.

British Regional Geology. Northern England, Pennines, Central England, South-West England, Bristol and Gloucester.

9

THE JURASSIC SYSTEM

Although the name of this system was derived, in the last years of the eighteenth century, from the Jura Mountains, the coast of England, and particularly that of Dorset, provides some of the best sections. Indeed many of the stages are named after English towns and districts. Moreover, of all the systems the Jurassic exhibits most clearly the principles and practice of stratigraphical palaeontology; it cannot compare with the Carboniferous System, for instance, in variety of rock-type nor with the Ordovician in the complex problems of stratigraphy, vulcanism and tectonics. But the relationship between sedimentary strata and their faunas, and the identification and correlation of beds by means of their contained fossils, were first and most clearly established in Jurassic rocks. Not all the Jurassic stages in Britain are well-provided with ammonites, but most of them are, and these are then incomparably the best zone fossils. Their value was already appreciated in the middle of the last century and many of the zones and stages defined then still survive, though some under slightly different names, as working units to the present day. It is thus not accidental that discussions on principles and methods in stratigraphical palaeontology include persistent references to Jurassic rocks and faunas, and the precision which is possible in their study may usefully form a kind of standard against which correlations of other ages can be measured.

The long history of Jurassic research does not mean that there are no problems left in the classification and correlation of this system in Britain. Although abundant in many formations the ammonites are not ubiquitous; even among marine rocks they are rare or absent from most of the bioclastic limestones, and particularly from the coral-reef facies. Naturally they are lacking from the Middle Jurassic non-marine rocks of Yorkshire and Scotland, and their similar absence from all the Purbeck formations and the succeeding Wealden of the Lower Cretaceous above is the prime cause of the problems attendant on the Jurassic-Cretaceous boundary. The major divisions adopted in this book are given in Table 43 and the stages and zones in the Appendix.

The Jurassic period is not one of the longest in Britain, nor is it outstanding in thickness, the known maximum being about 5,000 feet, in borings in southern England. The sediments are on the whole fine-grained and the most abundant are marine shales, clays, and mudstones; sandstones are relatively unimportant, though they include those of the Middle Jurassic deltaic facies. The marine limestones, though varied and conspicuous, are much less in total thickness than the clays and shales.

Table 43. *Principal Jurassic formations of southern England*

UPPER JURASSIC	Purbeck Beds Portland Beds Kimeridge Clay Corallian Beds Oxford Clay Kellaways Beds
MIDDLE JURASSIC	Great Oolite Inferior Oolite
LOWER JURASSIC	Upper Lias Middle Lias Lower Lias

The base of the Upper Jurassic is sometimes taken at the base of the Oxfordian stage (p. 297); its upper limit may be within the Purbeck Beds (p. 302).

Many of the limestones are oolitic, and this word has a confusing variety of usage. In addition to 'oolite' as a normal lithological type, employed for rocks of any age, the word was early incorporated in the title of certain stratigraphical units, rather similarly to 'Millstone Grit'. Thus there is the Inferior Oolite (Bajocian), and Great Oolite (Bathonian), and other smaller units in the Corallian Beds. A third, and virtually obsolete, usage is to employ Lower, Middle and Upper Oolites for all Jurassic strata above the Lias. This last may be ignored, but Inferior Oolite and Great Oolite are distinctly convenient. Ammonites are scarce or absent in these (the Middle Jurassic) formations, but they and their lithological subdivisions can nevertheless be defined and recognized in the field or drawn on a map. Such index fossils as they do contain (chiefly brachiopods, lamellibranchs and the rare ammonite) can then be used to place these rock units as closely as may be in the hierarchy of international stages and zones.

In its geographical setting Jurassic stratigraphy inherits the scene already opened by the Rhaetic marine invasion. To the west and north there were remnants of the earlier uplands. Of these Devon and Cornwall (Cornubia), Wales and some type of Scottish landmass are fairly certain in their general position, though not in detail. The form of north-western England, southern Scotland and most of Ireland is unknown, for we lack reliable evidence anywhere nearby. If, as already suggested, there is little firm reason to postulate a Permian Pennine ridge, its existence in Jurassic times is equally dubious; differing views are held on this topic, but the elevation of a north–south upland may have been a Tertiary feature. Within the area of the major outcrops, from Yorkshire to Dorset and their easterly and southerly extensions underground, there are two outstanding subsiding basins of deposition, one large positive element or island, and a number of smaller fringing elements. The island is the London Platform; all the major Jurassic (and Lower Cretaceous) divisions become thinner as this upland is approached and none overrides it. The Wessex basin lay to the south-west in Dorset, south Wiltshire and the Isle of Wight; it seems to have been partly separated from the basin of the Weald and both probably continued into northern France.

The northern margin of the London Platform was more irregular and the sediments on this side were thinner than those on the south, with a greater variation in facies and areas of local non-sequences, erosion and minor phases of emergence. In particular there was a belt of shallows centred on Oxfordshire, which perhaps extended across to the Midland coalfields over areas from which Jurassic rocks have now been removed. Another, somewhat similar, region lay to the west, around the limestone islands of the Mendips and northwards across the Radstock 'shelf' to Bath and Bristol; this belt may have continued westwards to South Wales to link up with the Welsh Massif. Between Bristol and Oxfordshire there was the basin of the Lower Severn and Gloucestershire Cotswolds, and north-east of the Oxfordshire shallows lay the Lincolnshire basin. The northern margin of the last just crossed the Humber; it was separated from the Yorkshire basin by the Market Weighton upwarp—a very persistent structure which produced thinning or absence in all formations from the Rhaetic to Lower Cretaceous.

In this summary of partly connected basins, belts of shallowing and actual islands, the term 'axis' has been avoided. Nevertheless, it has been used so much in descriptions of Jurassic environments in the past forty years that some mention is desirable. 'Axes' denoted positive areas of little sedimentation and occasional unconformity, separating basins from one another, but they were usually considered to be linear in plan, with a definite direction: in particular there have been described the east–west Mendip and Market Weighton axes, and (rather less precisely) an Oxfordshire or Moreton-in-the-Marsh axis with a dominantly north-

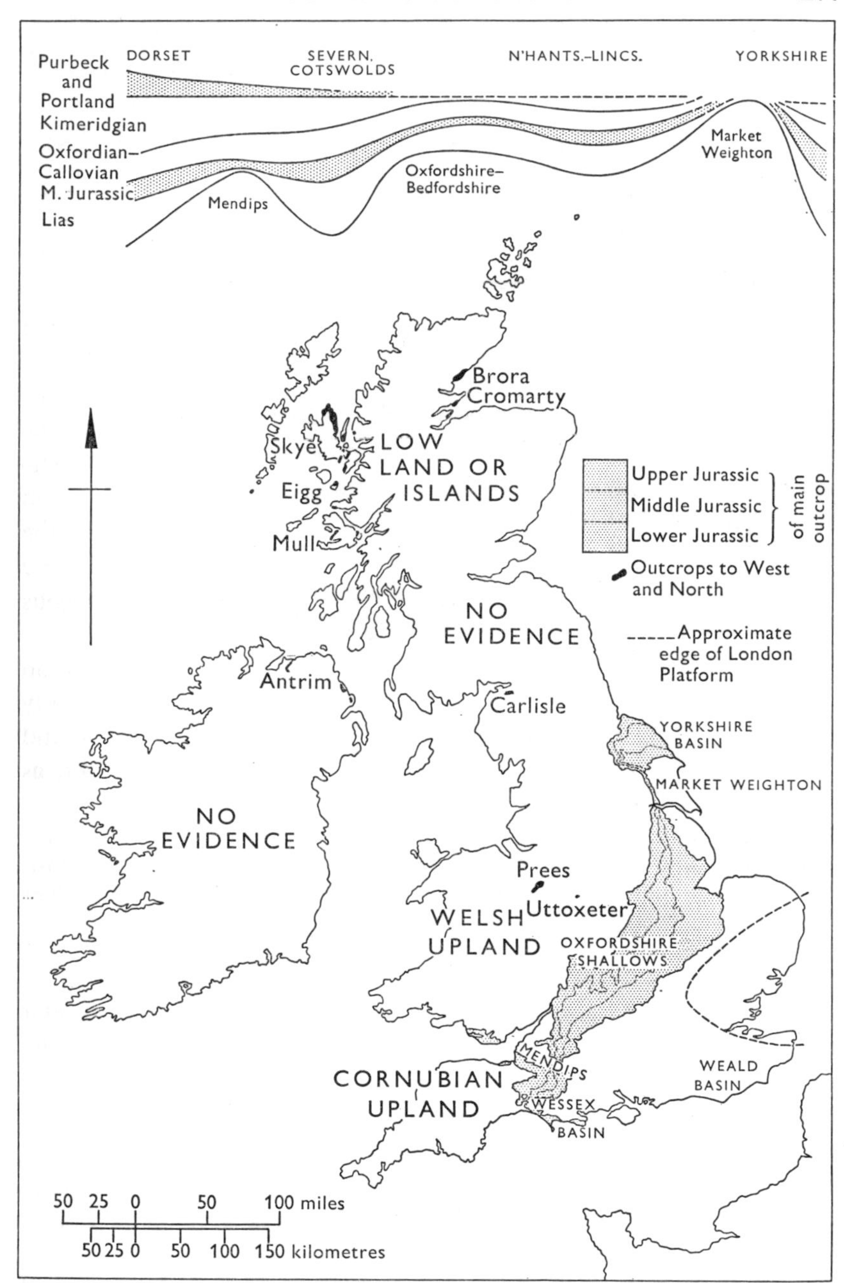

Fig. 55. The main outcrop of Jurassic rocks, together with much smaller inliers or outliers and the approximate outline of the London Platform. The section at the top shows broadly the variations in thickness (not to scale) along the main outcrop from Dorset to Yorkshire; for the Weald basin see Fig. 62.

westerly alignment. The idea of positive elements of lesser subsidence remains, but the concept of linear axes has been largely replaced by one of areas—variously called upwarps, structures, shallows or swells, the last after the German 'schwellen'. Their degree of emergence and their limits vary somewhat through the Jurassic period, as does their effectiveness in separating one type of facies from another. Nevertheless, by and large, the individual formations and their facies commonly maintain a unity within one sedimentary basin, and are more liable to change over the major upwarps.

From Yorkshire to the southern Midlands the Jurassic rocks have a gentle regional dip to the east or south-east, the many minor divergences having little effect on the broad crescentic sweep of the main outcrop. In southern England the Tertiary folds are more acute, and some of the outcrops are also interrupted by Upper Cretaceous overstep. In particular the latter conceals Upper Jurassic rocks over much of north Dorset and Wiltshire. The erosion of two Tertiary anticlines, cut into the western chalk scarp in Wiltshire, produces the Vales of Wardour and Pewsey, where the small inliers of Upper Jurassic and Lower Cretaceous rocks help to fill the gap.

In south Dorset there is again overstep and folding, but the situation is complicated by the acuteness of the anticlines (which have locally overturned northern limbs and are associated with reversed faulting) and by some intra-Cretaceous movements as well. The general effect, as seen, for instance, around Weymouth and Lulworth, is to produce asymmetric anticlines bringing up Jurassic beds, the overstepping Cretaceous outcrops being conspicuous on the flanks (Fig. 60). The monocline of the Isle of Wight belongs to the same group of folds, but does not expose Jurassic beds.

LOWER JURASSIC OR LIAS

The term Lias has long been used for the lowest of the three major Jurassic divisions, which is also the most uniform in lithology, being usually argillaceous. Even so there is more variation than appears at first sight, or in poor exposures, and what seems to be a monotonous series of shales proves on detailed examination to have many minor variations. These include coarser detrital additions, silt or fine sand, precipitated components such as impure limestones or ironstones, and more rarely an exceptional abundance of organic matter to form dark-grey or black bituminous shales, often accompanied by abundant iron pyrites.

Many Liassic outcrops are also seen to be rich in concretions, formed at an early stage in diagenesis by local concentrations, usually of calcium carbonate but sometimes of iron compounds as well. Fossils often form the centre of the concretion and these are locally plentiful in the Lias as a whole. In addition to the ammonites there are abundant lamellibranchs (especially oysters and gryphaeas), brachiopods, belemnites and many other invertebrates. Certain horizons are famous for fishes and reptiles, including the Lower Lias of Somerset and Dorset and the Upper Lias of Yorkshire. Trace fossils (tracks of invertebrates, such as worms or molluscs) are also found, as burrows or tubes, either parallel to the bedding planes or running vertically through them. Both lithology and fauna are those of a muddy, but normally well-inhabited sea of wide extent, deepening slightly after the first shallow invasions of the Rhaetic but never very deep. Local stagnant, anaerobic, conditions developed in some basins.

Variations in the main outcrop

Of the three Jurassic divisions the Lias shows best the marked variations in thickness in the different subsiding basins and across the intervening swells. In addition each major basin had its own variants in sedimentation and the way in which certain zones increase or decrease in thickness, are cut out entirely, or change in facies, from one basin to another is often extremely complicated.

In the Midlands and south of England there is a general tendency for the lowest Lias to be partly or entirely calcareous. In Lincolnshire, for instance, the pre-*planorbis* beds (i.e. the lower part of the *planorbis* zone, below the entrance of the ammonites) are called the Hydraulic Limestones and consist of 20–30 feet of impure calcareous rocks; farther up the Granby Limestones reach about 80 feet.

The Lias of the Wessex basin is very well exposed in the cliffs of west Dorset where the beds dip gently eastwards along the sweep of Lyme Regis bay. Most of the 800–900 feet is composed of Lower and Middle Lias, much of the latter being sandy and standing out as beds of calcareous sandstone. The lowest beds are the Blue Lias, a facies consisting of four principal rock-types: marls, limestones (calcite mudstones or calcilutites), very finely laminated limestones and bituminous shales. The last two are largely unfossiliferous, but the first two (which differ chiefly in the much lower carbonate content of the marls) are rich in fossils—both free swimming and bottom living forms, and as in other parts of the Lias ammonites and lamellibranchs are outstanding. The Blue Lias as a whole was laid down in shallow marine waters, but the laminated limestones and shales represent temporary periods of stagnation and poor aeration. The repetition of the four lithological variants and conditions responsible for them forms one of the many types of repetitive or cyclic sedimentation found in English Jurassic rocks.

The Blue Lias extends across the Bristol Channel into Glamorgan, but here there are also exceptional littoral facies in the most westerly outcrops. These abut against the Carboniferous islands in the same way that the underlying, abnormal Rhaetic beds do, but with a different type of lithology. The Liassic littoral beds consist of massive limestones with chert, sandy and conglomeratic beds, the last including chert gravels. Some of the limestones are rich in corals, which are otherwise rare, as might be expected, in the muddy type of Lias. No beds above the *semicostatum* zone are known in Glamorgan, and, in fact, no further Mesozoic or Tertiary sedimentary history in Wales at all, before the Pleistocene glaciation.

In the main outcrop, the Wessex basin stretched northward through Somerset to the Mendips, which formed islands or shallows of Carboniferous Limestone in the Lower Liassic sea. Beyond lay the shallows of the Radstock shelf, where sedimentation was slow and intermittent. This is well seen in the lowest zones of the Lower Lias, which are locally thin and conglomeratic with layers of phosphatic pebbles, and in places are absent altogether; the upper zones (*ibex* and *davoei*) are thicker and more normal in their clay facies but later there was a further regression of the sea, for the Middle Lias is missing. North of Bristol there is a marked expansion into the Lower Severn basin where the Lias is complete and thick; nearly 1,658 feet, mostly of shales, were found in the Stowell Park borehole of the northern Cotswolds.

There is a general reduction over the Oxfordshire shallows where the total thickness is under 550 feet; the effect is particularly conspicuous in the Upper Lias, which at its extreme attenuation is reduced to about 5 feet, including a basal limestone containing the rolled and eroded ammonites of several Whitbian zones. The expansion through Leicestershire, Northamptonshire and into Lincolnshire is rather irregular and the Upper Lias continues to show local erosion and overlap within its zones—evidence that even in the basins sedimentation was not always continuous. Moreover, there was a general emergence towards the end of the Yeovilian, the upper zones of which are missing and the lower overlain unconformably by the Bajocian stage; towards the north this gap widens and the latter lies on Whitbian beds.

All the Liassic divisions become thinner northwards through Lincolnshire and the long north–south outcrops become narrower in consequence. For instance, from the Lincolnshire basin to the Humber the Lower Lias is reduced from 700 to 300 feet; at Market Weighton it is about 100 feet and is the only Jurassic formation present, overlain by Albian (top of the Lower Cretaceous). The thinning and ultimate absence of all the later Jurassic stages is partly responsible for the marked attenuation of the outcrop over this structure. The effect is heightened, however, by the overstep of the Cretaceous beds.

Where the full sequence of Liassic beds reappears in the Yorkshire basin it amounts to over 1,400 feet; the excellent exposures in the cliff sections rival those of Dorset, except that the lowest zones (below *semicostatum*) and the Rhaetic junction are not visible, being concealed under the alluvium of the Tees. Compared with south-west England the sequence is more argillaceous and less variable, but there are many of the typical subsidiary components—beds rich in silt, impure limestones, organic matter and ironstones.

Liassic ironstones

By far the largest source of iron ore in Britain is the bedded ironstones of the Jurassic and in recent years slightly under half the total production has come from Liassic rocks and the remainder from the Bajocian Northamptonshire Ironstone. Many parts of the Lias contain iron compounds but only at a few levels and in a few places are they thick and rich enough to be economically important.

Table 44. *Distribution of Liassic iron ores*

UPPER LIAS	North-east Yorkshire, *dispansum* and *levesquei* zones; Raasay, Hebrides, *bifrons* zone
MIDDLE LIAS	*Spinatum* zone: Cleveland, Yorkshire, and the Marlstone Rock Bed of Midlands and Lincolnshire
LOWER LIAS	Frodingham, north Lincolnshire, *semicostatum* and *obtusum* zones

Petrologically the various ironstones have much in common. The iron is present as ferrous carbonate (siderite), hydrous alumino-silicate (chamosite) and hydrous oxides (limonite); magnetite is sometimes present but is much rarer than the other three. The non-ferrous components include calcite, an opaline form of silica and clay minerals. In texture the rocks consist of ooliths, fragmental debris (chiefly fossils, the ironstones often being distinctly fossiliferous) and matrix. Siderite, chamosite and calcite are important in the matrix and the first two may be altered to limonite. The ooliths were originally of chamosite and have locally been altered to siderite or limonite, or less frequently to calcite, magnetite or opal.

At the present time the Frodingham ore is the greatest producer among the Liassic ironstones. Whereas in south Lincolnshire there are some 150 feet of clays that are locally ferruginous, but not in a workable degree, in north Lincolnshire in the neighbourhood of Scunthorpe the equivalent beds are reduced to 18–32 feet, which consist almost entirely of a calcareous oolitic ironstone. Fossils are abundant, especially belemnites and species of *Gryphaea*, *Lima* and *Cardinia*.

A similar but more widespread lateral transition is found in the Middle Lias ironstones. Over much of the country from Yorkshire to Dorset there is a tendency for the lower part of this division to be sandy and the upper calcareous or ferruginous. The latter is also often relatively hard, forming a small scarp feature below that of the Middle Jurassic limestones, and is called the Marlstone. This bed is particularly rich in brachiopods (e.g. *Tetrarhynchia tetraedra*) and in three areas the iron content rises to economic proportions. In the Banbury and North Oxfordshire ironstone field the Marlstone forms the scarp of Edge Hill, and reaches a local maximum of 25 feet of workable ironstone. The absence of ironstone between this field and that of Leicestershire is largely attributable to intra-Liassic erosion—the Marlstone being locally eroded before the deposition of the Upper Lias. In the opposite direction, towards the Oxfordshire shallows, terrigenous detritus comes in and the iron content falls so that the rock becomes a sandy ferruginous limestone.

The Cleveland iron ore field in north-east Yorkshire is geologically more extensive than those in the Midlands, but economically it is non-existent. Several seams, interbedded with shales, reach a combined maximum of 70 feet; this is the Ironstone Series, which included part of the *margaritatus* as well as the *spinatum* zone. All the ironstone seams are thickest in the north of the outcrop, on the northern edge of the Cleveland Hills, where the uppermost Main Seam is 11 feet. As they are traced round to the east and south, in the coastal sections, they become impoverished through the reduction of seam thickness and splitting by shale beds. In the late nineteenth century the Cleveland iron ore was the foundation of Middlesbrough industry, as the Frodingham ore is at present that of Scunthorpe.

The Upper Lias ironstones of Yorkshire are very small in extent and occur chiefly in small synclines in the upper Yeovilian rocks of the inland dales, where these have escaped the pre-Bajocian erosion. The ores are no longer worked, but they include the 'magnetic ironstone' of Rosedale, which is unusual in that magnetite is one of the replacing minerals of the ooliths.

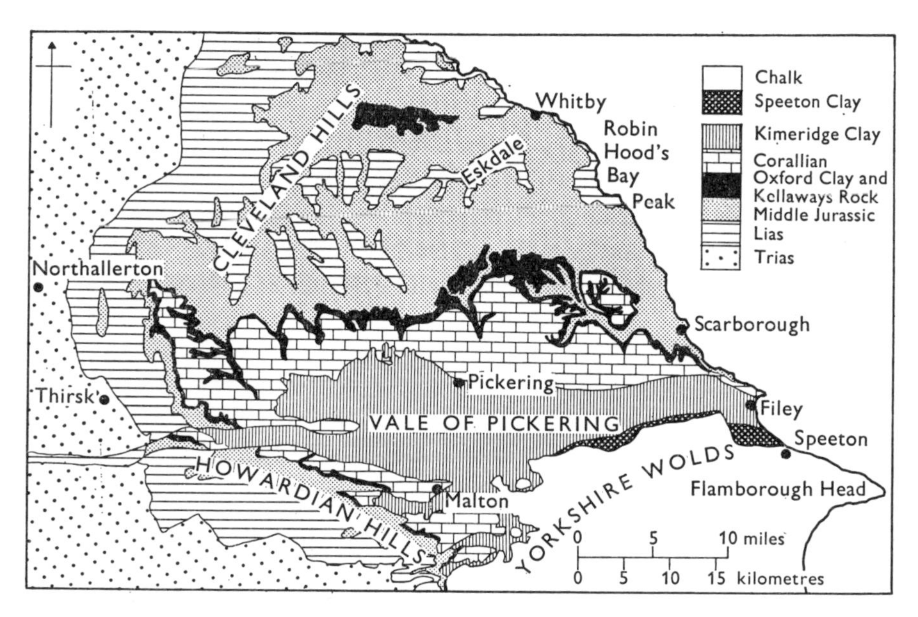

Fig. 56. Geological sketch map of north-east Yorkshire.

The Upper Lias of Yorkshire

At both ends of the main outcrop the Upper Liassic rocks illustrate problems that extend beyond their immediate counties. Over most of the Yorkshire outcrops the Whitbian alone is present, overlain unconformably by a relatively thin Dogger, the lowest Bajocian formation. In a few places in the inland valleys, however, there are isolated synclines (pre-Dogger basins) of Yeovilian beds, including the ironstones already mentioned, and along a short section of the coast between Whitby and Scarborough there is a much fuller Upper Lias succession, where thicker Dogger overlies Yeovilian beds with only a small non-sequence. This fuller sequence is cut off on the west by the Peak Fault, on the south side of Robin Hood's Bay, which is probably a transcurrent fault, bringing together areas differently affected by the pre-Bajocian erosion. Over the

Yorkshire basin this erosion resulted in the removal of almost all the Yeovilian, but in a few synclinal basins it survived.

The Upper Lias beds present east of the Peak Fault are shown in Table 45. From the Jet Rock Series upwards this sequence demonstrates a cycle of increasing shallowing and aeration. The Jet Rock Series is highly bituminous (the ammonite chambers sometimes contain oil), and pyrites is abundant in the dark-grey shales; the Top Jet Dogger is a 6-inch bed of laminated argillaceous limestone. At this period the basin was stagnant and the muds were laid down below wave action; bottom faunas are very rare or absent. Through the succeeding beds, up to the Alum Shales, the sediments become progressively a lighter grey, less pyritic and less bituminous. The calcareous concretions are accompanied by thin bands of sideritic mudstones. This shallowing and aeration culminates in the grey and, finally, yellow beds of Blea Wyke which contain an increasing proportion of sand and evidence of a thriving bottom fauna.

Table 45. *Upper Lias of Yorkshire (east of Peak Fault)*

Stages	Lithological formation	Maximum thickness (feet)
YEOVILIAN	Blea Wyke Beds (yellow and grey sandstones)	68½
	Striatulum Shales	52½
	Peak Shales	28
WHITBIAN	Alum Shales	65
	Hard Shales	21
	Bituminous Shales	65½
	Jet Rock Series	29
	Grey Shales	30

This is not the only cyclic sequence of the Lias (those of the Blue Lias have already been mentioned and others may be detected, rather less completely, in the Lower and Middle Lias of Yorkshire, and also in the Wessex basin) but it is one of the clearest and best known. Variations in depth, aeration, and in the chemical conditions of the bottom muds were clearly of great importance during the period. The many non-sequences and erosion levels from one basin to another also emphasize the way in which fine-grained sediments may, in certain circumstances, be laid down in quite shallow waters.

The Upper Lias of the south-west, the Cotswold Sands

From Gloucester southwards to the Dorset coast the most conspicuous formation between the Middle Lias and the Bajocian limestones is a thick mass of fine-grained yellow sands, sometimes loose and friable, sometimes rendered coherent by a calcareous cement. These sands go by different names in different parts of their outcrop: the Cotswold Sands of Gloucestershire, the Midford Sands near Bath, the Yeovil Sands of Somerset and the Bridport Sands on the Dorset coast. In the field they appear to be a single homogeneous formation, but their relation to the zones and subzones of the Upper Lias and lowest Bajocian show how, in fact, the facies boundaries cross the zonal boundaries, or 'time-lines'; in the north of their outcrop the sands are confined to the *variabilis* zone, lowest of the Yeovilian, but become younger and younger southwards and in Dorset overlap the Lower and Middle Jurassic junction (Fig. 57).

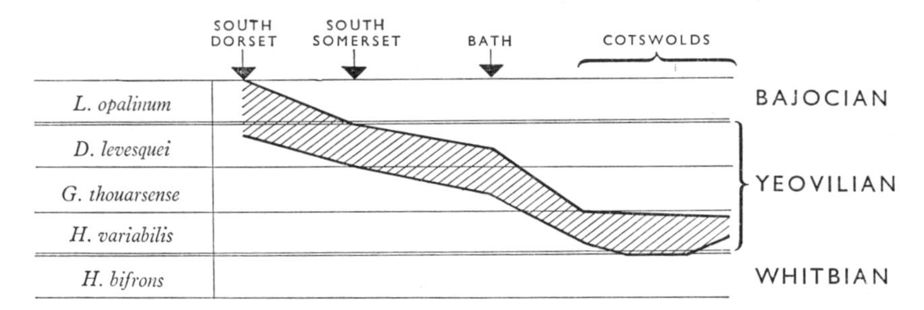

Fig. 57. Diagram to show the distribution and age of the Cotswold–Bridport sands, in relation to the zones of the Upper Lias and lowermost Bajocian. The thickness of the shaded band refers to the zones represented and is not related to the actual thickness of the sands. (After Wills, 1929, p. 139; zones from Dean *et al.* 1961.)

The 'cephalopod' or 'junction' beds which are found in various regions above or below the sands are thin layers of clay or limestone crowded with ammonites from more than one horizon. They were formed in current-swept conditions, so that little or no normal sediment was retained. True Upper Liassic clays in these outcrops are largely restricted to the coast where 70 feet underlie the Bridport Sands. Clays also reappear in north Gloucestershire, in the Lower Severn basin where the sands disappear. The origin of the sandy detritus is not very clear: the mineral content has little to link it with the nearest landmass, of Wales, or with any other possible source.

In their transgressive boundaries and false appearance of contemporaneity, the Cotswold Sands are a classic example of a diachronous deposit. They also show how, in order to establish the details of diachronism, some rigorous time-sequence is necessary, which in this case is supplied by the ammonite species. In systems less well supplied with zonal fossils a diachronous unit may be suspected, but is more difficult to prove conclusively.

MIDDLE JURASSIC: BAJOCIAN AND BATHONIAN

In the west of England the Middle Jurassic limestones are the most conspicuous rocks of the system because they form the fine scarps and long dip-slopes of the Cotswold Hills. Along part of that range, from Bath north to Oxfordshire, the scarp-forming beds are the Inferior Oolite, which is also responsible for the limestone uplands around Grantham and the north–south ridge of Lincolnshire. South of Bath the Inferior Oolite is less important, but the Great Oolite limestones are found in the relatively high ground of east Somerset and the borders of Wiltshire. The Inferior Oolite limestones become much thinner when they are traced southwards to the Dorset coast and those of the Great Oolite disappear in Somerset. The change from Lincolnshire to Yorkshire is even greater, for marine beds in the latter are restricted to relatively minor incursions in the Middle Jurassic Deltaic Series, the transition being associated with, but not rigidly governed by, the Market Weighton upwarp.

Table 46. *Middle Jurassic stages and rocks of the Cotswolds*

BATHONIAN	Upper Cornbrash Great Oolite Limestone etc.
BAJOCIAN	Upper Inferior Oolite Middle Inferior Oolite Lower Inferior Oolite

It follows that the Middle Jurassic rocks are characterized by marked variation in facies and in conditions. Moreover, with the exception of some of the Bajocian and late Bathonian faunas, ammonites are rather rare or absent; to some extent brachiopods and lamellibranchs are employed instead, but where there is strong lateral change among thin lithological units a large number of local names have had to be used for beds of not very great extent. This applies particularly to the limestones in the west of England.

INFERIOR OOLITE SOUTH OF MARKET WEIGHTON

Cotswold Hills and Oxfordshire

The Inferior Oolite, as a lithological group, is the approximate equivalent of the stage Bajocian, from Bayeux in Normandy. In the Cotswolds it is thickest in the basin of north Gloucestershire, which had earlier received a full complement of Liassic clays, and like the Lias it also becomes thinner over the Oxfordshire shallows. Even in the basin, however, Bajocian subsidence was not uniform or continuous, but was twice interrupted by uplift and gentle warping; thus arises the subdivision into Lower, Middle and Upper Inferior Oolite (Fig. 58). The second phase of movement was the more pronounced, producing a series

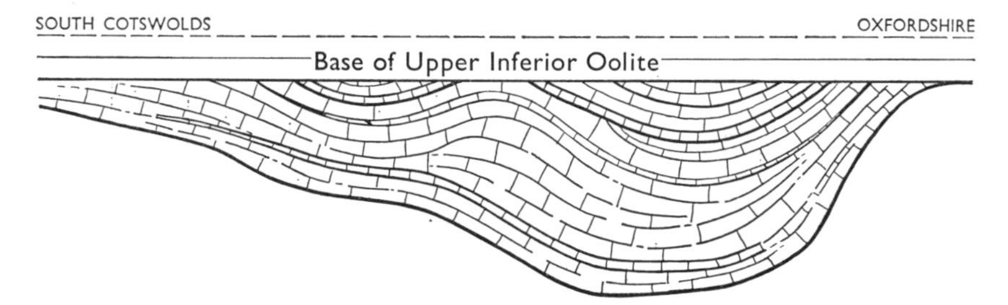

Fig. 58. Diagram to show gentle folding and unconformities in the Inferior Oolite of the Cotswold, the transgressive Upper Inferior Oolite being taken as horizontal. (After Kellaway and Welch, 1948, p. 68.)

of gentle flexures which were planed off and covered by the transgressive base of the Upper Inferior Oolite (Upper Trigonia Grit, or base of the *parkinsoni* zone). A further result is that certain of the underlying beds are confined to the synclines; the fullest succession is near Cheltenham, but even here the Inferior Oolite is only about 350 feet thick.

Almost all the Inferior Oolite is calcareous, including sandy and rubbly limestones, pisolites and oolitic freestones—the last term referring to the ease with which the rock can be cut in any direction. Some of the oolites are 'iron shot', or limonitic in a white or cream-coloured calcareous or clay matrix. Shells and banks of shell debris are abundant, together with echinoderms and corals; the last are common in certain beds but do not approach a true reef. Ammonites are rare but enough are known to establish the succession.

From the typical Cotswolds of the Cheltenham hills towards Oxfordshire, the Lower and Middle Inferior Oolite become thinner and ultimately disappear under the overlapping upper division. Western Oxfordshire (Evenlode Valley or Vale of Morton) thus acts as a further, major intra-Bajocian anticline. The Upper Inferior Oolite survives as 10 feet or so of limestones, the eroded remains of ammonites at the base lying on a similar condensed sequence of Upper Lias. This is the very meagre Bajocian link between the Cotswold and Lincolnshire basins.

Somerset and Dorset

Similar intra-Bajocian erosion was also active over the Mendips and only Upper Inferior Oolite is found here, as a thin capping of Doulting Stone (rubbly limestones) which overlie the steeply dipping Carboniferous Limestones. On the south side of the Mendip ridge the Doulting Stone thickens to 45 feet of building stone.

For the rest of the outcrop, down to Bridport on the coast, all three divisions of the limestones are present, but with many signs of non-sequence and minor erosion levels. As an example of lateral change in this shallow, variable belt, the thickest sequence, north-east of Sherborne, is some 90 feet (including 60 feet of the upper Bajocian Sherborne building stone) whereas only 6 miles away at Yeovil Junction the three divisions are present in 6 feet of beds. Nevertheless, in this region as a whole ammonites are more plentiful than in the Cotswolds; zones and subzones can be recognized in detail even though some may be represented only by thin layers of conglomerate with water-worn fossils.

On the Dorset cliffs the Inferior Oolite forms a thin capping to the Bridport Sands, the upper part of which is in the *opalinum* zone. Above there are a few feet of sandy limestone, forming a transition facies, and then an almost incredibly condensed sequence of oolitic and conglomeratic limestones, with lamellibranchs, ammonites and sponges. Of the 11-foot total the Upper Inferior Oolite forms more than half.

There is little in this exceedingly thin, variable development to suggest a Wessex 'basin' during Bajocian times. It has been suggested that the north–south bank of the Bridport and Yeovil Sands stood up as a submarine ridge, or an accentuated belt of shallows, so that sediments were largely swept off its top but were able to accumulate more successfully on the flanks; the Sherborne and other thicker outcrops would thus lie somewhat to the east of the submarine bank.

Northamptonshire and Lincolnshire

Where the Bajocian expands again from Oxfordshire into the basin of the east Midlands neither the lithological grouping nor the erosion levels correspond closely with those of the Cotswolds, although shallow water limestones are still important and most of the sequence is marine:

Lincolnshire Limestone
Lower Estuarine Series
Northampton Ironstone

There was a profound change in conditions towards the end of the Upper Lias with widespread emergence in this region. The base of the Northampton Ironstone (or Northampton Sand Ironstone) includes a layer of phosphatic pebbles and lies for the most part on Whitbian clays, the Yeovilian being absent. The ironstone contains ammonites of the *opalinum* zone and lithologically much resembles the Liassic ironstones; in recent years it has been the largest iron ore producing formation in the country and is worked extensively around Corby and Kettering.

The rock is oolitic, the ooliths being of chamosite, limonite and more rarely kaolinite, set in a matrix dominated by siderite and calcite. The proportion of sand grains is higher than in the Liassic ores and in north Lincolnshire the rock becomes a ferruginous sandstone of no commercial value. There is a similar transition in outliers to the west and south of the orefield. The maximum thickness, north-west of Northampton, is 70 feet, but this includes several minor variations in facies of which one (the Main Oolitic Limestone Group) is the principal ore producer. The complex lateral changes in the facies, erosion channels and non-sequences point to very shallow water conditions. Locally the whole ironstone group may be absent through channelling at the base of the beds above.

There was clearly a further broad recession of the seas during the Lower Estuarine Series. The name is misleading, and these beds were probably the product of very low-lying deltaic and coastal flats; the topmost beds of the ironstones are penetrated by the rootlets of the first, colonizing vegetation. The series is thin, with a maximum of 25 feet, and consists chiefly of light and dark-grey, fine sands, silts, and clays. There are thin beds of dark-grey carbonaceous clays and rootlet beds. Current-bedded sands and channels are not very common but occur towards the top. It is probable that these coastal flats extended northwards from the London Platform, and may have linked up temporarily with the somewhat similar deposits (though much thicker and bearing more signs of active distributary channels) in Yorkshire.

The coastal flats were submerged slightly before the end of the Lower Inferior Oolite period, by the seas of the Lincolnshire Limestone, which represents the rest of the Bajocian. The limestone is lenticular, with a maximum thickness of 132 feet in south Lincolnshire; in lithology it resembles the Cotswold limestones, the dominant rock being oolitic, but there are various other detrital types and the ooliths and grains are in a very fine calcite matrix. Some parts are rich in molluscs or brachiopods and others form a valuable freestone. The so-called Collyweston Slate at the base in Rutlandshire is a fissile sandy limestone, the sand grains of which resemble those of the beds below, but were incorporated in the basal calcareous facies during the flooding of the coastal flats.

The Lincolnshire Limestone is divided into upper and lower units and a band with the rhynchonellid *Acanthothiris crossi* is a useful marker at the base of the former; ammonites have been found occasionally but are rare. Towards the south, in the Northamptonshire orefield, the whole limestone becomes much thinner and finally disappears south of Kettering; here too the upper beds rest on an eroded and channelled surface of the lower limestones. These features and the various lithological types strongly resemble modern carbonate deposits in very shallow seas, and the channels were probably submarine, formed by currents on the shoals rather than by subaerial erosion. At the northern end of the outcrop the Lincolnshire Limestone partakes of the thinning that affects all the Jurassic rocks as they approach Market Weighton, but to a lesser extent than many. It is recognizable on the north of the Humber as 38 feet of limestone locally called the Cave Oolite.

A marked feature of Bajocian sedimentation in the Midlands and south of England is the somewhat paradoxical one of its paucity. Thin and condensed sequences, current action, winnowing and channelling of sediments, non-sequences and local unconformities—all these are characteristic, even in the basins. At certain periods there was actual emergence, for instance during the accumulation of the Lower Estuarine Series, and (probably) during the pre-*parkinsoni* erosion in the southwest; more frequently currents in the very shallow waters were effective. In such conditions the submarine shoals and the coastal flats could be transformed into one another by small changes in relative land and sea level, and any rigid or persistent distinction between the sea areas and the landmass of the London Platform, as palaeogeographic entities, is rather misleading.

GREAT OOLITE SOUTH OF MARKET WEIGHTON

The Great Oolite is almost but not quite equivalent to the Bathonian stage, since the uppermost zone (of *Clydoniceras*) is represented by the Lower Cornbrash above. The formation also supplies a good example of the problems that may still attend Jurassic stratigraphy even where the rocks are marine, fossiliferous, conformable and fairly well exposed; many of the beds in the type area around Bath have been known since the descriptions of William Smith. The difficulties arise not from erosions and transgressions like those that complicate Inferior Oolite stratigraphy but from the lack of ammonites, the bewildering variation of local facies, and the corresponding difficulty of identification and correlation of lithological sequences from one area to the next. Brachiopods and other groups supply index fossils to some extent, but being partly dependent on facies are not entirely satisfactory.

Cotswolds, Somerset and Dorset

The major change in south-west England is from the calcareous and clay facies found in the Cotswolds to the more uniform clays of south Somerset and Dorset. Where the limestones form the plateau country behind the Cotswold scarp and around Bath the exposures are fairly good, but they are much less satisfactory in the clay country. The successions and approximate correlations are shown in Table 47.

Table 47. *Great Oolite and correlatives in south-west England*

North Cotswolds	Bath	South Somerset and Dorset
Forest Marble or Wychwood Beds	Forest Marble Bradford Clay (10 feet)	Forest Marble *digona* bed 60 feet above base *boueti* bed at base (140 feet)
Great Oolite Limestones (150 feet)	Great Oolite Limestones (120 feet)	
Lower Fuller's Earth (60 feet)	Upper Fuller's Earth Fuller's Earth Rock Lower Fuller's Earth (150 feet)	Upper Fuller's Earth Fuller's Earth Rock or *wattonensis* beds Lower Fuller's Earth

In the type area the basal clays, which take their name from a thin seam of commercial fuller's earth south of Bath, are much thinner than those farther south, but nevertheless amount to 150 feet. The Great Oolite limestones mark a return to the current-swept, well-aerated limestone shoals. There are again local channels in the shelly and oolitic rocks, with coral beds and fine, white calcite mudstones; the different rock-types replace one another laterally in such profusion that their total thickness is difficult to estimate, but is over 100 feet. The Bath Stone, a typical, poorly fossiliferous, oolitic freestone was quarried and worked underground near that city; it is one of the great historic building stones and justifiably gives its name to the whole stage.

The Bradford Clay is much thinner and marks an incursion of black muds. The bed is famous for the well-preserved specimens of *Apiocrinus* which grew on the limestone floor and are found in place in the clays above; there is also an abundance of brachiopods. The Forest Marble—a name derived from Wychwood Forest in Oxfordshire—consists of shelly, current-bedded limestones with intercalated clays.

In the transition to the thicker, more argillaceous Bathonian sequence of south Somerset and Dorset a broad correlation can be made by a combination of lithological and fossil marker horizons. The Forest Marble at the top continues southward and thickens somewhat—a group of shelly and sandy limestones with clays. The '*boueti* bed', rich in the rhynchonellid *Goniorhynchia boueti* and *Apiocrinus* ossicles, recalls the Bradford Clay. The Fuller's Earth Rock of the south is an argillaceous limestone that can be traced northwards as far as Bath at the base of the Upper Fuller's Earth Clay. By these means it appears that the Great Oolite limestones of the Cotswolds correspond to the upper part of the Fuller's Earth Clay farther south. Thus during the Bathonian period the Wessex basin was once more a distinctive entity, in which was laid down a sequence both thicker and more muddy than that to the north, and which was in strong contrast to the more typical limestone shoal deposits that prevailed from Bath through the northern Cotswolds and into Oxfordshire.

Oxfordshire to Lincolnshire

The Great Oolite suffers less attenuation over the Oxfordshire shallows than do the earlier Jurassic formations. The last remnant of the Fuller's Earth Clay at the base disappears and is replaced by limestones. The Stonesfield Slate at the base of the Great Oolite Limestone is a facies largely peculiar to north Oxfordshire. It is a fissile, sandy, oolitic limestone with a local use as roofing material, and as such not unlike the Collyweston Slate; but in addition to a normal shelly fauna it contains a remarkable assemblage of fishes, bones of reptiles (including dinosaurs) and rare, early mammals. A few plants are also known. Presumably the terrestrial components were washed in gently from the neighbouring land, probably the London Platform.

In Northamptonshire and Lincolnshire the Great Oolite Limestone maintains its name and lithology but new formations appear below and above:

Great Oolite Clay
Great Oolite Limestone
Upper Estuarine Series

The Upper Estuarine Series is composed dominantly of grey or lavender clays and is thus finer in grain than the Lower Estuarine Series. The clays contain several rootlet beds and the whole represents the colonization of the banks of the Lincolnshire Limestone by the vegetation of coastal swamps and marshes. There is little evidence of channelling but a minor marine invasion allowed the temporary establishment of a marly oyster bed in the middle of the series.

The Great Oolite Limestone is a return to calcareous marine deposits, but possibly in a rather more open sea than that of the Lincolnshire Limestone. The rock is similar but with more skeletal debris and fine calcite mud, and fewer oolitic layers. This calcareous phase was terminated by a fresh influx of mud. The Great Oolite (or Blisworth) Clay is variegated in colour and poor in fossils, though a few oyster beds are found; rootlet beds are also lacking. It may represent an enclosed basin or lagoon, breached only temporarily by the open seas, bringing in the oysters. All the Bathonian is thin in this basin and three divisions described together only amount to about 100 feet, the Upper Estuarine Series being the thickest.

THE DELTAIC SERIES OF YORKSHIRE

Although there were several marine incursions among the deltaic beds of the Yorkshire basin, these only very rarely brought in ammonites and their main faunas, of lamellibranchs, are poor substitutes in matters of correlation. Consequently a detailed equivalence with the Bajocian and Bathonian formations of the Midlands and south is not possible. Ammonites of the *opalinum* zone are found in the Dogger and, very rarely, those of the *humphriesianum* zone in the Scarborough Beds; the Bathonian is probably represented by the Upper Deltaic Series alone, for the Lower Cornbrash is absent. The whole series is admirably exposed on the cliffs of north-east Yorkshire and the deltaic sandstones form the high ground of the moors inland. The Howardian Hills at the east end of the Vale of Pickering extend the Middle and Upper Jurassic outcrops farther south in the direction of Market Weighton.

Table 48. *Middle Jurassic rocks of Yorkshire*

Lithological formations	Maximum thickness (feet)
(Upper Cornbrash)	
Upper Deltaic Series	225
Scarborough Beds	104
Middle Deltaic Series (including Millepore Bed, etc., in the south)	253
Ellerbeck Bed	20
Lower Deltaic Series	170
Dogger	40
(Upper Lias)	

The Marine facies

As would be expected from their general geographical setting the marine intercalations tend to thicken to the south and west, where they become substantial marine invasions of the delta complex. In rock-type they are dominantly sandy, calcareous and ferruginous, and are locally rich in a shallow water fauna.

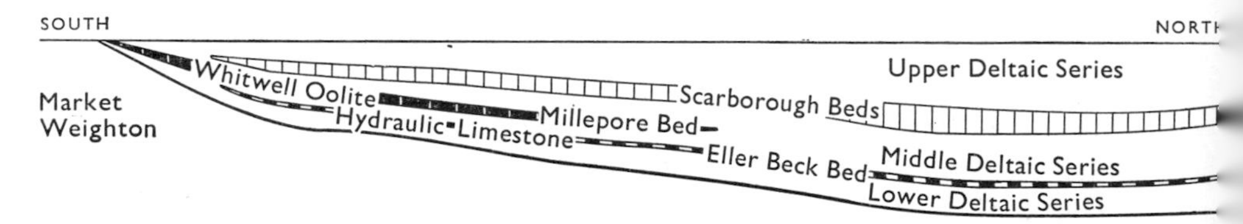

Fig. 59. The Deltaic Series of Yorkshire, showing the extent of the marine formations, excluding the Dogger. (Redrawn from Black, 1934, p. 263.)

The Dogger at the base is the most variable and its changes in thickness are related to the magnitude of the unconformity on the Upper Lias below. At Blea Wyke on the east, or downthrow, side of the Peak Fault (p. 283) the Dogger consists of 40 feet of ferruginous sandstones and chamosite oolites which contain an abundant shelly fauna and occasional corals. The junction with the Yeovilian beneath is an inconspicuous erosion level with pebbles and rolled fossils. On the west side of the fault and in the remainder of the coastal outcrops north-west to Whitby, not more than 12 feet are present, lying with a major unconformity on the Whitbian Alum Shales. Over north-east Yorkshire as a whole further facies of the Dogger are known, including sideritic sandstones, chamosite oolites and black shales. Some facies only occupy hollows and channels cut into the beds beneath and contain phosphatic pebbles at the base. It appears that the Upper Lias was gently uplifted, folded and locally eroded, and then the varied, but dominantly ferruginous, deposits of the Dogger were laid down on an uneven surface in a series of very shallow marine basins or lagoons.

The two succeeding marine incursions are less important. The Ellerbeck Bed of the central moorlands consists of ferruginous sandstones and shales; in some places it is crowded with the casts of lamellibranchs and occasionally includes thin ironstones. To the south and west, however, it passes into grey argillaceous limestones, the Hydraulic Limestone. The Millepore Bed, unlike the others, does not cover the full extent of the outcrop; in its typical, coastal sections, south of Scarborough, it is a calcareous sandstone, rich in the 'millepore' (i.e. polyzoan) *Haploecia*. The 30-foot Whitwell Oolite of the Howardian Hills is a lateral equivalent in a very different facies. It represents a northward extension of a normal Bajocian limestone, and in containing *Acanthothiris crossi* can be correlated with part of the Lincolnshire Limestone.

The Scarborough Beds are the thickest of these marine incursions, and reach 100 feet in the centre of the basin, between the Peak Fault and Scarborough itself. The rocks include sandstones, shales and impure limestones and are often richly fossiliferous.

The Deltaic facies

The Middle Jurassic deltaic beds have much in common with the Coal Measures, but although coal seams up to two feet in thickness are known they are always poor in grade. Clays and fine silts are dominant; they very rarely contain animal remains (the freshwater mussel *Unio* is an exception) but there are several rich plant beds. These include spreads of the horse-tail *Equisetum*, which were commonly the first to colonize the delta flats and which are often represented by rootlet beds even where the stems have been washed away. There are also drifted remains of ferns, conifers, gingkos and cycad-like plants. No stratigraphical sequence can be derived from the macroscopic plants but the spores do show some change through the succession.

The sandstones of the Deltaic Series are often lenticular and form washouts or channel-fillings which can be seen cutting down as much as 20 feet into the beds beneath. Some of those at the base locally replace the Dogger and even cut into the Whitbian shales below, and detritus from these forms a pebble bed of ironstone and calcareous nodules at the bottom of the channel. There is another group of washouts in the middle of the Upper Deltaic Series and isolated examples at other levels. The channels were formed by the delta distributaries as they wandered across the topset beds already formed, and which they dissected in times of gentle uplift. Broad spreads of current-bedded sandstones are not common in this deltaic facies, but the Moor Grit, a silica-cemented sandstone at the base of the Upper Deltaic Series, consists dominantly of foreset beds, building up the delta in its last phase, after the marine invasion of the Scarborough Beds.

The Middle Jurassic beds of this basin become thinner as a whole southwards, in the Howardian Hills, and disappear well north of Market Weighton. It is at this period in particular that the absence of a complete barrier over that upwarp is demonstrated. Just by what route the seas reached the Yorkshire delta we do not know, but intermittently there was a free passage from the south. The converse facies change, however, represented by the non-marine 'Estuarine Series' of Lincolnshire and Northamptonshire, need not have been a simple southward extension from Yorkshire; it is at least as likely that it was related to the London landmass. The palaeogeographic details are not decipherable as yet, particularly as we are largely restricted to relatively narrow north–south outcrops in eastern England.

Another embarrassing gap remains in our synthesis of Jurassic geology at this time—whence came the detrital components of the Yorkshire delta complex? There is no obvious source in Britain, and what little can be deduced of northern England and Scotland does not suggest a land mass capable of supplying this amount of detritus. On rather negative grounds a north-easterly source is possible, but little is known of Jurassic history in this part of the North Sea or in Scandinavia.

THE CORNBRASH, MIDDLE AND UPPER JURASSIC

'Cornbrash' was one of William Smith's original terms, for a thin, rubbly, brown-weathering bed of limestone and marl that overlies the heterogeneous components of the Great Oolite, from almost one end of the main outcrop to the other. It commonly forms a flattish belt of land suitable for arable crops.

The name therefore has a respectable antiquity as a formation or rock unit, but when the faunas and detailed stratigraphy are studied certain anomalies appear and the uniform outcrop of the small-scale map turns out to be rather misleading. Ammonites reappear in the Cornbrash after their extreme rarity in the Great Oolite beds, but those of the upper and lower divisions are widely dissimilar. *Clydoniceras* of the

Lower Cornbrash has Bathonian affinities, and a new, Callovian genus, *Macrocephalites*, enters in the Upper Cornbrash. As a result the lithological formation, either half of which may be only a few feet thick, overlaps two major Jurassic stages. In its slender way the Cornbrash illustrates neatly the progress of stratigraphical definition from lithological to palaeontological criteria.

A study of the variation in the Cornbrash across country resolves itself, in more detail than is appropriate here, into thickness and minor facies-changes of the lower and upper divisions. The whole must have been laid down in extremely shallow water conditions comparable with those of the earlier limestones, with much local erosion and non-sequence. The thickest succession is that of the Wessex basin, as it is exposed in the Dorset cliffs. Of the 30 feet here one-third belongs to the lower division and two-thirds to the upper. From a reduction in Bedfordshire to a total of three feet, the outcrop extends to the Humber as only a few feet of rubbly limestones, nearly all Upper Cornbrash. There is then a gap of 50 miles over the Market Weighton upwarp, and the upper division alone is known in the Yorkshire basin. It appears here in a slightly different lithology, as a grey or purple, hard, shelly limestone, which represents the final submergence of the Middle Jurassic delta by shallow seas.

CALLOVIAN AND OXFORDIAN: KELLAWAYS BEDS, OXFORD CLAY AND CORALLIAN BEDS

The boundaries both above and below the Upper Jurassic in Britain are awkward, and exemplify some of the troubles that follow on the necessary attempts to fit traditional or lithological groupings into an internationally agreed system. The rocks and faunas are described here under the English lithological headings, whose general relations to the international stages are as follows:

English lithological headings	International stages
Purbeck Beds (in part) Portland Beds	Portlandian
Kimeridge Clay	Kimeridgian
Corallian Beds	Upper Oxfordian
Oxford Clay	Middle Oxfordian Lower Oxfordian Upper Callovian
Kellaways Beds Upper Cornbrash	Lower Callovian

Kellaways Beds and Oxford Clay

After the widespread transgression of the Cornbrash, there was in southern and central England a relatively brief phase of clays and sands (the Kellaways Beds) and then a long period of muddy seas, producing the Oxford Clay. The Kellaways Beds commonly do not exceed 20 feet, though as much as 60 have been reported from Wiltshire. The lower part, the Kellaways Clay, is rather poor in fossils, and is succeeded by sands or calcareous sandstone, the Kellaways Rock, which is locally rich in the zonal ammonite, *Sigaloceras calloviense*. This facies is rather irregularly developed, however, and as a result, so is the base of the clay facies above.

The Oxford Clay is the lowest of the great clay strata that dominate the Upper Jurassic of southern and eastern England. Its relatively unresistant nature renders the outcrops often low-lying or marshy and it forms a belt of such country to the south and east of the Cotswold uplands. In the east Midlands, for instance around Peterborough, vast pits have been dug in the Oxford Clay for brick-making (particularly in the lower and middle parts), but even so the absence of complete sections makes the thicknesses over the outcrop as a whole difficult to estimate. There is probably about 500 feet in Dorset and an increase to 600 feet in Wiltshire; the customary reduction northwards into Lincolnshire (300 feet) follows and only 50 feet survive north of the Humber.

In Upper Jurassic times the Market Weighton upwarp was the only one of the earlier structures to exert a profound influence on sedimentation. Nevertheless, on the north-western edge of the London Platform there is a broad, positive area, centred on Bedfordshire and Buckinghamshire, where the Kellaways Beds are reduced and upper zones of the Oxford Clay missing. This element can be considered as part of the same, unstable region as the Oxfordshire shallows, but its manifestation inevitably shifts eastwards because the outcrops of the younger formations succeed one another in that direction.

The faunas of the Oxford Clay are well known because the brick pits have afforded exceptional opportunities for collecting, and it is interesting to note how they recall those of the previous great clay and shale formation, the Lias. Ammonites, lamellibranchs and belemnites are abundant, with the broader *Gryphaea dilatata* replacing *G. arcuata*. Marine vertebrates are similarly represented by ichthyosaurs and plesiosaurs and a variety of fishes. Terrestrial dinosaur bones have been found, a reminder that even the muddy formations were laid down not far from land.

In spite of the uninspiring drab appearance of the clays, a detailed study shows that they are not homogeneous throughout. Some beds are more sandy or pyritic or contain calcareous concretions; the fossils are often concentrated in certain layers and some of the clays are finely laminated. As in the Lias there were minor differences in the incoming detritus, in the gentle bottom currents and in chemical conditions of the bottom muds—all of which produced minor modifications in the environment, affected the deposits and faunas and influenced later diagenesis. In the Yorkshire basin there were further, local variations in sedimentation. On the coast the Cornbrash is overlain by sandstones which amount to 100 feet; the lower part corresponds to the Kellaways Rock but the ammonites of the upper part (the Hackness Rock) are those of the lower Oxford Clay of the Midlands and south of England. The clay facies above is substantially thinner in compensation and does not exceed 150 feet.

Corallian Beds, Dorset to Oxford

Although varied and interesting the Corallian limestones are a relatively thin intermission in the thick series of Upper Jurassic clays. In Dorset they amount to some 200 feet, intervening between the Oxford and Kimeridge clays which together are ten times that thickness. On the coast they are well exposed east of Weymouth and form three cycles, passing upwards from shales to calcareous sandstones and to shelly, marly or oolitic limestones. Of these the clays and the Osmington Oolite (60 feet) are the thickest.

3rd cycle	Ringstead Coral Bed Sandsfoot Grit Sandsfoot Clay
2nd cycle	*Trigonia clavellata* Beds Osmington Oolite Bencliffe Grit Nothe Clay
1st cycle	*Trigonia hudlestoni* Beds Nothe Grit (Oxford Clay, upper part)

The 'grits' in this succession are sandstones, sometimes ferruginous or calcareous and often current-bedded. The fossils include abundant lamellibranchs, as in the *Trigonia* beds, some ammonites, and corals at the top.

The group can be traced inland with minor variations until, like all the succeeding Jurassic strata, it disappears under the Cretaceous overstep of north Dorset. In Wiltshire in the Vales of Wardour and Pewsey some new facies appear, including the striking Steeple Ashton Coral Bed at the top of the second cycle. This grew on a foundation of current-bedded oolites and in addition to the corals includes a specialized echinoid fauna. Another coral bed develops at the top of the first cycle in north Wiltshire. The corals here, as in the reefs of Oxfordshire and Yorkshire, are principally *Isastraea* and *Thamnasteria*. In addition to the actual coral colonies the erosion of the reefs produced spreads of coral and shelly detritus in a loose rubbly limestone; the two facies combined are known as Coral Rag.

The Corallian beds are particularly well seen in the horseshoe of low hills that border Oxford on the west, south and east, the centre and north of the city being built on the Oxford Clay. There are three main facies, the Lower Calcareous Grit, resembling that of Dorset and Wiltshire, which is overlain by Coral Rag and the Wheatley Limestones. The last two are not in regular superposition, but replace one another laterally. The Coral Rag includes colonies growing in place, as patchy reefs, together with a specialized fauna of lamellibranchs and echinoids. Ammonites are conspicuously absent, as are the trigonias, which are common in other types of Corallian limestones. The detrital Wheatley Limestones were formed from the debris of corals and shells, together with some ooliths. They may overlie or underlie the reef spreads, or occupy channels between them; they also have an indigenous specialized fauna, different from that of the reefs, of echinoids, small oysters and *Exogyra*.

Despite its impressive richness and variety of facies the Oxford Corallian corresponds to only the middle and lower parts of the Wiltshire and Dorset successions, not rising above the Osmington Oolite. The Kimeridge Clay followed after a period of uplift and erosion.

The Ampthill Clay

The impressive exposures around Oxford are almost at the margin of the calcareous Corallian typical of southern England. Ten miles away to the north-east a totally different facies prevails—the black or dark-grey, tenacious Ampthill Clay. The Wheatley Limestones of Oxford diminish, become more clayey and then pass into the slightly calcareous basement facies of the clays. In Bedfordshire and Cambridgeshire this basal bed is the Elsworth Rock, a soft-weathering, cream-coloured, iron-shot oolitic limestone; in the former county it lies non-sequentially on the Oxford

Clay, the upper zones of which are missing, together with those of the Lower Calcareous Grit.

The main mass of the Ampthill Clay is very badly exposed and its thickness rather uncertain; 50 feet have been seen in Buckinghamshire, and the total in this region may be of the order of 100 feet. There is certainly some expansion in Lincolnshire where 200 feet are recorded. This incursion of muds into the Corallian seas has its effects on the present topography of the east Midlands. The slight belt of uplands formed by the limestones and sandstones around Oxford fails to the north-east, and the flat, ill-drained lowlands of the Oxford, Ampthill and Kimeridge clays have no such minor relieving features. Here lie the lower reaches of the rivers Ouse, Nene and Welland, and parts of the East Anglian and Lincolnshire Fens.

A very small exception in this widespread stretch of clays is the reef rock of Upware, ten miles north-east of Cambridge (Fig. 71), where a low mound of oolitic limestones and Coral Rag rises from the Elsworth Rock. It is overlain on the flanks by Ampthill Clay and presumably was a short-lived attempt to establish reef growth in unfavourable surroundings. The clay facies continues north to the Humber, but between that river and Market Weighton there is a mixed development of both limestones and clays. As in earlier periods the structure was no hard and fast line of facies demarcation.

Corallian limestones and reefs of Yorkshire

Coralliferous and other limestones, interbedded with sandstones, established themselves successfully in the shallow seas of the Yorkshire basin and their sequence can be described in terms comparable to those of southern England:

Upper Calcareous Grit
Osmington Oolite Series
Middle Calcareous Grit
Hambleton Oolite Series
Lower Calcareous Grit

The similarity does not imply strict contemporaneity, however. The Corallian beds of Yorkshire are nearly twice as thick as those in the south and the Lower Calcareous Grit belongs to the *mariae* zone, or well down in the Oxford Clay of Oxford. The base of the Corallian is thus diachronous, the initial sandy phase, and the clearing of the seas that it represents, appearing earlier in the north. The Osmington Oolite is about the same age as its Dorset namesake, but the Upper Calcareous Grit is probably older than the uppermost beds of the south.

The Corallian beds dip southwards off the north-east Yorkshire moors, under the Kimeridge Clay and alluvium of the Vale of Pickering (Fig. 56). The Lower Calcareous Grit forms a conspicuous scarp above the softer Oxford Clay. It is rich in the ovoid, siliceous spicules of the sponge *Rhaxella*, which in places were sufficiently abundant to be an important component in the rock; their replacement by calcite released substantial amounts of silica contributing to the formation of chert. The Lower Calcareous Grit is also the most extensive of the three arenaceous phases, and on this sandy substratum there developed the coral-sponge and oolite associations of the Hambleton beds.

The Osmington Oolite is the more widespread of the two limestone groups and in the Coral Rag facies at the top there is discernible a reef-belt in the north-westerly parts of the outcrop, with associated fore-reef and back-reef developments. In the Howardian Hills to the south these are replaced by bioclastic limestones. The faunas of both oolitic and reef sediments are alike in Dorset, Oxford and Yorkshire, but the facies of the last are on a more massive scale; they nevertheless show much lithological variation and are comparable to the shallow water carbonate associations of the present day.

KIMERIDGIAN AND PORTLANDIAN: KIMERIDGE CLAY, PORTLAND AND PURBECK BEDS

During late Jurassic times, or approximately at the beginning of the Portlandian stage, the minor uplifts that had already caused reduced or intermittent sedimentation in the southern Midlands swelled to form a broader shallow region that extended north-westwards from the London Platform. Deposition was concentrated in a southern province, the Wessex and Wealden basins, and in a north-eastern, from Norfolk to Yorkshire. Marine conditions continued in the latter, though their exact history is not very clear. In the south, however, the seas retreated, and during a long period, which spans the Jurassic-Cretaceous boundary, non-marine facies predominated. Ammonites are consequently absent from the Purbeck and succeeding Wealden beds and the supplementary groups (ostracods, lamellibranchs and plant spores) have not as yet provided the same authoritative correlation.

In this book the Purbeck Beds are included in the Jurassic System—a time-honoured practice but one probably needing revision. It is more likely, by analogy with the Lincolnshire sequence and those on the continent, that the base of the Cretaceous should be taken lower down, perhaps at the Middle Purbeck 'Cinder Bed' (p. 307).

Kimeridge Clay

The onset of marine muds at the beginning of Kimeridgian times brought in a period of exceptionally uniform sedimentation in the Jurassic shelf seas of Great Britain, a period perhaps associated with slight downwarping of the sea floor and neighbouring lands. Even so it is highly improbable that the term 'deep water' can be applied to this or any other British Jurassic formation and certain positive regions, notably Market Weighton and the south-east Midlands, still exercised some effect and there were many minor facies variations.

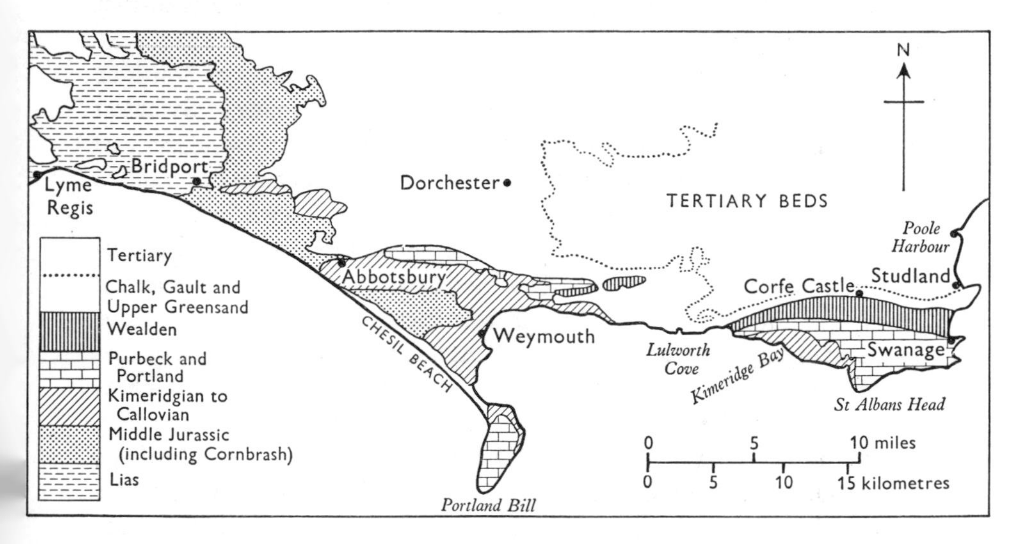

Fig. 60. Sketch map showing Jurassic outcrops of south Dorset and the Cretaceous overstep. Detailed outcrops affected by the Purbeck folding (e.g. of Lulworth Cove, Worbarrow Bay, etc.) cannot be shown, but some appear in Pl. 6. The small amount of Lower Greensand in the east is included with the Wealden Beds.

The thickest sequences of Kimeridge Clay come from Wessex and the Weald; 1,773 feet are recorded from the latter and near the type locality of Kimeridge Bay in east Dorset 1,650 feet are known from combined cliff and borehole sections. Even in the Wessex basin this figure is exceptional, however, and 10 miles away to the west, at Ringstead, there is less than half this amount. Kimeridge Clay also appears between Weymouth and Portland and in its relative softness has been eroded in the stretch underlying the Chesil Beach that separates the 'island' from the mainland.

Ammonites are abundant throughout the clays, together with lamellibranchs (especially oysters), *Lingula* and rhynchonellids. Reptiles and fishes are known, though not so commonly as in the Oxford Clay, but this may largely reflect the fewer artificial exposures. Much of the clay is bituminous and on the Dorset coast an oil shale seam of 3 feet was once known as Kimeridge Coal. There are several thin impure limestones, known as 'white bands' or 'white stone bands'. At the western end of the Dorset outcrops the Abbotsbury ironstone at the base is an oolitic rock with a sandy matrix. The greater variability of the Kimeridge Clay, and in particular the carbonate components, make it less suitable for brick-making than the Oxford Clay and it lacks the large inland pits of that formation.

There are broad outcrops at the western ends of the Vales of Wardour and Pewsey and the Kimeridge Clay reappears again south of Swindon and continues through Oxfordshire into Bedfordshire. At Swindon there is only about 300 feet of Kimeridgian beds, nearly all clays, but a series of sands and sandstone interrupts the upper part. At Oxford the thickness is reduced to half and the lowest clays lie unconformably on the Corallian limestones. The Shotover Grit Sands of the Upper Kimeridgian (*pectinatus* zone) are outstanding in this succession—yellow calcareous sands with large concretions—but from Bedfordshire north-eastwards the formation is for a stretch overlapped by the Cretaceous Lower Greensand and Gault. Borings in Cambridgeshire have proved over 125 feet, but the upper zones are missing. The clays also underlie much of the fens on either side of the Wash; although soft as 'solid' rocks go they are more resistant than the peat and alluvium and are responsible for many of the fen islands, which were the sites of early habitation.

The clays in Lincolnshire are thicker, and though the break at the top persists, 320 feet are known in the south of the county, including thin seams of oil shale in the upper part. The last traces are found on the northern shores of the Humber under the Chalk. North of the very long gap over Market Weighton, the Kimeridgian of Yorkshire is very badly exposed indeed, its broad outcrop having been planed down to form the Vale of Pickering. From such information as can be gained it appears that the clays are mainly of Lower Kimeridge age. In spite of this they are thicker (400 feet or more) than anywhere in the Midlands or east of England.

The distribution and thickness of the Kimeridge Clay follows the general picture of Upper Jurassic times in spite of the broadly uniform muddy facies. In the north there is the Yorkshire basin, and the lesser one of Lincolnshire; in the south those of Wessex and the Weald. Between lay an unstable tract from north Wiltshire to Cambridgeshire with reduced thicknesses, non-sequences and some sandy invasions.

Portland Beds of the Dorset coast

All the inland outcrops of the Portlandian strata are incomplete and meagre compared with the superb coastal exposures on the Dorset coast. If a geological map of that county shows a lesser area of Portland Beds than of any other formation, a view of the cliffs gives a very different impression. Of all the Jurassic limestones the Portland Stone is the most massive and resistant to wave action. It builds the bastion of Portland Bill, the cliffs of the Isle of Purbeck, and the two breakwaters, or 'horns', that almost enclose Lulworth Bay. In the first two the beds form part of the gently dipping southern limb of the Weymouth (or Purbeck) anticline; at Lulworth they are part of the very steep northern limb and are in places almost vertical.

Like the clays below, the Portland Beds are thickest in the Isle of Purbeck and also best exposed. Here there are 240 feet, comprising the Portland Sand below and the slightly thicker Portland Stone above. Lamellibranchs and ammonites are the chief fossil groups, the giant ammonites of the limestones being especially famous (*Titanites* itself comes from the freestone at the top). Rather surprisingly corals are absent, in contrast to the earlier Jurassic limestones.

There is a gradual transition in facies from marly beds at the top of the Kimeridge Clay. The sand proportion increases upwards, but some mud remains; the Portland Sand when fresh is dark-grey to black and there is a repetition of the cementstone bands. The Portland Stone comprises a considerable range of limestones: shelly rocks, calcite mudstones, white limestones with large chert nodules and seams, the Roach (a massive rock made porous by the hollow moulds of molluscs), and a fine oolitic freestone with comminuted shells. Only the last is the Portland Stone of commerce; the lower beds are called the Cherty Series and have no economic value; the upper beds are the Freestone Series. The quarries on Portland Bill in particular have supplied vast quantities of stone for solid masonry in southern England since Wren used it in the rebuilding of St Paul's Cathedral.

The threefold sequence—Kimeridge Clay, Portland Sand and Portland Stone—is one of the best examples of this type of repetition (clays—sands—limestones) among the Jurassic formations. Comparable examples have long been recognized, such as those in the Corallian beds of the south or the sequence from Upper Lias to Northampton Sands and Lincolnshire Limestone. The scale, and hence duration, of the successions varies enormously and some are only local, so that perhaps the term cyclic is not wholly appropriate. Nevertheless, it does appear that during the clearing of the seas and the establishment of carbonate shoals there was commonly an intervening arenaceous phase; or that if no distinct sandstone is present the basal beds of the limestone are decidedly sandy. The process is perhaps related to a decrease in depth (caused by lesser subsidence?) and an increase in currents.

Purbeck Beds of Dorset

These beds are exposed in the cliffs of the Isles of Purbeck and Portland and overlie the Portland Stone inland. They also contain building stones, in the Middle Purbeck, and a decorative limestone (Purbeck Marble) near the top. Both were worked in shallow shafts near Swanage. Although not so impressive in exposures as the Portland Beds beneath they are somewhat thicker, reaching 400 feet near Swanage. There is a profusion of rock types, mostly fine-grained, argillaceous or calcareous; the majority were laid down in very shallow waters, occasionally marine but more frequently of lakes, marshes or coastal lagoons, both fresh and saline. There were also periods of emergence which left their traces in a terrestrial flora and fauna and in remnants of soils. No ammonites are found, not even in the fully marine incursions, and the Lower, Middle and Upper Purbeck divisions are based upon the ostracod faunas—these small crustacea fortunately inhabiting both fresh and salt waters.

In Dorset the change from the marine freestones at the top of the Portland to the basal Purbeck beds is conformable, but the seas retreated and a swamp forest grew on the new land surface. The dark brown 'dirt beds' are interpreted as fossil soils and sometimes contain the remains of trees—cycads and conifers; in the 'fossil forest', east of Lulworth Cove, tree stumps and fallen logs are surrounded by tufa and the wood fibres are preserved in silica. In the rather curious, traditional terminology of the Purbeck rocks the Broken Beds above refer to the collapse structures following the solution of thin evaporite beds and the Caps are algal and tufaceous limestones; the early Purbeck land surface was thus re-submerged by muddy, saline lakes. The succeeding clays have traces of gypsum and halite, but also contain shells. Another local emergence is marked by the Mammal Bed, at the base of the Middle Purbeck, which is a lenticular dirt bed, famous for its bones of early mammals and reptiles, though these are no longer found. The former were small insecti-

Table 49. *Purbeck Beds near Swanage*

UPPER PURBECK	*Viviparus* Clays Marble, Shell and Ostracod Beds *Unio* Beds Broken Shell Limestone	72 feet
MIDDLE PURBECK	Chief Beef Beds *Corbula* Beds Upper Building Stones Cinder Bed Lower Building Stones Mammal Bed	157 feet
LOWER PURBECK	Marls with gypsum and insect beds Broken Beds and *Cypris* Freestone Caps and Dirt Beds	169 feet

vores or arboreal forms; the dinosaurs, turtles and crocodiles suggest a climate distinctly warmer than that of the present day, a deduction reinforced by the late Jurassic and early Cretaceous floras.

The Middle Purbeck Beds represent the most extensive marine invasion. The Building Stones are compact grey limestones with occasional well-preserved fishes and a variety of molluscs, some marine. The Cinder Bed (a rather absurd term) is a thin dark-grey layer that is crowded with small oysters and also contains, much more rarely, the echinoid *Hemicidaris*. Limestones and shales above with *Neomiodon* mark a return to fresher waters; here and elsewhere 'shales with beef' refer to thin layers of fibrous calcite. The Upper Purbeck Beds consist very largely of freshwater clays and limestones, containing *Unio* and the small gastropod *Viviparus*, the latter being conspicuous in the mottled limestones of the Purbeck Marble. At the top the *Viviparus* Clays are divided only arbitrarily from similar rocks at the base of the Wealden beds.

Portland and Purbeck Beds, Wiltshire to Buckinghamshire

Along this stretch, round the base of the Chalk scarp bordering the Wiltshire and Berkshire downs and the edge of the Chilterns, the highest Jurassic rocks are usually the Kimeridge Clay. Later beds appear, however, in the Vales of Wardour and Pewsey, in a small syncline near Swindon, and in a more extensive series of outliers (chiefly Portland but some capped with Purbeck) from the eastern outskirts of Oxford to Whitchurch, north of Aylesbury. Throughout these outcrops the beds are much thinner than in Dorset, as would be expected from their position on the northern edge of the Wessex basin, or, to put it another way, on the western fringes of the London Platform. Together with the reduction in thickness there is often a relative increase in the littoral sandy or pebbly components.

The nearest, and most similar, succession to that of Dorset appears in the Vale of Wardour where the 115 feet of Portland Beds contain two building stones. All three divisions of the Purbeck Beds, which total 65 feet, are recognizable from their ostracod faunas, as are the marine beds in the middle, and dirt beds and fossil wood of the lower division. In the Swindon syncline there are 40 feet of Portland Beds, but unequally divided, the equivalent of the Portland Sand being reduced to 4 feet of sandy, glauconitic limestones with a basal bed containing black chert ('lydite') pebbles and phosphatic fossils from the Kimeridge Clay, on which it rests non-sequentially. The beds above consist of two limestones separated by the 20 feet of the Swindon Sands. There is locally an erosion level beneath the Purbeck Beds, which comprise 20 feet or so of marls, limestone and further pebble beds and include both marine and freshwater fossils.

From Oxford to Aylesbury the Portland succession is similar to that at Swindon, interbedded limestones and sandstones, but thinner. Purbeck beds are rarer, since only a few capping outliers survive. Twelve feet have been recorded near Aylesbury of clays, marls and thin beds of limestone. Although they contain ostracods their age is not very certain.

The various outliers in this belt from Swindon eastwards owe their varied successions to a complex combination of events, in a region of reduced and intermittent sedimentation between northern and southern basins. After the relatively tranquil period of the Kimeridge Clay there was local uplift and erosion. Then followed the deposition of Portland Sand and Portland Stone; then another erosive phase and the deposition of a rather uncertain amount of Purbeck Beds. The outliers are overlain, at one place or another, by different divisions of Lower Cretaceous rocks, themselves also mutually unconformable. For instance the Shotover Sands that overlie the Portland Beds near Oxford are usually taken to be Wealden; Lower Greensand is found overlying several earlier formations, and Gault in places oversteps it on to the Jurassic rocks. These complicated relations among beds that are themselves attenuated, marginal and poorly fossiliferous make the stratigraphy of these outliers debated and difficult. We may recognize the problem, however, as one of many resulting from the abnormal, reduced sedimentation over this south-eastern Midland region.

THE JURASSIC BASIN OF THE WEALD

The underground structure of the Weald has long been known as the reverse of the surface anticlinorium displayed by the Cretaceous formations, for the Jurassic (which are entirely hidden except for some small Purbeck inliers) are synclinal in form. The Weald was one of the major subsiding basins of Jurassic sedimentation. Not only do all the beds become thinner to the north and north-east, and the younger overlap

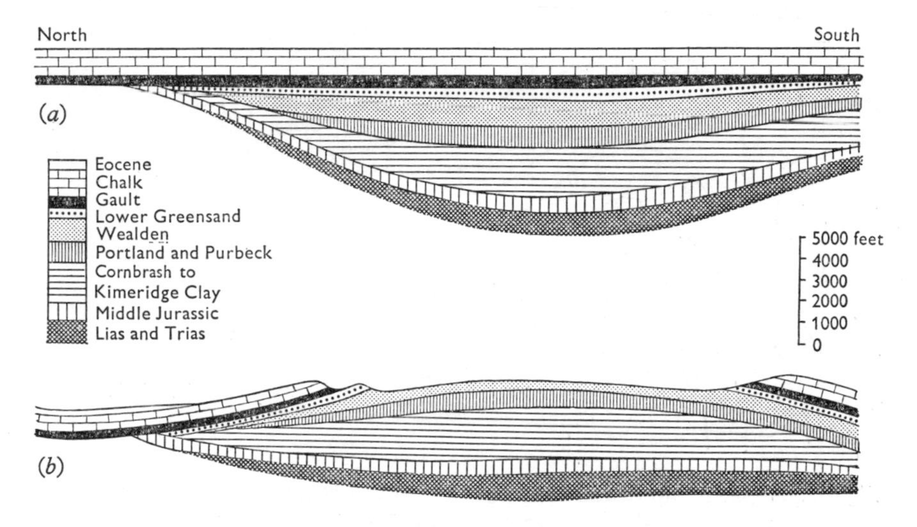

Fig. 61. Sections through the Weald. (*a*) The sagging basin of Jurassic and Lower Cretaceous beds is covered by Gault and Chalk, which transgress northwards over the London Platform. (*b*) Mid-Tertiary folding and subsequent erosion produce the present structure and topography (minor folds and faults omitted). (Redrawn from Gallois, 1965, p. 52.)

the older, as they approach the London Platform and the Kent Coalfield, but there is some indication of thinning to the south as well, towards a ridge or feature somewhere near the present coastline. A similar feature, the Portsdown ridge, partly separated the Weald and Wessex basins. Our knowledge of the underground Jurassic beds is drawn from the shafts and borings of the Kent Coalfield and the more widespread borings for oil exploration in the central and southern Weald. An isopach map for all southern England is shown in Fig. 62.

In general the Jurassic facies of the Weald are similar to those already described in the southern outcrops; from the Lias to the Kimeridge Clay the resemblances are nearest to the Cotswold successions and we need only note the main differences. The Lias tends to be more sandy to the east and south, suggesting a source of detritus in these directions. A substantial limestone formation in the Inferior Oolite reaches as far as east Sussex, but in east Kent a more sandy facies again prevails. From the Great Oolite to the Oxford Clay the rocks again resemble those of Gloucestershire, and the Great Oolite Limestone maintains a thickness of 150–200 feet. The expansion of the Kimeridge Clay to 1,773 feet in the central Weald (boring at Ashdown) is one of the clearest contributions to the greater thickness in the centre of the basin.

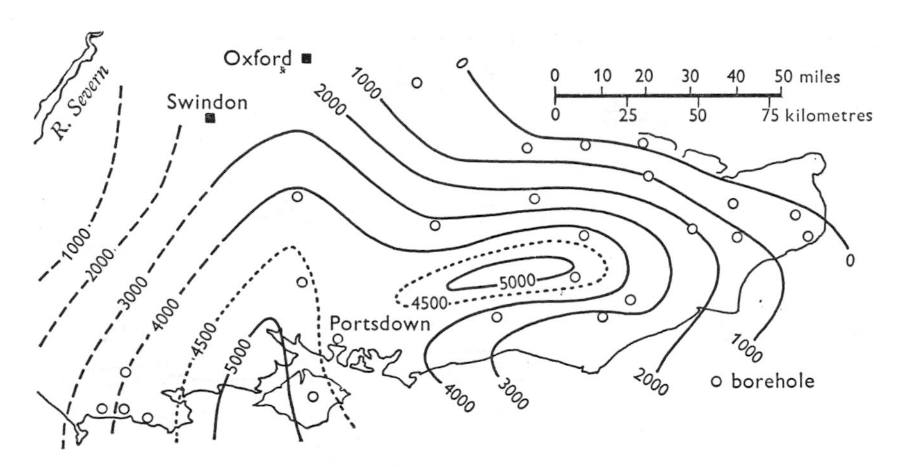

Fig. 62. Isopach map of the Jurassic rocks in the Weald and Wessex basins; thickness interval 1,000 feet. (Redrawn from Howitt, 1964, p. 97.)

The Corallian Beds exhibit thickness and facies changes comparable to those of the main outcrop and could probably be interpreted similarly if more detail were available. Along the northern edge of the basin there is a belt of limestone, including the 148 feet of coral-bearing rocks penetrated in the Warlingham borehole in Surrey; but with the expansion southwards into the central Weald the calcareous facies diminish and most of the 480 feet at Ashdown (an exceptional amount for any English Corallian succession) are argillaceous. There is a reduction to 111 feet over the Portsdown ridge and in late Jurassic and early Cretaceous times thinner sequences are characteristic here, suggesting a positive upwarp, comparable to the better known structures of the main outcrop.

A major change appears in the Portlandian stage, for only the lower, sandy, beds have been recognized and there appears to have been a late Portlandian hiatus or non-sequence over all or most of the Weald basin.

The Lower, Middle and Upper Purbeck beds can be broadly correlated with those of Dorset, with the marine invasion of the Cinder Bed again as the most widespread and reliable marker horizon. The evaporites of the Lower Purbeck are well developed, and exploited, in east Sussex and there is a further sandy influx among the Upper Purbeck clays and limestones.

The Purbeck Beds of southern England reflect a new phase in the geological history of the area. Hitherto a shallow sea had extended southwards from the edge of the London Platform and covered much of northern and central France. The Purbeck emergence transformed the fringes of this sea into coastal marshes and saline lagoons, still subject to occasional marine invasions. In Wealden times a heavier and coarser load of detritus was carried into this downwarp and so developed the deltaic and lagoonal Wealden, or more broadly Anglo-Parisian, Basin—to be an important geographical entity for a long time to come.

SCOTLAND AND IRELAND

The shallow Jurassic seas extended far beyond the main outcrop, into western and north-eastern Scotland and northern Ireland. The relics of their sediments are scanty, but have an importance out of all proportion to their size in showing that beds of all stages from Hettangian to Kimeridgian were deposited at some place and were probably fairly extensive. Between the Midlands and Scotland there are a few outliers of Lias, in Cheshire, Staffordshire and on the Carlisle plain; down-faulted fragments are known from Arran. An outstanding thickness (4200 feet, ? part of the Irish Sea basin) was found in the Mochras borehole, drilled on the coast near Harlech, N. Wales.

The Lias of Antrim is almost co-extensive with the Rhaetic beneath and is found in many narrow outcrops around the Antrim Basalts, especially on the south and east (Fig. 53). For the most part it is thin, not more than 100 feet of shales with thin impure limestones, but at one place or another zones up to *davoei* have been found, representing most of the Lower Lias. Rolled remains of Middle and Upper Lias fossils have been recovered from the basal Cretaceous conglomerates above, and presumably these beds were also fairly widespread. How much of the succeeding Jurassic strata was laid down and later removed is not determinable, but the much more complete sequence in west Scotland suggests that sedimentation may have continued longer in Ireland as well.

In Scotland the ammonite faunas are sufficiently like those of England to indicate free migration between north and south and there are some broad similarities in lithology. As would be expected it is the clay formations, which are the most uniform in England, that extend least changed into the Hebrides and Moray Firth. Oolites and other limestones are diminished and sandstones increased, both suggesting a nearer approach to the sources of detritus, although with one conspicuous exception (in the Middle Jurassic) the formations in the north are still marine.

The Hebridean province

Outcrops of Jurassic rocks are found on many islands of the Inner Hebrides (Fig. 8) and are extensive in northern and eastern Skye and the neighbouring island of Raasay. At the southern end of the latter there are fine sections from Lower Lias shales up to the Great Estuarine Series (Bathonian). The low-lying belt of Strath, between the Red Hills (Tertiary Granites) of Skye and Torridonian country, is largely formed of Lias. Higher beds occur on the peninsula of Strathaird to the west and in Trotternish in north Skye. There is also a small patch of Lias across on the mainland at Applecross.

These comprise the larger northern Hebridean outcrops and there are minor ones on the islands of Rhum, Eigg and Muck. A southern group includes two coastal strips on Mull, others across the Sound of Mull, in Morvern, and yet other small patches on the coast of Ardnamurchan. Nearly all the Jurassic rocks of west Scotland essentially owe their preservation to a cover of Tertiary basalts, which in many places is still present (Pl. 7). Some are also downfaulted against Cambrian or Pre-Cambrian rocks and it is clear that their present outcrops bear no relation to their original extent.

The Jurassic succession falls naturally into three unequal divisions: a thick, marine, lower group (Lias to Inferior Oolite); the Great Estuarine Series; and a thin, incomplete Upper Jurassic group. A composite sequence is given in Table 50, but over the 90-mile stretch from north Skye to Mull the facies and thicknesses show a good deal of variation.

The Broadford Beds of the Lower Lias are dominantly sandy shales, with calcareous sandstones and impure limestones in the lower part. On Skye they contain corals and an oyster bed and rest with a probable non-sequence on red Trias; elsewhere they are also broken by non-sequences and the *planorbis* zone is found only in Mull, and perhaps at Applecross. The Pabba Shales form the small island of that name, east of Skye, and are found at the base of the sections on Raasay, where there are also good exposures in the Middle Liassic Scalpa Sandstone above.

Table 50. *Composite Jurassic succession of the Inner Hebrides*

Formations and lithology	Maximum thickness (feet)
Kimeridge Clay, lowest beds only	20
Corallian Shales	180
Oxford Clay, shales or sandstones	100
Kellaways Beds (?), calcareous sandstones	25
Cornbrash (Lower)	23
Great Estuarine Series (chiefly Bathonian) shales, thin limestones and sandstones	600
Inferior Oolite (Bearreraig Sandstone)	600
Upper Lias, shales	77
Middle Lias, Scalpa Sandstone	245
Lower Lias { Pabba Shales	700
Lower Lias { Broadford Beds	420
(Rhaetic, Trias or older rocks)	

The Upper Lias bears a greater resemblance to the Yorkshire facies than do the lower divisions, and includes shales with jet, calcareous concretions and pyrites. The oolitic Raasay ironstone also resembles the southern ores in that it contains chamosite and siderite, but it lies at a slightly different horizon, in the *bifrons* zone. Only Whitbian, not Yeovilian, zones have been found in this region and there is probably a non-sequence below the rather condensed basal zones of the Inferior Oolite.

The Inferior Oolite is particularly thick in the north; on Raasay the 600 feet consist largely of sandstones, which are coarse and current-bedded, forming some of the finest Jurassic coastal scenery in Britain. There is a marked southward reduction and limestones appear in Mull and Ardnamurchan. Despite the difference in facies the ammonite faunas are distinctly similar to those of the Cotswolds, and include forms from the *opalinum* to the *garantiana* zones.

The name 'Great Estuarine Series' is no more appropriate in Scotland than is 'Estuarine Series' in Lincolnshire, nor is there any clear link with the nearer contemporary non-marine facies of the Upper Deltaic Series of Yorkshire. It is true that the beds contain sandstones with plants, but there are also limestones and shales with freshwater shells and ostracods, fishes, reptiles, *Estheria* and algae. In the upper part marine incursions produced oyster beds. The whole is much more varied in sediments and fauna than the Deltaic Series. The beds are thickest (600 feet) in north Skye and have a poor coal at the base, a horizon represented on Raasay by an oil shale. There is again a southward reduction, seen for instance on Raasay itself, Eigg and Muck. The mineral assemblage of the Great Estuarine sandstones is consistent with derivation from the rocks exposed at present in the central and northern Highlands.

Cornbrash fossils, of the lower division, have been recognized only on Raasay, in beds that are probably equivalent to the top of the Great Estuarine Series of north Skye. Kellaways Beds and Oxford Clay are also seen in that peninsula, but the former lack the ammonites for certain correlation. The latter is followed by more shales with cementstones; most of these are Corallian but the uppermost contain basal Kimeridgian fossils. The Tertiary lavas lie above and there is no further information about Jurassic history in the west of Scotland.

In spite of non-sequences, variations in facies and difficulties of correlation, some generalizations can be made about this fascinating northern province of Jurassic sedimentation. The thickest and most continuous sequences are not towards the south, in the direction of the English basins of deposition, but northwards towards the Minch and the entirely unknown. In north Skye over 3,000 feet were laid down, up to early Kimeridgian times. In England the Kimeridge Clay is the thickest of the Upper Jurassic formations and the most widespread; moreover, several hundred feet are known in eastern Scotland so the Jurassic total in the Hebrides may have been nearer 4,000 feet. This is comparable with the Wessex basin, but the facies responsible for the thickening in Skye and Raasay are significantly different and are chiefly the Middle Jurassic sandstones, both marine and non-marine. It is possible therefore that we ought not to think of yet another Jurassic 'basin'—of northern Skye or the Minch—but rather of a marginal downwarp, nearer a persistent source of arenaceous detritus than any area in England, even Yorkshire. On such a view the Great Estuarine Series would represent a temporary enlargement of the bordering lands, even supporting enough local vegetation to form peat at the outset. Later there were widespread coastal flats and lagoons, but still receiving more sediment in the north. In late Bathonian times the flats were resubmerged, and the Upper Jurassic muds spread thinly over the basin.

Even so we must admit that any palaeogeographical deductions that can be made in western Scotland are very tenuous. The data are slim beside the impressive array that builds up the pattern of basins, shoals, swells and islands of England.

North-Eastern Scotland

The outcrops here are smaller than on the opposite side of the Highlands and are obscured by faulting and poor exposure. The chief area is a 20-mile coastal strip in south-east Sutherland, on either side of the small town of Brora; it is supplemented by two very small patches farther south on the shores of Cromarty, at Ethie and Port an Righ. The sequence at Brora is shown in Table 51. No reliable estimate of thickness can be given for most of the divisions. The Oxford Clay is best exposed, being seen in the Brora River as well as on the foreshore. The northern half of the coastal strip is occupied by Kimeridge beds alone, and even there estimates have varied between 700 and 1,500 feet.

Table 51. *Jurassic succession at Brora, Sutherland*

Stages	Formations and lithology
KIMERIDGIAN	Shales and sandstones, with boulder beds
OXFORDIAN	'Corallian' sandstones and limestones Brora Arenaceous Series, upper part
CALLOVIAN	Brora Arenaceous Series, lower part Brora Shales Brora Roof Bed
BATHONIAN	Great Estuarine Series, coal seam, shales and sandstone
	(fault gap)
LOWER LIAS	Shales, sandstones and limestones

Exposures in the Lias are poor, but probably very little is present, partly owing to faulting; ammonites from the *raricostatum* and *jamesoni* zones have been recorded. The lithology is also abnormal; the rocks are dominantly clays, but also contain minor sandstones, limestones and thin seams of coal. Above the fault gap, estimated to cut out beds from Middle Lias to Bajocian, the Estuarine Series includes a 50-foot massive sandstone at the base and above that increasingly finer beds, the clays of which have yielded several plant species also found in Yorkshire. Higher yet there are black bituminous shales with, apparently, a mixture of marine and freshwater shells and the series terminates with the Brora Coal. Although small in extent this is over 3 feet thick and has been worth working in a region naturally poor in fuel.

From the Brora Roof Bed above, a hard ferruginous sandy limestone, there succeeds a varied series of Upper Jurassic sandstones, shales and

CHRIST'S COLLEGE
LIBRARY

occasional thin limestones. Their fauna includes lamellibranchs, belemnites and ammonites, and in the Kimeridgian shales several species of plants. Here also are certain famous boulder beds. The most impressive is some 200 feet thick and consists of very large angular blocks of Middle Old Red Sandstone which have yielded fishes. The blocks are now visible on the foreshore and the Kimeridge shales are squeezed up and distorted around them. Colonies of the coral *Isastraea*, littoral shells and plants have also been collected from the shales. The jumbled mass is deduced to have accumulated as an underwater breccia, at the foot of a submarine scarp, possibly a fault scarp, as the result of an earthquake, the littoral fauna and the corals at the same time being swept down by the accompanying tsunami or earthquake wave.

The influence of nearby land and actual emergence is even more marked here than on the west of Scotland. This is apparent not only in the relatively thick Brora coal, but also in the thin Liassic seams (remarkable intercalations in beds of this age) and in the numerous and well-preserved drifted plants. The general succession, however, the ammonite faunas and the persistent influx of sandy detritus are alike on both sides of Scotland, and we should probably do well to remove from our imaginary view any long-continued or substantial Scottish Highlands. As in the English landscape, coastal flats, shallow seas and intermittently emergent lands are more probable, supplying extensive detritus only in Middle Jurassic times. Modern, mountainous Scotland is a Tertiary product, not to be superimposed on the Mesozoic scene.

REFERENCES

The classic work on British Jurassic stratigraphy is Arkell (1933); the references given below cover more recent work or specialized aspects.

Arkell, 1933, 1947*a*, *b*, 1956; Davies, 1956; Dean *et al.* 1961; Donovan and Hemingway, 1963; Hallam, 1958, 1960*a*, 1965; Hemingway, 1951; Hollingworth and Taylor, 1951; Howitt, 1964; Hudson, 1964; Kent, 1955; Lee and Pringle, 1932; Swinnerton and Kent, 1949; Taylor, 1949, 1963; Whitehead *et al.* 1952; Wilson *et al.* 1958.

British Regional Geology. East Yorkshire and Lincolnshire, East Anglia, Bristol and Gloucester, Hampshire Basin, Wealden District, Tertiary Volcanic Districts.

10

THE CRETACEOUS SYSTEM

In north-western Europe Cretaceous history follows on from Jurassic as a record of shallow seas, contrasted with deltaic and other non-marine environments. In southern England the rocks also follow conformably and there is again a large main outcrop, in this case occupying much of eastern and southern England, and isolated but valuable outliers in Scotland and Northern Ireland. But in addition the period closes on a valedictory note: it is the last in which sedimentation extended far and wide over the area of these islands. After the unprecedented expansion of late Cretaceous seas there was a contraction, which, with fluctuations, has lasted till the present day.

In facies the Cretaceous rocks are both like and unlike those of the preceding period. There are the marine clays still replete with ammonites, of which the Gault is the outstanding example; there are marine sands, such as the Lower Greensand of southern England and lesser formations in the north; there is the remarkable deltaic and lagoonal complex of the Wealden. But the detritus-free seas of late Cretaceous times did not give rise to oolites, shelly or coralliferous limestones. Instead there were deposited over 1,600 feet of that most peculiar of calcareous rocks, the Chalk.

Ammonites continue to be the most important zone fossils where they are available, but in addition to their absence in the Wealden facies they are not plentiful at all levels in the marine rocks of the north-east. In the Chalk seas they gradually declined, family after family becoming extinct, until all were gone shortly before the close of the period. In the non-marine rocks, as in those of the Jurassic, ostracods are the best guide fossils, with some assistance from plant spores and pollen.

The Cretaceous stages of western Europe are chiefly named from France and the Swiss Jura mountains. The English traditional lithological groups agree with them fairly well, but unfortunately the base of the Gault Clay, sometimes taken in this country (though not in this book) as the base of the Upper Cretaceous is not at the base of the Lower Albian stage, but nearer its top. The relations between the lithological and cyclic divisions of the Wealden and the marine stages are not precisely determined. 'Neocomian' (from Neuchâtel) has been used for Wealden

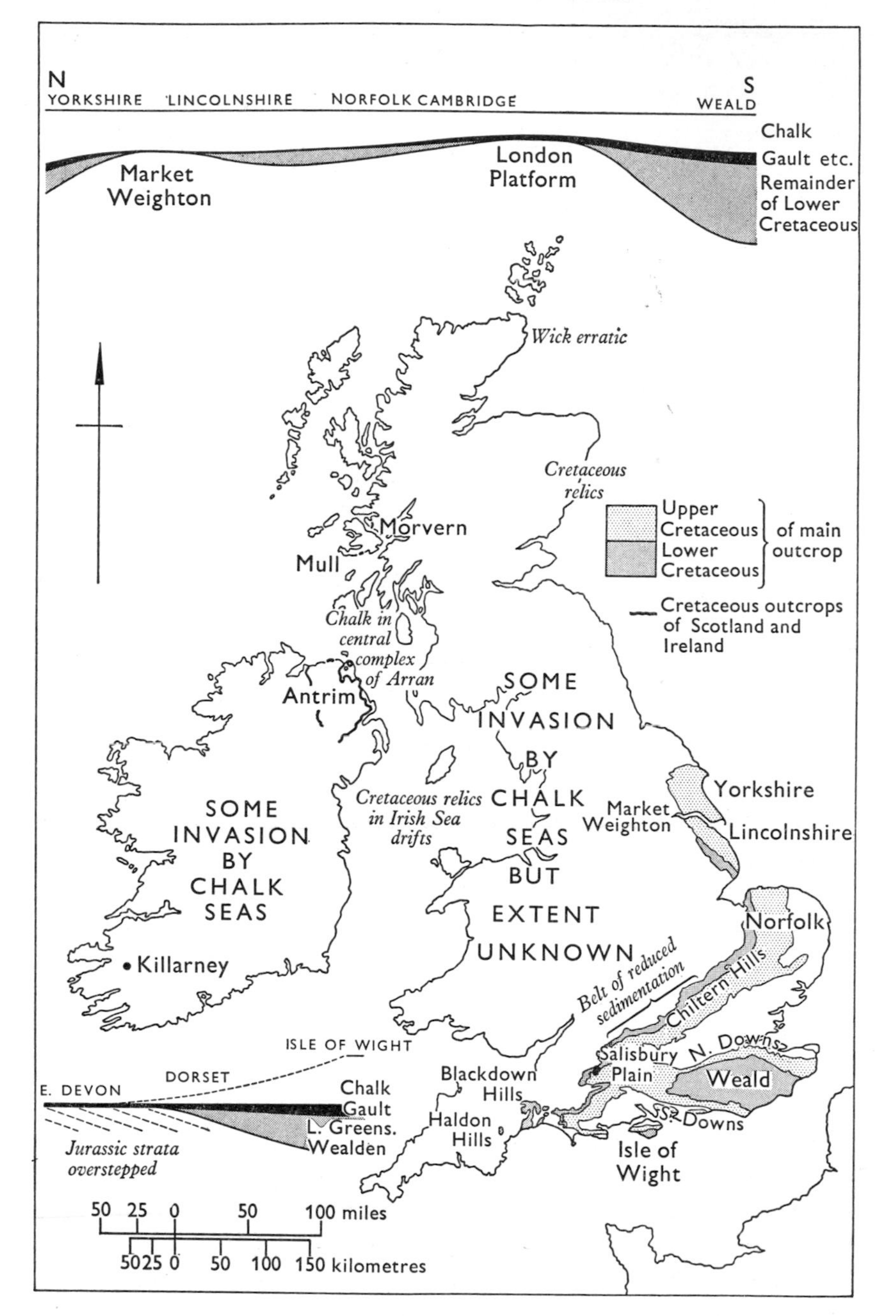

Fig. 63. The main outcrops of Cretaceous rocks together with smaller areas of Ireland and west Scotland. The section at the top shows the approximate thickness variations from the Weald basin to Yorkshire; that at the bottom the decrease in thickness, and Albian overstep, from the Weald to east Devon. (Not to scale.)

facies in Britain and elsewhere, but not being based on a standard marine succession it does not form a satisfactory stage name. Table 52 follows traditional usage in that the Jurassic–Cretaceous boundary is equated with that between Purbeck and Wealden beds. But, as noted in the last chapter, this is only a common assumption and most probably the Upper Purbeck and most of the Middle is lowest Cretaceous in age.

Table 52. *Divisions of the Cretaceous System in the British Isles*

	Stages	Principal groups, etc.	
UPPER CRETACEOUS	MAASTRICHTIAN SENONIAN TURONIAN CENOMANIAN	Upper, Middle and Lower Chalk	
LOWER CRETACEOUS	UPPER ALBIAN MIDDLE ALBIAN	Upper Greensand and Gault	
	LOWER ALBIAN APTIAN	Lower Greensand	
	BARREMIAN HAUTERIVIAN VALANGINIAN BERRIASIAN (RYAZANIAN)	Marine rocks in north-east England	Approximately equivalent to Wealden of south-east

Zonal Table in Appendix (p. 408).

The uplift on the northern margins of the London Platform, in the eastern Midlands and East Anglia, which restricted late Jurassic deposits largely to the Weald and Wessex, was effective through much of Lower Cretaceous times. These two basins were still the main foci of sedimentation in the south, but there was a gradual and irregular encroachment northwards during Aptian and Albian stages. A Lower Greensand transgression, for instance, is conspicuous around the edge of the Berkshire Downs, over the feather edge of the Purbeck and Portland littoral facies, and Gault Clay is the main overlapping formation on most of the London Platform. The northern seas, with their boreal ammonite faunas reminiscent of north Germany and Russia, were established in Yorkshire and Lincolnshire and spread southwards into Norfolk. By Middle Albian times the marine invasion of south-east England was largely complete.

No such history can be made out in the western and northern parts of the British Isles, but the few outcrops that remain also suggest an increasing marine expansion. There are some traces of littoral or shoreward facies to the Chalk, but for the most part the sea seems to have encroached on a country of little relief, which had suffered no major upheaval since the Hercynian orogeny. The extraordinary absence of detritus from the bulk of the Chalk is a testimony, negative yet eloquent, to the lack of transport from such low-lying lands as may have survived.

WEALDEN DELTAS

In its geographical sense the Weald lies between the outwardly dipping Chalk ridges of the North and South Downs. These dips conform with the broadly anticlinal structure, but there are a number of minor folds

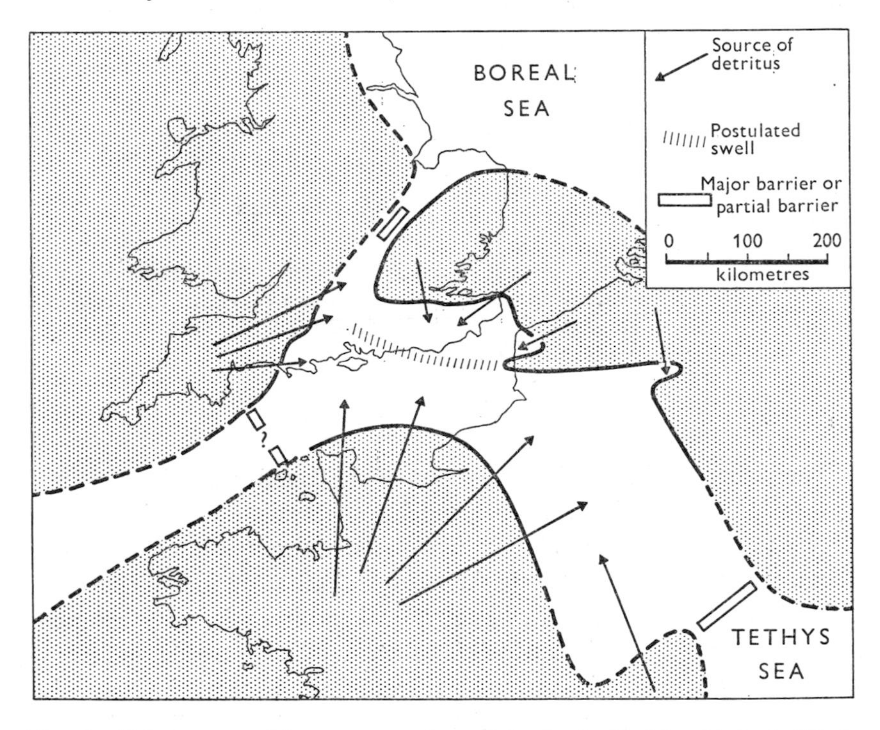

Fig. 64. Palaeogeographic reconstruction of the Anglo-Parisian Basin in Wealden times, showing the barriers separating the basin from the northern and Tethys seas and the sources of detritus. A low swell or ridge is deduced to extend eastwards from Portsdown to the French coast. Within the British sector the boundaries of the Cornubian uplands and London Platform are fairly certain but those in the midlands and north are much less so. (From P. Allen, 1965, p. 324.)

with the same east–west trend, that makes the term anticlinorium more appropriate. Similar folds affect the Chalk of Salisbury Plain, and the erosion of two to form the Vales of Wardour and Pewsey has already

6. Mupe Bay and Worbarrow Bay, Dorset, from the south-west. These two coalescent bays have been eroded into Wealden Beds. Steeply dipping chalk forms the ridge on the left. In the foreground and on the extreme right there are cliffs of Portland Stone capped by Purbeck Beds.

7. Tertiary and Jurassic cliffs on the north-east coast of Skye. A columnar sill of dolerite overlies Middle Jurassic (Inferior Oolite) sandstones, the lower beds of which crop out in the foreground at sea level.

been mentioned. Chalk is brought up by the Portsdown Anticline among the Tertiary outcrops of Hampshire at approximately the boundary between sedimentary basins of the Weald and Wessex.

The Wealden cyclothems of the Weald

On the north the Wealden basin was bordered by the gentle uplands of the London Platform; to the south-east a gulf extended far into central France, the whole representing a restricted phase in the long history of the Anglo-Parisian Basin (Fig. 64), which during Lower Cretaceous times was separated from the southern European seas in south-eastern France by an oscillating, transitional belt. The facies in France and Wessex, however, are not exactly like that in the Weald of Kent, Sussex and Surrey, and it is possible that the latter was limited by some ridge or swell stretching eastwards from Portsdown to the French coast south of Boulogne.

The English Wealden is thickest north of the present coast line, and consists of over 2,000 feet of clays, silts and sands with many signs of very shallow water, plant growth and thin but important spreads of gravel.

The traditional divisions are lithological:

Weald Clay	Upper Weald Clay Horsham Stone Lower Weald Clay
Hastings Beds	Upper Tunbridge Wells Sand Grinstead Clay Lower Tunbridge Wells Sand Wadhurst Clay Ashdown Beds, silts and sands, with Fairlight Clay at base

The cyclothems resulting from the interaction of deltaic and lagoonal conditions chiefly affect the Hastings Beds; the third is less complete and less well known, and is capped by the Horsham Stone, a thin calcareous sandstone:

3rd cycle	Lower Weald Clay Upper Tunbridge Wells Sand
2nd cycle	Grinstead Clay Lower Tunbridge Wells Sand
1st cycle	Wadhurst Clay Ashdown Beds

Each cyclothem represents the southward advance of a delta into a large lagoonal area, fresh to brackish, and its submergence by muds. The lower part consists dominantly of silts and sands, the latter increasing upwards to terminate in current-bedded sandstones with channels and plant remains. These are capped by a thin pebble bed formed on a retreating strand-line as the delta surface was gradually covered by the deepening waters. In early stages the horse-tails (*Equisetum*) were able to maintain themselves in the shoreward shallows, but later they were replaced by a fauna of shells (*Neomiodon* and *Viviparus*) and ostracods. The next cyclothem begins with the renewed building out of deltaic silts.

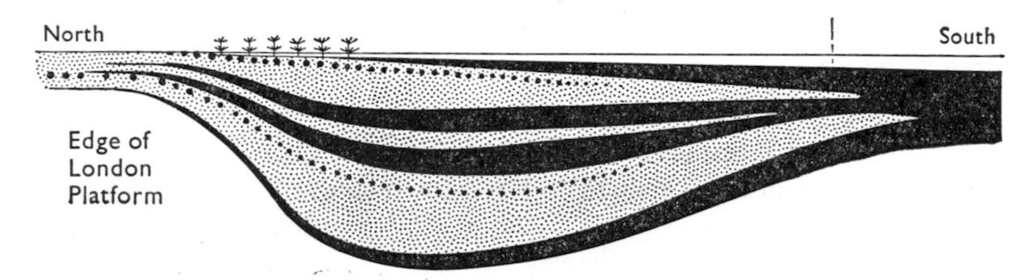

Fig. 65. Wealden deltas. Two major deltaic advances are shown with a minor one in between. They are separated by periods of transgression when muds advance northwards. The pebble beds near the top of the sands are those of a northward retreating strand-line. The water-level at the end of the latest transgression is shown at the top and a belt of *Equisetum* growing in very shallow water. (Redrawn from Allen, 1959, fig. 21 D.)

The north–south oscillation of the two principal facies means that the clays (Fairlight, Wadhurst and Grinstead) tend to be thicker in the south, near the Sussex coast, and the delta silts and sands swell out in the central weald. All the Wealden becomes much thinner towards the edge of the London Platform, from which the sediments were largely derived. The pebbles show that Devonian, Carboniferous and Jurassic rocks were being eroded, and there is evidence of one river at least draining the London area, and another from north-east Kent. At times of maximum inundation these rivers became temporary estuaries. The Wealden Beds also overlap older strata in this direction, and in East Kent the Hastings Beds and then Weald Clay come to lie on the Carboniferous rocks of the Kent Coalfield.

The flora and fauna of the Hastings Beds are appropriate to this deltaic-lagoonal environment and are restricted almost entirely to the clays. Apart from the horse-tail beds at several levels the Fairlight Clay contains many terrestrial plants, such as ferns and cycads, drifted out into the shallow lake. Reptiles, amphibia and fish are known, especially from the Wadhurst Clay, and the widespread footprints of *Iguanodon* show how shallow the water must have been.

The deltaic incursions ceased by the middle of the Weald Clay, and a more uniform facies prevailed. This was dominantly freshwater muds, colonized from time to time by swarms of shells, especially *Viviparus*; these calcareous shell banks form a local ornamental stone, 'Sussex Marble'. Towards the top brackish shells appear (*Paraglauconia*), and, as further signs of increasing salinity, oyster beds; the Wealden formation was finally brought to a close by the widespread Aptian invasion.

A boring on the Portsdown ridge recorded only 785 feet of Wealden Beds, but there is a marked expansion into the Wessex basin. Here no division is possible into Weald Clay and Hastings Sands, and the cyclothems are absent. In the east the Wealden is dominantly argillaceous, with a rather arbitrary division into the Wealden Marls (variegated marls and sandstones) below and Wealden Shales above. The latter are grey, with local layers of shelly limestones, clay ironstone and sandstones, and, as in the Weald itself, brackish and marine shells appear near the top.

In the Isle of Wight the two divisions amount to 750 feet, and the base is not seen. From here westwards there is a general increase in the sandy proportion, and, in Dorset, a marked decrease in total thickness. In Swanage Bay the marls with sandstone make up almost the whole 2,300 feet, with only 25 feet of Wealden Shales at the top, while 10 miles farther west at Worbarrow Bay the latter have disappeared entirely. West of Lulworth 215 feet of sandy beds include nearly 30 feet of conglomerate. In west Dorset no Wealden remains, having been overstepped by Albian; it is probable that the upper beds disappear first, so that the residual sandy facies represents Hastings Sands. Its westerly source is evident from the pebbles which include quartzites, radiolarian chert and tourmaline rocks, all recognizable in Devon and Cornwall. Moreover, the mineral assemblage is characteristic of the Dartmoor Granite and its aureole, so that the intrusion must have been unroofed by Wealden times.

APTIAN AND ALBIAN SEDIMENTS IN THE SOUTH

The Lower Greensand, Gault Clay and Upper Greensand have an irregular curved outcrop inside the chalk downs from Folkestone on the Kent coast westwards to Surrey and Hampshire, and back to Eastbourne on the Sussex coast (Fig. 66). They also form much of the southern part of the Isle of Wight, and extend into Dorset, but only the higher Albian beds reach beyond the Jurassic and Wealden outcrops into east Devonshire. The relationship between the stages and lithological grouping of the formations is shown in Table 53. Some of the lithological boundaries are diachronous; thus the Gault Clay facies in the south of the Weald ranges down into the Folkestone Beds, and, much more important, the Upper Greensand increases in proportion westwards at the expense of the Gault. Furthermore there is no simple comparison between the Lower Greensand formations in the Weald and Isle of Wight.

Table 53. *Aptian and Albian formations of southern England*

Stages	Weald (e.g. east Kent)		Isle of Wight
UPPER ALBIAN	Upper Greensand and Upper Gault Clay		Upper Greensand and Gault Clay
MIDDLE ALBIAN	Lower Gault Clay		
LOWER ALBIAN UPPER APTIAN	Lower Greensand	Folkestone Beds	Carstone Sandrock
		Sandgate Beds Hythe Beds	Ferruginous Sands
LOWER APTIAN		Atherfield Clay	Atherfield Clay

Lower Greensand

This name arose through a series of misconceptions, principally an old confusion between these beds, below the Gault Clay, with others above it, which in the westerly outcrops are sometimes faintly green. Surface exposures inland range in colour from yellow to rusty red and brown, owing to the oxidation of the iron compounds; in cliff sections, or underground, however, the glauconite is fresh enough to give a greenish tinge to some of the beds.

Fossils are usually rare in the sandy facies, probably because these have been leached and the shells dissolved. They may be fairly abundant in the harder or calcareous bands and nodules, and from such rocks enough ammonites have been collected to allow a correlation between the lithological formations and the system of stages and zones. Other molluscs and brachiopods appear in certain beds and together with sporadic sponges, corals and echinoids make up a representative marine Mesozoic assemblage.

The Atherfield Clay is grey to reddish, silty and the most fossiliferous of the Aptian formations. Several ammonite species (especially of *Deshayesites*) are found, and the '*Perna* Bed' at the base is a harder band, cemented with carbonate, that contains the conspicuous large valves of the lamellibranch *Mulletia mulleti*, once called *Perna*. The formation is

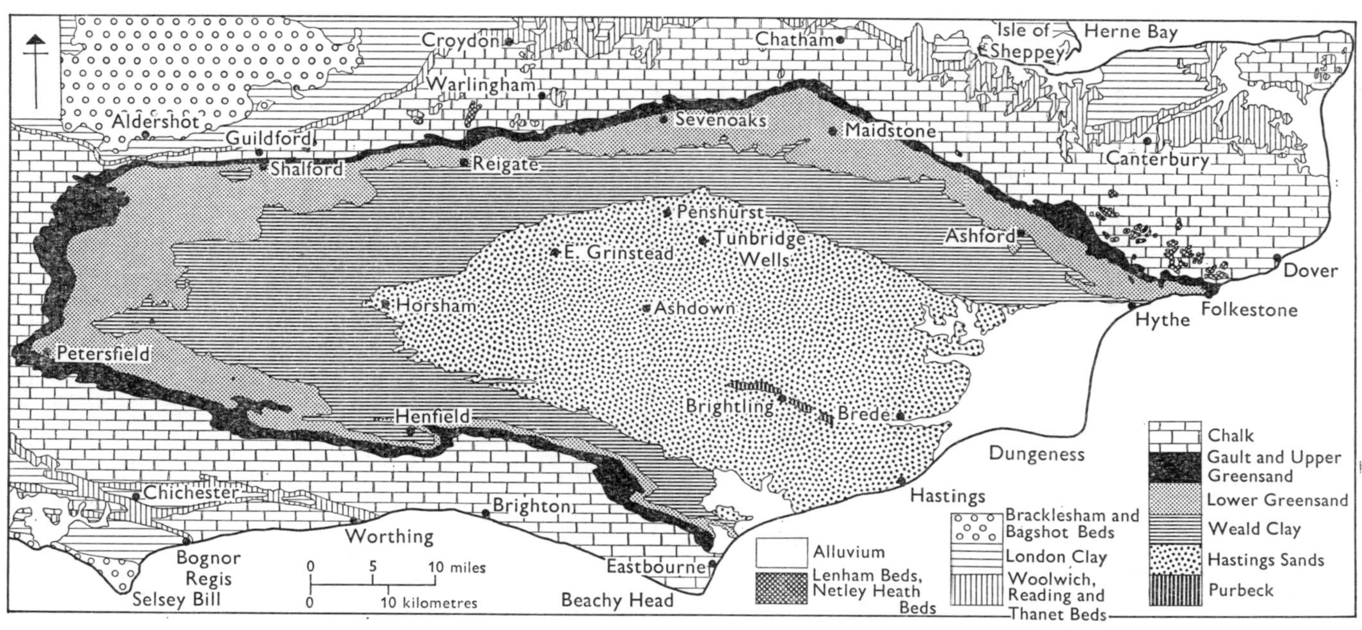

Fig. 66. Geological sketch map of the Weald and neighbouring areas. (Adapted from Gallois, 1965, pl. 1.)

thickest in the north-western part of the Weald where it amounts to 60 feet; the reduction to the south-east is very marked in Sussex where there is only a thin remanié deposit of nodules and rolled fossils, while in east Kent the *Perna* Bed is missing through a non-sequence.

The sagging basin of the western Weald is again shown in the three sandy formations; their combined thickness at Folkestone is under 200 feet, on the Surrey–Hampshire borders it expands to 600 feet, and at Eastbourne is less than 100 feet of sands, with some Gault-like clays in the uppermost beds. The sands are normally poorly cemented, often mixed with silt or clay and sometimes current-bedded or with seams of pebbles and phosphatic nodules. The harder and better cemented beds are usually calcareous, or more rarely siliceous, and form local sandy limestones or chert, the former being known as 'Ragstone' or 'Kentish Rag'. 'Hassock' refers to a greenish silty sandstone with specks of glauconite. Both types are also exceptional in their abundant fossils, which in the calcareous rocks include corals. The silicified variants of the Hythe Beds in particular are responsible for the upstanding scarps in the north-west, such as the Leith Hill range of Surrey which reaches 900 feet and overtops the chalk downs.

The Sandgate Beds often contain an admixture of silt and clay and lack fossils except in calcareous bands; in the north they have yielded derived Jurassic fossils, particularly from the Oxford Clay, washed in from the London uplands. At their type area the Folkestone Beds are yellowish greensands with a rich Upper Aptian and Lower Albian fauna, but farther west thicken into current-bedded ferruginous sands, largely unfossiliferous.

Both palaeontological and lithological evidence point to non-deposition and possibly erosion at several periods in the shallow seas of the Lower Greensand. Local non-sequences, with the absence of certain subzonal ammonites, are found for instance at the base of the Atherfield Clay in east Kent, below the Sandgate Beds in Surrey and where they overlap on to the Kent Coalfield, and—most widespread—in the middle of the Folkestone Beds. The breaks are commonly accompanied by seams of rolled phosphatic nodules containing fossils derived from the beds below. Such phosphatic lumps and rolled remains frequently mark periods of little or no sedimentation when the sands and silts were carried away by gentle currents.

Over the Portsdown ridge the Lower Greensand is reduced to 100 feet, but as in earlier times the eastern part of the Wessex basin was a further subsiding region, and there are 800 feet of beds in the south-west of the Isle of Wight, where there are exceptionally fine cliff sections. This expansion is accompanied by a thicker and more sandy Atherfield 'Clay', succeeded by a varied series of sands—often glauconitic, sometimes ferruginous or argillaceous, and with an abundant fauna. Phosphatic and calcareous nodules are found here too, but for the most part sedimentation seems to have been more continuous than in the Weald, the only marked break being at the bottom of the Carstone. The latter is a term often applied to Cretaceous rocks, a dark-brown, coarse ferruginous sandstone with quartz pebbles and phosphatic nodules. Here it represents the Lower Albian zone of *Douvilleiceras mammillatum*, which is reduced to a remanié nodule bed in much of the Weald.

In Dorset the Lower Greensand shows a more marked westerly reduction than the Wealden below or the Gault above. It amounts to only 200 feet in Swanage Bay, where ammonites are rare; some of the lamellibranchs are apparently brackish types, and the sands contain lignite. These littoral characters are enhanced farther west, the beds being much thinner in Worbarrow Bay and at Lulworth Cove are reduced to a 6-inch ironstone band with a brackish Wealden-like fauna. Nevertheless, this bed is probably equivalent to the upper part of the Atherfield Clay in the Isle of Wight, so that the overstep of the Gault Clay is combined with the absence of all the higher Aptian beds. West of Lulworth the Gault lies on Wealden.

The Gault Clay and Upper Greensand

Where both these facies are well developed the greensand lies above the clays, as for instance in the western Weald; but over southern England as a whole there is also a lateral transition, and the sand replaces an increasing proportion of the clay westward. Typical Gault Clay is dark-grey, stiff and tenacious, and best seen in the lowest beds; the higher levels become paler and more calcareous. Phosphatic nodules are the main lithological variant, and a particularly persistent band marks the junction between Lower and Upper Gault. The clay fauna is a rich one; in addition to ammonites and other mollusca there are brachiopods, echinoids and crustacea.

The term Upper Greensand covers a considerable lithological range from glauconitic sandstones and siltstones to calcareous marls and cherts. Ammonites are not so plentiful as in the clays and the chief fossils are lamellibranchs and sponges. The upper and lower limits of the greensand are not very precise and in places there may be a lithological passage from the marly Upper Gault below to the impure, slightly glauconitic, Lower Chalk above. The facies reflects the same westerly source of sandy detritus that is a persistent feature of Cretaceous sedimentation in southern England, but in Middle and Upper Albian times the effect is unusually striking because of the contrast with the muddy seas of the Gault Clay.

The type sequence of ammonite-bearing Gault is at Folkestone, where almost all the 118 feet of Middle and Upper Albian consists of clays; there is only a small amount of glauconitic sands in the upper beds and nothing recognizable as an Upper Greensand formation. As the beds are traced westwards along the foot of the North Downs there is an increase in total thickness and also in the greensand; the latter reaches 80 feet in west Sussex out of a total of 220 feet, and the two combined are 300 feet in Surrey. Borehole evidence on the Gault cover to the London Platform suggests that it varies from somewhat under 200 feet to somewhat over that figure, and consists largely or entirely of the clay facies.

In the Isle of Wight both clay and greensand facies are again present and their difference is accentuated. In the south of the island 160 feet of 'greensand', including massive sandstones with seams of chert, overlie 100 feet of clay. It is this superposition, and the thick-bedded sandstone sliding on the impervious clay surface, that is responsible for land-slipped 'undercliff' on the south coast. The westerly trend for sands or sandstone to increase at the expense of the underlying clay is continued in Dorset, and the overstepping base of the Gault from Swanage to Devon is one of the most impressive transgressions in British stratigraphy. In east Dorset a small thickness of clays overlies Lower Greensand, though probably with unconformity and the absence of the Lower Albian, Carstone and upper part of the Sandrock; the latter is reduced to 10 feet compared with the 184 in the Isle of Wight. From here westward there is a downward transgression on to the remnants of the Wealden, the Purbeck, Portland and Kimeridge beds, the last being the underlying formation on the coast east of Weymouth. Along the sweep of Lyme Regis Bay there are small, outlying Albian outcrops; at Abbotsbury in the east on Oxford Clay, at Bridport in the centre on Bathonian, and at Lyme Regis in the west on Lias (Fig. 60). In east Devon the underlying formation is Keuper Marl, and on the Haldon Hills it is the breccias at the base of the New Red Sandstone.

In south Dorset this overstep is complicated by intra-Cretaceous folding, which produced small, acute, east–west folds and some major faults. These structures were planed off by the mid-Albian transgression, and so, north of Weymouth for instance, the underlying rocks vary in short distances from Oxfordian to Wealden. Through south and west Dorset the overlapping Albian is mostly formed of Upper Greensand, but some 20 feet of dark-grey silty clays persist at the base as far west as Lyme Regis. The Middle Albian is probably overlapped and disappears in east Devon, for although ammonites are few, it is apparently absent on the Blackdown Hills, the most northerly of the Devonshire outliers.

The Upper Greensand in this county consists of glauconitic sands which weather rusty brown or yellow, and contain *Exogyra*; large calcareous concretions abound at some horizons and chert is common at the top; the shells in these beds are beautifully preserved in chalcedony. The pebble and mineral contents, like those of the Wealden sands, testify to the erosion of the Hercynian granites and surrounding country rock.

The last outpost of all Cretaceous rocks in south-west England lies 800 feet up on the Haldon Hills, east of Dartmoor (Fig. 32). Here there are 100 feet of glauconitic and pebbly greensands, which contain abundant *Exogyra* and a few corals. This is not the actual shore facies, but near it. At present, after all the further denudation of late Cretaceous and Tertiary times the Dartmoor Granite rises to over 1,900 feet, and is thus most unlikely to have been submerged by Upper Albian seas. Of all the Cretaceous areas in the British Isles, this in the south-west preserves best the evidence of neighbouring lands, both in the change of facies (nearly always to some type of sands) in the shoreward direction, and in the recognizable detritus brought into the shallow seas.

The western and northern fringes

Over the London Platform and around its western and northern edges the Gault Clay is the most conspicuous overlapping formation. But in places underlying the clay or standing as outliers beyond it, there are outcrops of sands: many are unfossiliferous or contain only long-range forms, but a few have a sparse Lower Greensand fauna.

In the Vale of Wardour, and at Dilton in north Wiltshire there are a few feet of ferruginous sands and gravels; they form a basement bed to the Gault Clay and contain rare fossils of the *mammillatum* zone, at the top of the Lower Greensand. The Seend 'Iron-Sands', near Devizes, were once worked for their iron content, and produced a large fauna of *nutfieldensis* zone (Upper Aptian). The best known of these outliers is that formed by 150 feet of sands and gravels at Farringdon in Berkshire; within a small area these beds lie on Corallian and Kimeridgian and are famous for the sponge-gravels at the base; in addition to the many species of sponge the deposits are rich in polyzoa, brachiopods and echinoids, but relatively poor in lamellibranchs. The sands and gravels are current-bedded and accumulated in channels on the floor of a clear, shallow sea.

From the hills south of Oxford eastwards to the Aylesbury district of Buckinghamshire there are several small sandy outliers, lying on Kimeridgian, Portland or Purbeck beds, many of which are uncertain in age. The Shotover Ironsands, 80 feet thick, south of the city of Oxford, have been called Wealden on a basis of lithology and freshwater molluscs, but the latter are only long-range forms. In lithology they are fine sands and silts, with seams of clay. Similar beds are found filling a shallow channel at Stone, 2 miles west of Aylesbury, but again diagnostic fossils are lacking. Yet another example at Whitchurch, north-east of Aylesbury, has been found to contain Middle Purbeck shells, so the age of all this group is doubtful.

Marine Aptian fossils have been collected from ferruginous sands near Aylesbury and south of Oxford, and there are several intervening outcrops of similar type. Commonly they are distinguishable from the supposed Wealden beds in being coarser-grained, reddish, ill sorted, and sometimes pebbly.

Although the ages of this group of outliers are rather dubious, and some may long remain so, their geographical position is of some importance. They represent the fringe-edge of deposition, thin and interrupted, which persisted from late Kimeridgian to Middle Albian times, in the shallow strait between the southern basins of Wessex and the Weald, and that of eastern and northern England. Ten miles away to the north-east lay the margin of the latter, and the much thicker and continuous Aptian beds of Bedfordshire and Cambridgeshire.

THE NORTHERN SEA

North of the London Platform, from Cambridgeshire to Yorkshire, all the Lower Cretaceous rocks are marine. In their several outcrops they overlie Upper Jurassic beds (usually Kimeridge Clay) unconformably and are overlain conformably by the Lower Chalk. But between these limits there is much variation, both in facies and in the stages present. Even the Gault, so widespread in the south-east, does not maintain its thick clay character to the north. The most diverse, best exposed and probably most extensive sequence is in Lincolnshire. Slightly later the seas spread into Yorkshire and Norfolk, and then after a longer interval into Cambridgeshire and Bedfordshire. At last, after the fluctuating earlier conditions, a union of northern and southern seas was achieved in Upper Aptian times.

The Lincolnshire succession is thickest and most complete in the south of the county, since, as in the Jurassic period, there was a marked reduction in deposition towards Market Weighton. The principal sediments in this northern sea were sands, often with pebbles, muds and oolitic ironstones. The last include the Roach as well as the Claxby Ironstone. The Tealby Limestone is a thin impure bed intercalated between thicker clays and in places is no more than a sandy rock with a calcareous cement. Stratigraphical fossils include belemnites and, more rarely, ammonites—the latter for instance in the Spilsby Sandstone, Claxby Ironstone and Sutterby Marl. In the last they provide a valuable correlation with the Upper Aptian faunas of the Weald.

Table 54. *Lower Cretaceous succession in Lincolnshire*

Stages	Lithological formations	Thickness (feet)
UPPER ALBIAN	Red Chalk	24
LOWER ALBIAN (?)	Carstone	24
UPPER APTIAN	Sutterby Marl	11
	(non-sequence)	
UPPER BARREMIAN	Fulletby Series: Roach, Ironstone	62
LOWER BARREMIAN and HAUTERIVIAN	Tealby Series: Upper Clay, Tealby Limestone, Lower Clay	92
HAUTERIVIAN and VALANGINIAN	Claxby Ironstone and clays	21
RYAZANIAN and ? PORTLANDIAN (BERRIASIAN)	Spilsby Sandstone	76

The Spilsby Sandstone, though dominantly arenaceous, includes some silts, pebbly bands and phosphatic nodules. Some levels are incoherent, but others are hardened with a calcareous cement. The problems raised by the ammonites of this bed have already been mentioned briefly (p. 302). At the bottom the lowest, glauconitic, sand includes some phosphatized fragments of Kimeridgian forms and the clay floor has been penetrated by boring molluscs. Indigenous Ryazanian species are known from the upper part, and it is those of the lower that are crucial. Though long held to be earliest Cretaceous they may (by comparison with

Russian faunas) be Upper Jurassic, and equivalent to the Purbeck and top part of the Portland Beds. If this be so, there would still have been a break after the deposition of much of the Kimeridge Clay though not so long as supposed hitherto; and later the Spilsby Sandstone would have accumulated slowly, and perhaps intermittently, with pebble beds and layers of nodules, over a considerable span of late Jurassic and early Cretaceous time.

A broad cyclic pattern can be detected in the succeeding sedimentation of this shallow northern sea: a transition from sandstones to ironstones, and then to clays, up to the thin Tealby Limestone, and then the same facies in reverse order until the non-sequence and erosion at the base of the Sutterby Marl. Here more than one Lower Aptian zone is represented in the layers of phosphatic nodules at the base.

The Carstone is a ferruginous and glauconitic sandstone with pebbles and broken shells, but no indigenous fossils. It passes up gradually into the Red Chalk—a curious facies confined to this northern province. It is a fine impure calcareous rock, reddened with a detrital component of iron oxide (hematite) and there are also scattered quartz pebbles. Typically it contains plentiful specimens of the little belemnite, *Neohibolites minimus*, and also lamellibranchs and terebratulids.

In Lincolnshire and Norfolk the Red Chalk is probably all Upper (or Middle and Upper) Albian in age—that is, it corresponds to much or all of the Gault Clay of the south-east. In Yorkshire a paler reddish or pink colour is continued up into the lowest Cenomanian beds where it extends over some 50 feet. The extreme thinness of the Red Chalk suggests very slow deposition, and though the cause of the colour is not known for certain, it may have been derived from a laterized land surface or from the Trias, unconformably below further east (p. 400).

As the Lincolnshire beds approach the Humber nearly all become thinner and then disappear (Fig. 67), but the Red Chalk remains, together with a small amount of Carstone below. Near Market Weighton there is a small section that shows these remnants lying unconformably on Lower Lias (*angulatum* zone)—the only formations to cross this formidable upwarp since early Jurassic times.

On the northern side, in the Yorkshire basin, the rocks between the Kimeridge Clay and the Red Chalk are surprisingly different and are all comprised in the Speeton Clay. The name place is a coastal village, north of the chalk of Flamborough Head, where a succession of Lower Cretaceous zones has been patiently built up from a series of badly slipped and intermittent exposures. The bulk of the Speeton beds consists of grey clays with seams of cementstone, ferruginous and phosphatic nodules; glauconite is found at many levels and pyrite particularly in the lower clays. The stratigraphical sequence is based chiefly on the belemnites and crushed ammonites; other fossils are ostracods, lamellibranchs and brachiopods (including *Lingula*); one bed is rich in crustacea (the 'shrimp bed') and another in *Pentacrinus* ossicles.

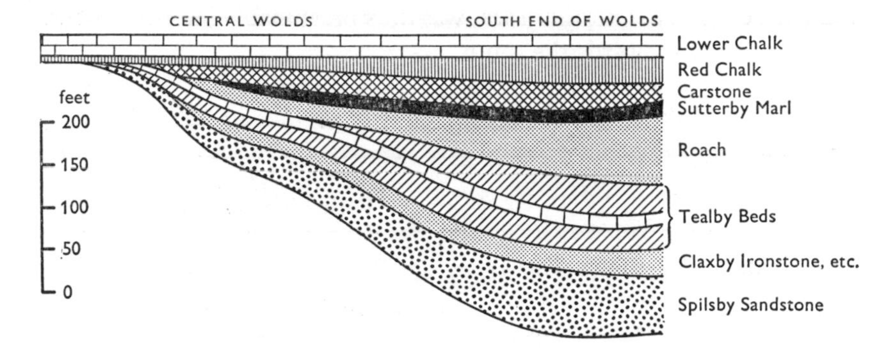

Fig. 67. Diagrammatic section through the Lower Cretaceous rocks of Lincolnshire. All the beds below the Red Chalk disappear some 12 miles south of the River Humber. (Redrawn from Wilson, 1948, p. 54.)

Derived Kimeridgian fossils are again found in the basal layer of phosphatic nodules; the lowest indigenous faunas are here Berriasian, and representatives of all the succeeding stages are known. Towards the top of the Aptian a marly layer with glauconite and nodules is probably equivalent to the Carstone, and above that grey and pink marls and impure chalk are succeeded by the Red Chalk. The thickness of clay is difficult to determine at Speeton itself, but has been estimated at just over 300 feet. In the borehole at Fordon, however, a few miles inland, nearly 700 feet were recorded. Thus in spite of their finer grain it seems that the Lower Cretaceous deposits were substantially thicker north of Market Weighton than were the more variable facies of Lincolnshire. Certainly this increase agrees with the general history and greater subsidence of the Yorkshire basin since early Liassic times.

At the southern end of the Lincolnshire outcrops the Cretaceous rocks disappear under the alluvium, peat and modern deposits of the Fens and Wash. They reappear in the low cliffs of the Norfolk coast, at Hunstanton, and there are isolated exposures in that county for some 25 miles southward (Fig. 71). The fullest succession, however, is in the north and is given here with such ages as are known (Table 55).

The Sandringham Sands rest on an eroded surface of Kimeridge Clay and for the most part consist of current-bedded, cream or white quartzose sands which have been worked for glass-making. Farther south where they are thinner these beds include coarse brown, glauconitic sands with nodules, and the latter have yielded moulds of Berriasian ammonites. Since the Snettisham Clay is here very thin or absent, there is a large gap below the Carstone above. Pre-Aptian deposition, on this southern edge of the northern sea, was thus presumably intermittent and became more so towards the south.

Table 55. *Lower Cretaceous rocks of north-west Norfolk*

Stages	Formation	Estimated thickness (feet)
UPPER ALBIAN and ? MIDDLE ALBIAN	Red Chalk or Hunstanton Red Rock	4
LOWER ALBIAN (?)	Carstone	40
	——— non-sequence ———	
BARREMIAN	Snettisham Clay	30
	——— ? non-sequence ———	
BERRIASIAN	Sandringham Sands	100

The Carstone resembles that of Lincolnshire in lithology and lack of an indigenous fauna; it does, however, carry derived fossils of at least two Lower Aptian zones. Again there is a transition upwards to the thin Red Chalk, which nevertheless includes faunas of both Lower and Upper Gault. It is worth noting that the thickness variations of the Red Chalk and Gault in Norfolk run contrary to those in the beds below, for the thin red rock at Hunstanton passes southwards into something nearer normal Gault, to reach 60 feet of grey, marly clay in the south-west of the county.

The southernmost outcrops of the northern province are those of Cambridgeshire and Bedfordshire, which extend from Ely to Leighton Buzzard. The Gault is of normal facies in this belt, and gradually increases southwards from 90 to 200 feet. More attention has been given to the beds beneath, grouped together under the southern term of Lower Greensand.

This is a region of exceptionally effective pre-Aptian erosion—yet another effect of the unstable northern margin of the London Platform. At either end of the outcrop the Lower Greensand lies on Kimeridge Clay, but in the middle at Sandy (Bedfordshire), a low arching of the

Jurassic clays has resulted in erosion of the Kimeridgian and Ampthill and the underlying bed is Oxford Clay. At Upware a few feet of Lower Greensand lie banked up against Kimeridge Clay and the small Corallian reef. Derived fossils, of many ages, are found here and elsewhere in the sands, and range from Oxfordian, Corallian and Kimeridgian forms to blocks of supposed Wealden Sandstone with iguanodon bones and earlier Aptian fossils from sediments which are no longer found.

In thickness there is a marked increase south-westwards from the few feet of Lower Greensand that cap several 'islands' (including the Isle of Ely) in the fens to 70 feet in south Cambridgeshire and over 200 feet of the 'Woburn Sands' near Leighton Buzzard. An exceptional flora of conifers, cycadophytes and early angiosperms has been collected from the last, which chiefly consist of silver sands. At the top they contain the well-known lenses of Shenley Limestone. In these masses, up to 10 feet long, the sand grains and pebbles are scattered in a pink or yellow calcareous rock; there is an exceptional fauna of brachiopods and other shells which grew on an uneven surface of the sands beneath, probably in very shallow water. The conditions appear to have been unsuitable for ammonites, since these are very rare. They do, however, include specimens of *Leymeriella*, showing the limestone to belong to the uppermost Lower Greensand (Lower Albian).

In review, the history of Lower Cretaceous sedimentation from Devon to Yorkshire can be epitomized by marking the successive marine invasions and increasing continuity of the seas. From approximately Berriasian to Hauterivian times the cyclic delta and lake régime held sway in the Weald and the less distinctive freshwater marls, clays and westerly littoral sands in Wessex. These contrast with the steady accumulation of Speeton Clay in the Yorkshire basin and a rather thinner more varied sequence in Lincolnshire, whence extended the Sandringham Sands into Norfolk.

Towards the end of the Wealden regime (?Barremian), first brackish and then minor marine incursions appeared in the Weald and Wessex, precursors of the full-scale Aptian invasion, which resulted in the *Perna* Bed in both basins. About of the same age is the Snettisham Clay on the southerly fringe of the northern basin.

Marine sands characterize the Aptian and Lower Albian in nearly all areas south of Market Weighton, but there is a great difference between the heavy sedimentation of the Lower Greensand of the south and the thin or intermittent deposition in the north. The Upper Aptian is a further period of marine extension and in Middle and Upper Albian times the Gault Clay covered the London Platform; Market Weighton subsided under a thin but lasting cover of Carstone and Red Chalk, and the seas encroached far to the west, within a few miles of the Dartmoor shore.

In this history the south-east Midlands—the 'fringe area' between north and south—are the most difficult to interpret. It may be that there was never any persistent barrier, but that the thin spreads (chiefly sands) were persistently laid down, and nearly as persistently removed. Whatever the detailed history may have been, from Upper Aptian times the differences between north and south were much diminished, heralding the greater, striking uniformity that developed in late Cretaceous conditions.

THE CHALK

In its extent and thickness the Chalk is a formation unique not only in the British Isles but in north-western Europe. It is found in the French as well as the English outcrops of the Anglo-Parisian Basin, occurs underground in Holland and reappears in typical lithology on the shores of Denmark, north Germany and south Sweden. Some marginal facies, commonly sands, are found near certain of the bordering land masses, such as Cornubia, Scotland, Brittany, the Ardennes and the Bohemian Massif. But most of these belong to the Cenomanian and Turonian stages, and how far the Senonian seas may have spread is largely conjectural though their sediments are known in Greenland; the present limits of Senonian chalk are largely those of erosion. During latest Cretaceous times there was probably a shrinking of this northern sea for the highest chalk, and the earliest Tertiary beds, are not widespread.

In England the Chalk outcrop has the greatest area of any single formation, and usually gives rise to gently rolling downs and uplands with a scarp-edge overlooking lower land formed of Jurassic or Lower Cretaceous rocks. Chalk also underlies the Pleistocene of East Anglia and Tertiary beds of the London Basin, where it is the source of artesian water. Rocks of Chalk age, though not all of that lithology, are the most important of the few Cretaceous outcrops in Ireland (Fig. 53) and Scotland.

Our familiarity with the Chalk, in quarries and cliffs from Yorkshire to Dorset, is apt to obscure its true singularity. Pure white chalk, typical of the Senonian stage, is a very fine-grained rock with a few principal constituents, all organic in origin. The coarser fraction, up to 0·1 mm, consists of the comminuted debris of shells, frequently the prisms of calcite from the valves of *Inoceramus*; this fraction also includes the less

8. The Isle of Wight, with the Needles, from the west. The steeply dipping chalk is seen in the foreground and along the central ridge of the island. On the left are the vertical Eocene beds of Alum Bay; the lower ground in the distance on the right is formed of Gault Clay, Lower Greensand and Wealden Beds.

abundant foraminifera and the calcareous 'spheres' which were also micro-organisms but of unknown affinity. An even finer fraction is derived from coccoliths, minute planktonic algae that secrete rings and groups of plates, each plate being a single calcite crystal. It is this matrix of very fine coccolith debris that particularly distinguishes the Chalk, and precludes a close comparison with either the modern deep sea oozes or the calcareous deposits of shoals such as the Bahama Banks. A sea of moderate depth (200–400 metres) is usually postulated, with periods of shallowing and accentuated current action. The various 'rocks' of the Chalk in particular probably reflect such phases.

The absence of mud and sand from typical white chalk suggests that it was laid down far from the contemporary shores, which, as we have seen, were gradually receding during earlier Cretaceous times. It is also probable that such lands as did remain unsubmerged were low-lying, and so very little detritus was carried into the later Chalk seas. The exceptional uniformity of conditions is also reflected in the lesser contemporary variations in thickness compared with beds beneath. The Wessex and Yorkshire basins still show some greater degree of subsidence. There is a slightly thinner sequence over the now submerged London Platform, and a more pronounced one south-westwards towards the Cornubian uplands; the present variations in thickness, however, are much more closely connected with the erosion that preceded Palaeocene and Eocene deposition. Maastrichtian chalk is rare in Britain but small remnants of the lower zone (*Belemnitella lanceolata*) are known from the Norfolk coast and north-eastern Antrim. The uppermost Senonian zone (*B. mucronata*) is widespread in East Anglia and parts of Wessex, but over several other outcrops erosion has cut down well into lower Senonian beds.

Chalk is not normally rich in fossils and the very large numbers in museum collections, often beautifully preserved, are rather misleading and are due rather to long collecting from quarries and cliffs than to an original abundance. Apart from the planktonic microflora and foraminifera, lamellibranchs, particularly *Inoceramus*, are the most plentiful, together with echinoderms, belemnites and brachiopods. Ammonites are rarer and in southern England and as zone fossils do not extend far up into the Chalk. The standard zones of the British Chalk are based on the successions of southern England and are given in the Appendix; in Lincolnshire and Yorkshire there are minor faunal differences which result in some replacements of southern species by northern ones.

Certain outstanding lithological bands are found in the lower part of the succession, especially in the south-east Midlands. Most of them are harder than the average chalk and these are called 'rocks'; they tend to be conglomeratic ('nodular') and to have an unusually high proportion of foraminifera, shell fragments and chalk spheres:

Top Rock	Top of *planus* zone
Chalk Rock	Base of *planus* zone
Melbourn Rock	Low in *labiatus* zone
Plenus or Belemnite Marls	Base of *labiatus* zone
Totternhoe Stone	Base of *subglobosus* zone

Lower Chalk, Cenomanian

There is nearly always a lithological transition from the upper Albian facies to the Lower Chalk, and the latter retains some appreciable percentage of detrital components. What these are—glauconitic sand, silt or clay—varies from one region to another, and is related to the facies below, whether Gault Clay, Upper Greensand or Red Chalk. The stratigraphical terms used for these Cenomanian impure chalks are old fashioned and a little misleading. Chalk Marl usually refers to a soft rock with a substantial addition of silt and clay; Chloritic Marl is a misnomer, for the scattered green mineral is glauconite and not chlorite; glauconitic chalk may contain a quartz sand component as well. Even where there is no very striking detrital addition the Lower Chalk is nearly always slightly muddy, and grey rather than white. Phosphatic nodules are found at the base in some places and also concentrations of sponges or sponge spicules, but flint is not common at this level.

The two distinctive lithological bands, the Totternhoe Stone and the *Plenus* Marls, are fairly widespread. The former is found from Berkshire to Yorkshire and is 20 feet thick at its name-place in Bedfordshire; here it contains an abnormal amount of *Inoceramus* fragments, which result in a gritty texture, and also has green coated 'nodules' (that is rolled, contemporary, pebbles redeposited in a chalk matrix) at the base. In Lincolnshire and Yorkshire it is a thin bed, up to 4 feet thick, rich in shell fragments and glauconite grains. The *Plenus* Marls take their name from *Actinocamax plenus* which is confined to this horizon. The band, which varies from 1 to 6 feet, is found in all the principal English outcrops as a layer of grey marl; in Yorkshire it is dark enough to be called the Black Band. This was a brief period of exceptional mud influx into the Cenomanian sea.

Besides these widespread lithological variations there are a few others, more local. In particular the outcrops along the south coast become increasingly sandy towards the west, a transition comparable to the development of Upper Greensand below and similarly related to the neighbouring Cornubian uplands. The present Chalk outcrops in the south-west have, however, been more depleted by erosion than those of the underlying beds; outliers are few in west Dorset and none extends beyond Sidmouth in the extreme south-east of Devon. In this region the Lower Chalk is thin and irregular, and for a short stretch absent, where Middle Chalk rests on Upper Greensand. It is usually sandy and glauconitic, and in some places inland is reduced to a few feet of calcareous sandstone; the basal beds are frequently conglomeratic and phosphatic.

Another unusual local basement bed is the Cambridge Greensand, a thin, rather irregular bed of calcareous clay with grains of glauconite. It lies on an uneven surface of Gault Clay, in the hollows of which are the phosphatic nodules that once were extensively worked for agriculture. They include the rolled remains of fossils from at least two zones of the Upper Gault which have been locally eroded leaving only this remanié bed behind. In addition there is a rather mysterious assemblage of boulders and pebbles, or 'erratics'. Far-travelled pebbles have occasionally been found at other levels and places in the Chalk but are most abundant and best known from the Cambridge Greensand. The rocks include igneous and sedimentary types whose nearest recognizable source is in the Midlands and Wales, and some of no known origin. It is possible that they are secondarily derived from some intermediate formation, such as the New Red Sandstone, but even so how they got into the early Chalk sea is a puzzle; transport in the roots of trees is a possible solution, though one not easily proved.

The erosion of the Upper Gault in Cambridgeshire can also be seen in a wider context: it is the latest of the long sequence of abnormally reduced strata or of actual unconformity, which intermittently affected Jurassic and Lower Cretaceous sedimentation in this region, and which is broadly related to the northern, unstable, edge of the London Platform.

There are variations in thickness over the Cenomanian outcrops as a whole though they are not spectacular. Two hundred feet is a normal figure in south-east England, but the accentuated subsidence of the Wessex basin is shown in the maximum of 350 feet in the Portsdown boring. The effects in the south-west have already been mentioned, but there is another, less marked, reduction in East Anglia, from 160 feet at Cambridge to 125 feet in south Norfolk and only 56 feet at Hunstanton. At the last most of this thinning takes place in the lowest bed (the Chalk

Marl) which rests with a basal layer of nodules on a comparably reduced amount of Red Chalk. The Lower Chalk remains relatively thin in Lincolnshire and there is only a modest expansion, to 125 feet, in Yorkshire. The increasing uniformity of Cretaceous conditions is further shown in this region, for the Cenomanian stage is the last in which any reduction can be traced over the Market Weighton upwarp.

Middle and Upper Chalk (Turonian and Senonian)

As would be expected there is a greater uniformity in these two divisions than in the lower one with the proviso, already noted, that the thickness of the Upper Chalk is strongly affected by later erosion. The Middle Chalk has the same order of thickness as the Lower, and reaches 220 feet in Bedfordshire. It consists dominantly of white chalk, often nodular, with flints appearing near the top. The Melbourn Rock is a 10-foot yellow band crowded with nodules and is found from Wiltshire to East Anglia. The detrital origin of nodules in the *lata* zone is well seen in the Isle of Wight, where the pebbles have been bored while lying on the sea floor and are coated with a green crust. The westerly facies is again accompanied by irregular thinning; the Middle Chalk at Beer, on the east Devon border, still amounts to 130 feet but is only 80 feet inland. The Beer Freestone is a 13-foot local marginal facies composed very largely of comminuted shells.

The Upper Chalk of the south-west has naturally suffered the most erosion and is represented by two small outliers. Neither rises above the *cortestudinarium* zone nor shows any marked sign of westerly detritus. There is, however, one further relic of Senonian sediments in the flint gravels that cap the Upper Greensand of the Haldon Hills, for these have yielded casts of chalk fossils of several ages, up to the *Marsupites* zone.

Over most of southern and eastern England the Upper Chalk is a relatively soft, white rock with flints, and the two hard bands are confined to the *planus* zone at the base. The 15-foot Chalk Rock stretches from Wiltshire to Cambridge and the thinner Top Rock has a slightly more restricted distribution in the same region. Where fairly complete the Upper Chalk is by far the thickest of the three divisions. It reaches a maximum of 1,360 feet in the Isle of Wight and is about 1200 feet in East Anglia. Despite its greater stratigraphical range in the latter, the total thickness of the Chalk (about 1,350 feet in north-east Norfolk) does not equal that in the former, so that though the 'Wessex basin' was no longer a recognizable unit in facies or palaeogeography, it was still an area of slightly accentuated subsidence.

The distinction of the Yorkshire basin was more clearly maintained into the Senonian. The position of Lincolnshire is uncertain because only 60 feet of Upper Chalk have escaped erosion there, but there are about 1,000 feet in the Yorkshire Wolds and the total complement of the Chalk there is over 1,500 feet. Moreover, there are faunal and lithological distinctions between north and south. The white chalk of Yorkshire is harder, better cemented and originally included a small proportion of aragonite crystals; the flints so characteristic of all the southern Senonian are restricted to the three lowest zones. The echinoderm index fossils, such as *Micraster coranguinum* and *Offaster pilula* are replaced by others, and similarly ammonites, which are rare in the south above the Chalk Rock, continue later in the north in such genera as '*Scaphites*' and *Hamites*. Some Yorkshire characters, in all three Chalk divisions, are found in the northern outcrops of Norfolk, and this county appears to be in an intermediate position between the northern and southern facies and faunas. The Maastrichtian chalk occurs only in the contorted, glacially transported masses found on the north Norfolk coast at Trimingham (Fig. 71).

TRANSGRESSIONS IN IRELAND AND SCOTLAND

Only Upper Cretaceous beds are found in Northern Ireland and the Hebridean Province and their outcrops follow the distribution of earlier Mesozoic rocks; the relative importance of the two countries is reversed, however, and the Irish outcrops are the more extensive and informative. They underlie the Antrim Basalts and form a nearly continuous strip on the south-east and north-east, and are seen at many places in the west. Uplift and erosion succeeded the Chalk deposition and not very much remains, the greatest thickness known being 479 feet in a boring west of Lough Neagh.

There is a fairly simple lithological distinction into the greensands, below and chalk above, but both rock-types vary in thickness, and to some extent in age, over the long, strip-like outcrops. The succession can be exemplified from the three regions of Table 56. The Cenomanian seas first invaded the south-eastern part of the area and laid down greensands, followed by more argillaceous sediments, with a combined thickness of perhaps 40 or 50 feet. The ridge of Dalradian schists in north-east Antrim was not submerged at this time, nor apparently was much of the country to the north and west. There was a minor phase of uplift and erosion within the Cenomanian, and a further series of greensands was deposited before the end of the period.

Table 56. *Cretaceous rocks of Northern Ireland*

Stages	North Antrim	North-east Antrim	East Coast (e.g. Larne)
MAASTRICHTIAN	(Pellet Chalk)		
SENONIAN (*mucronata* to *coranguinum* zones)	White Chalk Glauconitic Chalk and thin greensand	White Chalk Thin basal conglomerate	White Chalk Glauconitic Chalk
UPPER TURONIAN	(Lias)	(Dalradian Schists)	Glauconitic Sandstone
CENOMANIAN			Glauconitic Sandstone Sandy and argillaceous limestone
			(Lias)

Then followed a major regression of the sea, and Turonian beds are almost entirely absent. In late Turonian and early Senonian times there was a second major transgression with a gradual decrease in the sandy influx, greensands in the south-east being succeeded by glauconitic chalk. The Senonian seas spread a little later into the north, and still later submerged the Dalradian ridge. All the White Chalk is thus Senonian in age and much of it belongs to the *mucronata* zone. This was probably a further period of marine expansions; the small outlier or pocket of chalk far to the south, near Killarney in Co. Kerry, appears to belong to this age, and flints are known from the glacial deposits on both the east and west coasts of Ireland. That chalk deposition continued, in places at least, into Maastrichtian times, is suggested by the microfaunas of the residual clay-with-flints and the pellet-chalk. The last fills in a small erosion hollow at Ballycastle, overlying the White Chalk and underlying the Antrim Basalts.

In lithology the Irish white chalk bears some resemblance to that of Yorkshire in that it is a hard white limestone, well cemented with secondary calcite; the term Hibernian Greensands has been applied to the lower beds, but the foregoing summary and table shows this is a diachronous facies, developed at more than one horizon, as is common in basal sands laid down on an irregular land surface. Although much of the greensands are Cenomanian some sandy facies followed the Turonian regression, and in the north there is a thin basal Senonian bed.

The particular importance of the Irish outcrops is that they exhibit better than anywhere else in the British Isles the effects of the Upper Cretaceous marine invasions. There is much evidence of Cenomanian transgressions in France, but in England it is the earlier ones, Aptian and Albian, that are most impressive. The outcrops of Cenomanian and later rocks are now too far removed from their original shore-lines, whose position is chiefly deduced from the sandy influx in Dorset and Devon. In Antrim, however, not only are there the near-shore glauconitic sands, but also the overlap of Cenomanian by Senonian deposits and the gradual submergence of minor features such as the Dalradian ridge.

The same sort of history probably applies to the west side of Scotland, for though the outcrops are small and scattered there is again a lower sandy group with a thin remnant of white chalk above and Turonian beds are absent.

(Tertiary lavas and sediments)

~~~~~~~~~~~~~~~~~~~~

Silicified Chalk, probably *mucronata* zone
Cenomanian:
Thin brown clays with brachiopods
White quartzose sandstone (40 feet)
Glauconitic sandstone with shelly calcareous bands (*c.* 30 feet)

~~~~~~~~~~~~~~~~~~~~

(Lias)

This sequence is found in Morvern, opposite the north coast of Mull, and there are further representatives on that island, on Eigg and on Skye. The strongly transgressive Cenomanian overlies several different formations in its various outcrops—including all three Jurassic divisions, the Trias and locally, on the Morvern uplands, the Moine Schists.

The White Sandstone is an unusual Cenomanian facies; it is virtually unfossiliferous and in its very well-rounded, well-sorted grains and lack of accessory minerals, probably represents the long-continued working of sand on the shoals of a shallow sea. That the Senonian seas had a wide extent beyond these fragmentary outcrops is suggested not only by the general geography of the period, but by numerous remnants in the drifts. Masses of chalk are also included with the other blocks of Mesozoic rocks in the central ring complex of Arran.

No Cretaceous rocks are found in place on the east side of Scotland, but there are several relics scattered among the Pleistocene and superficial deposits and these come from a wider stratigraphical range than on the west. Fossiliferous blocks have been found in Fife and Aberdeenshire; in the latter the facies is a greensand, but it appears that some of the fossils have affinities with the Speeton faunas of Yorkshire. A more spectacular find is a large mass of Lower Cretaceous sandstone near Wick, on the east coast of Caithness; several species of lamellibranchs and ammonites have been collected from this mass and the age may be Valanginian or Hauterivian. The sandstone is considered to be a very large erratic, apparently transported bodily from some place on the floor of the Moray Firth. Upper Cretaceous rocks, in the form of flints and chalk blocks, are locally abundant in the boulder clays of the east coast, from Aberdeenshire to Caithness, and most of the Senonian zones, up to *mucronata*, appear to be represented. Again their origin is considered to be the floor of the North Sea.

The extent to which the later Chalk seas submerged the uplands of Wales, Ireland and Scotland is a subject of much speculation but virtually no decisive evidence. A broad westward extension across the Midlands and much of northern England is implicit in the foregoing description of the Upper Chalk of the main outcrop, its lack of detrital facies, and the known Senonian transgressions in Ireland and Scotland. But whether the present mountainous regions of Older Britain were ever covered by a substantial layer of chalk is a very different matter.

Arguments for such a cover follow chiefly from geomorphological studies. It is well known that the present drainage of the uplands shows in places a marked disregard for the structures of the rocks over which the rivers flow, and many of the major streams flow eastward or southeastward—that is, the drainage systems have been inherited. By some authorities they are accordingly deduced to have formed on a cover of newer rocks and have been gradually let down on to the older rocks during Tertiary erosion. By far the most likely formation for such a cover is the Chalk, and according to some views the pre-Cretaceous surface survives in certain of the existing summit plateaux.

On the other hand, no Cretaceous residual deposits have been recorded from these upland areas, and especially no flints, despite the vast volume of chalk that on this view must have been removed. Flints are known from Pleistocene deposits on the borders of Wales, in areas of drift or in gravels, but where they may reasonably be attributed to the ice sheets from the Irish Sea. In the Weald, which forms a useful comparison because an earlier chalk cover is certain, flints are not very conspicuous but do occur occasionally on the higher ground.

It is also possible to form alternative views on the drainage history, in that the streams may have been established on later, Tertiary, plateaux. Some at least of the upland surfaces must be later than the Tertiary dykes which crop out on them, otherwise the magma emerging would have spilled out in the form of lavas. This certainly applies to the Southern Uplands of Scotland and probably to North Wales. Such chalk relics as are known in the north-west at present are all at a fairly low topographic level, and any theory that entails the widespread uplift of a chalk cover and the development of new consequent rivers must be accompanied by some down-dropping mechanism to account, for instance, for the chalk near sea-level in Antrim or below the Irish Sea. Accordingly it is considered here that the chalk cover of upland Britain remains an interesting hypothesis, but one as yet lacking in much positive evidence.

REFERENCES

Allen, 1949, 1954, 1955, 1959, 1965; Arkell, 1947*b*; Black, 1953; Casey, 1961, 1963; Charlesworth, 1963, ch. 14; Hallam, 1965; Hancock, 1961; Hughes, 1958; Kirkaldy, 1939; Larwood, 1961; Lee and Pringle, 1932; Peake and Hancock, 1961; Swinnerton and Kent, 1949; Wright and Wright, 1951.

British Regional Geology. East Yorkshire and Lincolnshire, East Anglia, Hampshire Basin, Wealden District, Tertiary Volcanic Districts.

11

PALAEOGENE: PALAEOCENE, EOCENE AND OLIGOCENE BEDS

The Cretaceous period ended some 70 million years ago, and the time-span since then has commonly been called the Tertiary (or Cainozoic) Era, to which was added a quite disproportionately small Quaternary Era or Pleistocene System. The Tertiary systems were named Eocene, Oligocene, Miocene and Pliocene, terms derived from the increasing proportion of living forms. It has become apparent over the years, however, that these divisions are not comparable to the earlier systems, heterogeneous though these be, and so there has developed a common practice of referring to a Tertiary or Cainozoic System, and using the four lesser names for epochs or series. The Palaeocene, a division based on the mammalian faunas of North America, has gradually received more universal acceptance; and in many parts of the world there is a convenient distinction in Tertiary history between the three lower subdivisions, the Palaeogene, and the two upper—or three if the Pleistocene is included—the Neogene. Palaeogene and Neogene fit British stratigraphy well. The Mid-Tertiary was a time of uplift and folding in Britain, and very few, if any, Miocene outcrops remain; Pliocene deposits are small and scattered, but form an introduction to the complex history of Pleistocene refrigeration and glaciation. Much of this chapter is concerned with the history of sedimentation in the main outcrops of southern England, but certain aspects of Tertiary vulcanicity in Scotland and Ireland also lie within its compass.

THE ANGLO-PARISIAN-BELGIAN BASIN, SEAS AND SHORES

After the recession of the Chalk seas the British Isles were subjected to uplift, warping and, in places, gentle folding; when sedimentation was re-established it presented a very different picture. In some degree this change was also a return, for the Anglo-Parisian Basin of Jurassic and Lower Cretaceous times, whose boundaries were largely submerged by Senonian seas, on the succeeding uplift became a major sedimentary basin in north-western Europe, extending from Belgium on the east to southern England on the west. The earlier connexion with the Tethys through southern France, however, does not seem to have reappeared, and the basin was largely surrounded by land except for a wide opening to a northern sea which lay across Denmark and into North Germany. Nevertheless, marine faunas from the Mediterranean, especially the large foraminifera, the nummulites, intermittently found their way into the northern basin along a route in the western part of the English Channel, where outliers of Lower Tertiary beds still survive on the sea floor.

The principal Palaeogene outcrops within this semi-enclosed basin comprise those of the Paris region; Belgium, from Brussels westward to the coast; the London synclinorium and parts of East Anglia; and the Hampshire synclinorium, including the north of the Isle of Wight. The successions in all four have much in common, and are particularly linked by the successive marine invasions. Since these came in from the north-east, the Belgian and easterly London outcrops are the most persistently marine, whereas to the west and south the Parisian and Hampshire out-crops have a greater proportion of fluviatile, lacustrine or littoral beds. In the Paris region a full complement of Oligocene beds is succeeded by a small amount of Miocene sands; but in the Hampshire basin nothing is known above Middle Oligocene and erosion in the London basin has left even the Middle and Upper Eocene poorly represented.

Although the Palaeogene history of this northern basin can be worked out fairly satisfactorily, correlation on an international basis presents difficulties. The two principal groups of stratigraphical importance in Tertiary times are the mammals and foraminifera. The mammalian sequence is based on the continental beds of North America, and within our region can be employed to some extent in France but very little elsewhere. The larger foraminifera flourished in the warmer waters of the Tethys and are found in the countries around the Mediterranean; but they only entered the Anglo-Parisian-Belgian Basin occasionally and are helpful chiefly in Middle and Upper Eocene beds. This does not mean that the marine beds are unfossiliferous; on the contrary, they are often exceptionally rich in shells, chiefly molluscs, which in the Paris outcrops formed the basis of the earliest studies in stratigraphical palaeontology; there are also some important plant-bearing formations. Then the essence of Palaeogene history in this basin is a series of marine transgressions and regressions, with all the oscillations of facies and diachronous boundaries that such a régime entails. The total result is that the 'stages' in this province are less universal than those of the

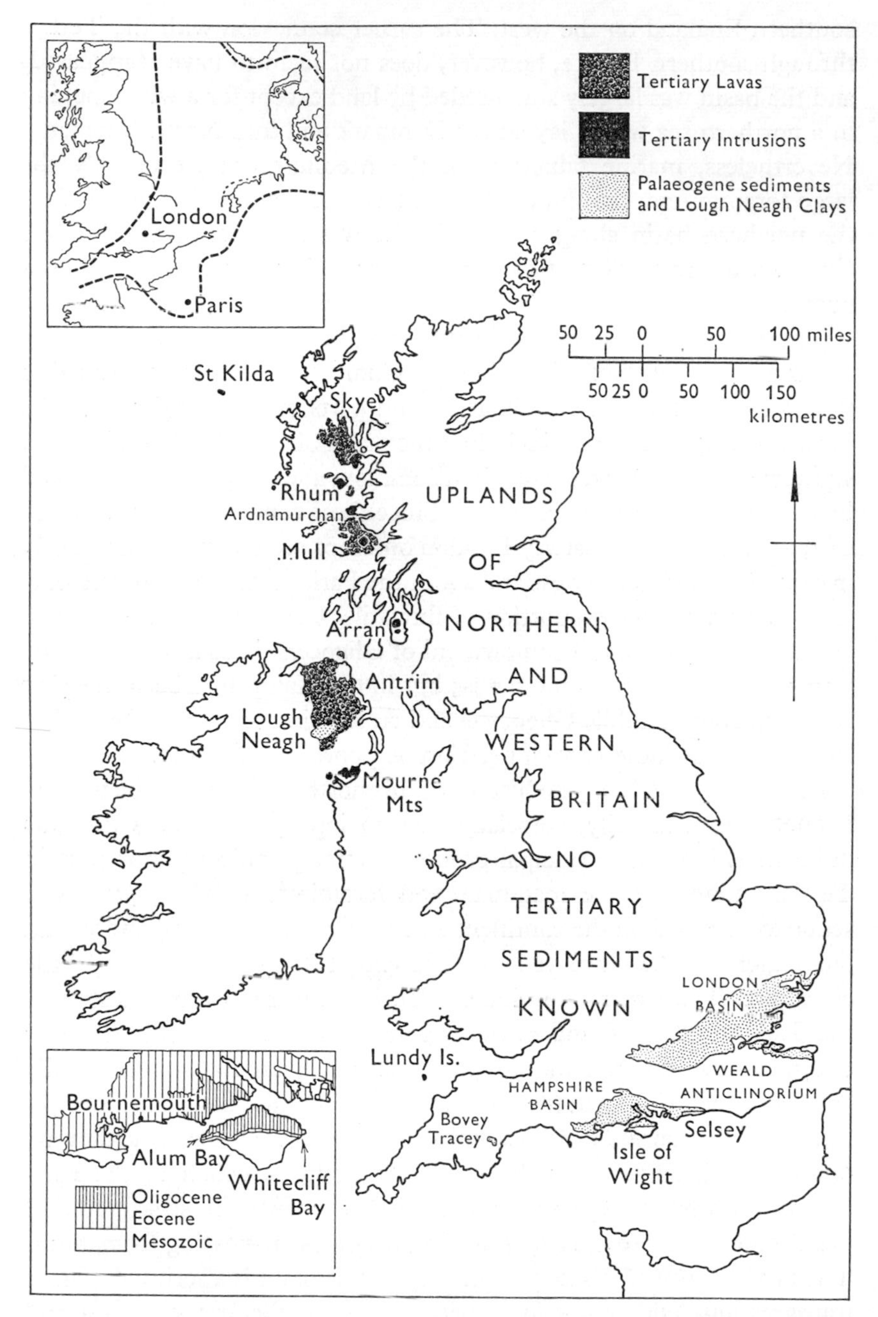

Fig. 68. Outcrops of Tertiary sediments in the south of England and of the Tertiary Volcanic Province. Insets: upper, relation of the Anglo–Parisian–Belgian Basin to the North Sea Basin; lower, outcrops of the Isle of Wight and neighbouring mainland.

preceding systems, except in so far as they can be translated into a foraminiferal or mammalian sequence, and those shown in Table 57 should be considered only as a general guide. The three Oligocene divisions can also be termed Lower, Middle and Upper; sometimes the Lutetian has been termed Middle Eocene and the stages above and below, Upper and Lower Eocene.

Table 57. *Palaeogene succession in southern England*

	Stages	Hampshire Basin	London Basin
	(Upper Oligocene, Chattian and Aquitanian, absent)		
OLIGOCENE	RUPELIAN	*Absent* Upper Hamstead Beds	*Absent*
	LATTORFIAN	Lower Hamstead Beds Bembridge and Osborne Beds Upper and Middle Headon Beds	
EOCENE	BARTONIAN	Lower Headon Beds Barton Beds	
	AUVERSIAN	Upper Bracklesham Beds	Upper Bagshot Beds*
	LUTETIAN	Lower Bracklesham Beds	Middle Bagshot Beds
	CUISIAN		Lower Bagshot Beds*
	YPRESIAN	London Clay	London Clay
PALAEOCENE	SPARNACIAN	Woolwich and Reading Beds	Woolwich and Reading Beds
	THANETIAN	*Absent*	Thanet Sands
	MONTIAN	*Absent*	*Absent*
		(Upper Chalk)	

* The limits and exact correlation of these divisions are uncertain; the Middle Bagshot Beds are Lutetian.

At present the two main English outcrops are separated by the Wealden anticlinorium and the chalk of north Hampshire. The earth movements responsible for this separation are, however, largely Mid-Tertiary and during Palaeogene times the deposits were broadly continuous from Essex to Dorset and possibly beyond.

The two major basins are structurally synclinal and like the Weald are diversified by minor flexures; these are especially conspicuous where they bring up anticlinal ridges of chalk, such as that on which Windsor Castle is built, or the longer one from Portsdown to Dean Hill near Salisbury. A further anticline is responsible for chalk outcrops along the line of the Lower Thames. Both synclinoria have much steeper dips on their southern limbs, that of the Hampshire basin being the monocline of the Isle of Wight, where the Eocene beds are almost vertical, while the narrow ridge of chalk along the Hog's Back, west of Guildford, combines dips up to 45° and thrusting.

Palaeocene and Eocene Beds, the Hampshire Basin

The principal sections here are on the east and west coasts of the Isle of Wight and along the cliffs of Christchurch and Bournemouth bays, from Milford westwards to the Dorset outcrops of Studland and inland. Some of the lower and middle beds also crop out on the Sussex coast around Selsey Bill. Owing to the gently synclinal structure the outcrops of Sussex and Dorset do not include the highest beds (Bartonian) and in these only a small degree of facies-change is detectable. In the Bracklesham Beds and below, however, the total extent of the outcrops is enough to demonstrate the greater proportion of marine beds to the east, and what we may group together as 'non-marine' on the west (Fig. 69).

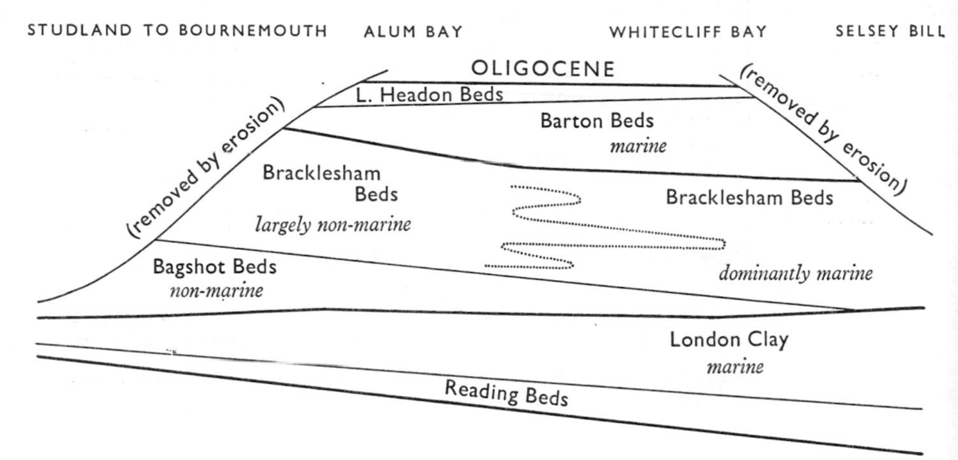

Fig. 69. Diagram to show the principal formations and facies changes in the Eocene beds of the Hampshire basin; the relative thicknesses are suggested but the diagram is not to scale. In the central outcrops Oligocene beds follow conformably.

The section at Whitecliff Bay at the east of the Isle of Wight will serve as an introduction to the stratigraphy in this basin (Table 58). The only facies that are not clearly marine are the Lower Headon Beds, the Reading Beds (which have a marine bottom bed with oysters in Sussex) and a relatively small amount of Bagshot Sands which intervene between the London Clay and Lower Bracklesham Beds. A remarkable change has taken place in the middle part of the succession when it reappears in Alum Bay 20 miles away on the west coast of the island. The Reading Beds and the London Clay are similar, though the latter has an increased proportion of sand in the upper part, but above these there are 770 feet of beds with no marine fossils, though glauconite is not uncommon near the middle of the sequence. The sediments are red and white sands and white and chocolate-coloured clays, the latter containing abundant vegetable remains and at least three distinct plant beds. Their original conditions are not very clear, but most probably they were largely non-marine. The lower 80 feet or so are conventionally referred to the Bagshot Beds and the remainder to the Bracklesham Beds.

Table 58. *Eocene succession at Whitecliff Bay, Isle of Wight*

Formations and lithology	Thickness (feet)
Lower Headon Beds	
Freshwater sands	28
Barton Beds	
Glauconitic clays and sands, molluscs	368
Upper Bracklesham Beds	
Glauconitic clays with sands, molluscs and *Nummulites variolarius*	
Lower Bracklesham Beds	585
Glauconitic clays with sands, molluscs throughout; *N. laevigatus* in the higher beds and *N. planulatus* lower down	
Bagshot Sands	138
London Clay	
Grey clays with silts	320
Reading Beds	
Red and mottled clays with silts and sands	150

Above there follows the basal pebble bed of the succeeding Bartonian marine transgression. The Barton Beds are dominantly clays at both ends of the Isle of Wight, and have yielded a rich fauna, chiefly of molluscs, though nummulites are found in the lower part. The facies is entirely marine, except towards the top, where marine fossils become fewer, being progressively replaced by freshwater forms, showing that the water became less and less saline. The uppermost Bartonian sediments, which were laid down wholly in fresh, or almost fresh water, are the Lower Headon Beds—clays, shelly marls and thin limestones. Mammal, reptile and plant remains are known from the Hordle cliffs of the mainland. The division also thickens westwards from 28 feet at Whitecliff Bay to 81 feet at Hordle.

Westward, round Bournemouth Bay to Studland, and thence inland into east Dorset, there is a further accentuation of the facies-change in beds of Bracklesham age and below. The Reading Beds become thinner and coarser through the Dorset outcrops, and their gravels contain flints and pebbles of chert from the Cretaceous greensand. In the same region the London Clay is reduced to about 100 feet and becomes rather more sandy. Conversely the Bagshot facies expands to 450 feet of bright red sands, pipe-clays and coarse, pebbly current-bedded sandstone; the pebbles become relatively more important westward and the outliers south of Dorchester consist of sandy gravels. The remainder of the Bracklesham Beds (?Lutetian and Auversian), about 700 feet, comprise current-bedded sands with some clays and pebble beds. These are locally known as the Bournemouth Marine and Freshwater Beds and the Boscombe Sands, but the truly marine elements of the first are few, and elsewhere there are leaves of palms, conifers, deciduous trees and ferns.

The Hampshire basin thus provides evidence of at least three marine invasions, from Sparnacian to Bartonian times, with intervening periods of regression. Thus far the succession conforms to the concept of a cyclic series of oscillations between marine and non-marine conditions. But the process was decidedly irregular and the marine sequences, as they may be interpreted from the surviving outcrops, varied considerably in extent and duration. The only sign of the first is the small thickness of marine oyster-bearing clays at the base of the Reading Beds, which are found in Sussex and at isolated places in Dorset and Hampshire. In contrast the succeeding Ypresian invasion took the London Clay right across into Dorset, to be covered by the Bagshot sands and gravels which gradually encroached eastward as the sea retreated. The third, Bracklesham, invasion produced a thicker series of sandy clays in the east (Whitecliff Bay) but did not penetrate nearly so far and gives rather the impression of a series of advances and minor retreats. There are some lignite seams and rootlets in the Bracklesham Beds themselves, and very low-lying coastal flats of muds and silts, occasionally colonized by shore plants, may be envisaged. In Bartonian times the transgression extended throughout Hampshire, as far as the limited exposures allow us to trace it, and was followed by gradual freshening of the waters of the Headon Beds, which continued on into Lower and Middle Oligocene times.

A serious gap in this reconstruction is a full understanding of the non-marine facies, which are poorly fossiliferous but show a great variation in lithology. Till their petrology has been subjected to the same rigorous scrutiny as the faunas of the marine beds it will not be possible to fill in the picture on the landward side of the successive oscillations. Pollen analysis of the carbonaceous beds would probably assist also. The pebbles and mineral assemblages of certain beds (especially Bagshot) show that the Dartmoor highlands were still providing a detrital element as in earlier periods, and the plants indicate a climate much warmer than that of modern England. This aspect, however, is better known from the flora and fauna of the London Clay (p. 355).

The London Basin

The much smaller vertical extent of the Eocene Beds here is shown in the table of strata (p. 349) but there are compensations in the earlier marine beds of the Thanetian and Sparnacian stages. These, and the typical London Clay, are best seen in the easterly outcrops around the Thames Estuary. In the west, at Reading and beyond, the facies more resemble parts of the Hampshire basin.

After the late Cretaceous and Montian emergence, the seas at first invaded the London basin for only a short distance, and the Thanet Sands are confined to the east, not quite reaching Colchester in Essex or Guildford in Surrey. Forty feet of unfossiliferous glauconitic sand is a normal development, but 80 feet are found at Herne Bay in Kent and shell beds also appear. Over most of the outcrop the Thanet Sands are underlain by a basal layer of green-coated flints, the Bullhead Bed. Unlike those in many of the later flint conglomerates these pebbles are largely unworn, and probably represent a clay-with-flints residue (such as occurs on existing chalk surfaces) which was incorporated in the incoming Thanetian sea without much transport and erosion.

The succeeding Woolwich and Reading Beds are much more extensive, and their double name covers three rather than two facies. The marine Bishopstone facies of glauconitic shelly sands, seen in north Kent, is again restricted to the east, and the Woolwich sea extended little, if at all, farther than the Thanetian. The non-marine Reading facies—sands and mottled clays—on the other hand, are found over almost all the western Tertiary outcrops, and so rest for a large part on the chalk. A 'bottom bed' of somewhat worn flints underlies both facies. In the London region, and forming the Woolwich Beds of the type area, there is an intermediate facies containing brackish and estuarine shells.

The Ypresian invasion is even more impressive in the London than in the Hampshire basin; London Clay is not only the thickest formation

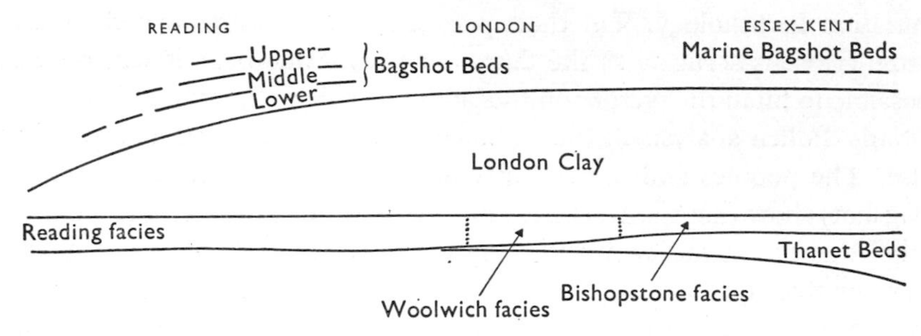

Fig. 70. Diagram, similar to Fig. 69, to show the principal Palaeogene formations of the London basin. The relative thicknesses are suggested but the diagram is not to scale.

that remains but has much the largest outcrop, covering most of the triangle from Ramsgate in Kent to Newbury in Berkshire and northwards (under Pleistocene) as far as Yarmouth. Typical London Clay is a brown, dark-grey or bluish homogeneous deposit diversified only by occasional cementstone layers. As such it is most characteristically displayed in the east, where it is also thickest, being over 500 feet in south Essex. Westward it becomes thinner and increasingly silty, especially in the upper and lower layers. The term 'Claygate Beds' has been used for alternating clays and sands in the upper part in Surrey and Greater London. South-west of Reading the London Clay is only 200 feet thick and includes sand near the top.

The westerly thinning probably represents partly a genuine reduction in contemporary sedimentation, and partly a diachronous boundary with the Bagshot Beds above (and possibly with the Reading Beds below), so that sands were advancing from the west while the London Clay sea still persisted in the east. As noted in the Introduction, however, diachronism is difficult to demonstrate clearly without a detailed, reliable series of zones, and these are largely lacking in the London Clay.

There is more than one type of pebbly or sandy basement bed beneath the clay. The best known is the Blackheath Beds of the London district, the pebbles being quantities of well-rounded flints; the Oldhaven Beds are a more sandy variant in east Kent. Both include marine faunas and there are brackish water elements as well in the Blackheath Beds. Both transgress southwards in places over Woolwich and Thanet beds against the rising chalk of the North Downs and show strong channelling at the base. It has sometimes been thought that an early 'Wealden island' was already uplifted, but there is little positive evidence of land at this time, and the existence at Newhaven and Dieppe of an estuarine Woolwich facies is somewhat against it. Where the London Clay itself transgresses beyond the Blackheath Beds, it has its own basement of rolled flints, overlying Reading Beds.

Fossils are not abundant throughout the clay, but a large and varied fauna has been compiled from various places and horizons, notably from the upper beds exposed in the cliffs of the Isle of Sheppey. Mollusca are the chief components, but there are also polyzoa, brachiopods, annelid worms, fishes and foraminifera (though the last have not yet proved to have stratigraphical value). Among the rarer invertebrates are crabs and cirripedes, the latter perhaps brought in on floating logs. So far there is nothing very exceptional for a Tertiary marine clay, but the plants are outstanding. Evidently the fragments of a coastal flora were swept out by currents and, finally becoming water-logged, sank on to the offshore muds. Apart from drifted wood, the assemblage consists very largely of the fruits and seeds of dicotyledons, palms and conifers. Of these the most abundant are the coconut-sized fruits of the stemless palm *Nipa*, a modern genus of Indian deltas and swamps. As a whole the flora gives a vivid picture of tropical or near-tropical flora living on the strand and in mangrove-like swamps on the lowland shores of this northern Eocene basin. Moreover, this picture is reinforced by many of the invertebrates and fishes, which have warm-water relatives at the present day, and the existence of such reptiles as crocodiles.

The Bagshot Beds of the London Basin succeed the London Clay over wide areas in the west, particularly in the heath country of west Surrey, north Hampshire and south Berkshire—the type area where the beds have a maximum thickness of 400 feet. Farther east they only form outliers, capping hills in south Essex and the heights of north London such as Hampstead and Highgate. The exact ages of the Lower, Middle and Upper Bagshot Beds are not clear because they contain very few marine fossils, but, like their counterparts in the west of the Hampshire basin they probably correspond to much of the Bracklesham Beds, with the possibility that the Upper Bagshot Beds may partly correspond to the lowest Barton Beds. The typical Lower Bagshot Beds are very like those of the Hampshire basin: fine, current-bedded, variegated sands with flint pebble beds, pipe-clay and lignite. Even so we should beware of identifying these sands too closely with environment, for in Essex some parts at least contain rare marine molluscs. Some marine forms are also found in the middle and upper beds, and in the former the mollusca and *Nummulites laevigatus* indicate a Lutetian age. Upper Bagshot Beds are restricted to a number of outliers on the borders of Berkshire and

Surrey; their basal pebble beds contain, in addition to the usual flints, rare fragments of chert. If, as commonly supposed, the last really are derived from the Lower Greensand of the Weald, then that structure must have been uplifted, and the Chalk considerably eroded by this time. But the identification of the chert is crucial.

Oligocene Beds, the Hampshire Basin

Oligocene rocks occupy the northern part of the Isle of Wight (Table 59), where, owing to the marked decrease in dip away from the central monocline, they cover a much larger area than the nearly vertical Eocene beds. The Headon Beds also crop out across the Solent on the Hampshire coast and in the New Forest. There is a much smaller proportion of marine beds than in the Eocene, marine horizons occurring only at the base of the Middle Headon Beds, in the Bembridge Marls and at the top of the Upper Hamstead Beds. The facies also are commonly finer with much clay, calcareous marl and beds of impure limestone; non-marine sands of Bagshot type are scarce. Thus the 700 feet or so were laid down in quieter conditions, with less current action and the local precipitation of lake marls.

Table 59. *Oligocene succession in the Isle of Wight*

Stages	Formations	Approximate thickness (feet)
RUPELIAN	Upper Hamstead Beds	30
LATTORFIAN	Lower Hamstead Beds	225
	Bembridge Marls	120
	Bembridge Limestone	25
	Osborne Beds	80
	Upper Headon Beds	60
	Middle Headon Beds	up to 130
	(Lower Headon Beds, Bartonian)	

The Middle Headon Beds consist of a variable group of sandy clays and sands laid down on the fringes of the sea; for the most part their fossils indicate brackish conditions. However, a thin bed with a purely marine fauna, including corals, is to be found at their base in the east (but not in the west) of the Isle of Wight. Somewhat surprisingly this bed is also found in certain north-westerly outcrops, as the Brockenhurst Beds of the New Forest, but is never more than 2–3 feet thick. A marine regression is indicated in the Upper Headon Beds, which contain marls and limestones rich in the freshwater shells, *Lymnaea* and *Planorbis*.

The same rock-types, freshwater marls and limestones, persist through the Osborne and Bembridge beds; a minor marine incursion is represented by oyster beds at the base of the Bembridge Marls in the east of the Isle of Wight. Many terrestrial fossils are found in the Bembridge Beds, land shells, mammals, and reptiles, and the 'Insect Bed' contains plants as well as insects. The flora includes some of the tropical forms already known from the London Clay, but they are mixed with more familiar, temperate trees such as the oak and the beech. The limestones yield the small oogonia (reproductive bodies) of the alga *Chara*, which then, as now, was an active agent in the precipitation of lime muds in shallow meres; *Chara* is also known from the Upper and Lower Headon limestones.

The freshwater régime continued throughout the deposition of much of the Hamstead clays and sands. A dark-grey carbonaceous clay at the base contains the remains of aquatic plants whose roots penetrated the Bembridge Marls beneath; higher up there are marls rich in shells and a layer of water-lily roots and leaves. The latest record of Palaeogene sedimentation is the beginning of the Rupelian marine incursion, represented by the 30 feet of clays and sands with marine shells at the top of the succession.

Outlying Tertiary deposits

Beyond the main outcrops just described, there are a very few isolated outcrops or basins to suggest some extension of Tertiary beds to the west. The first example is one of the minor curiosities of Dorset geology. Creechbarrow is a small upstanding hill north of the chalk ridge a few miles west of Corfe Castle. It thus lies on the westerly facies of the Bagshot Beds, and these form the foundations of the hill. The top is capped, however, by sands of a different type, containing weathered but unworn flints, and these by the Creechbarrow Limestone, a soft tufaceous rock which has been interpreted as an isolated, marginal remnant of Bembridge Limestone.

Outliers of unfossiliferous sands and gravels, usually ascribed to the Bagshot Beds, are found much farther west than any of the other Tertiary formations, rather in the same way that the earlier, Cretaceous, sandy facies extended farther west than the clays. There is also a similar overlap, in this case, over Reading Beds and London Clay, and south-west of Dorchester, coarse gravels fill solution hollows in the chalk. A

Bagshot age, on grounds of lithological similarity to the Dorset gravels, has been claimed for a few further outcrops. The Haldon Gravels, which cap the greensand of the Haldon Hills (p. 239) contain no fossils, except those in their derived Cretaceous flints; there are also patches of flint gravels, which have been compared to both the Haldon and Dorset examples, at Marazion in Cornwall, and near Bideford in north Devon, but whether these are really Eocene or later is doubtful.

A much more extensive Tertiary outcrop is found in the Bovey Tracey basin on the south-east slopes of Dartmoor (Fig. 54). It is a depression in the Palaeozoic rocks and though the Bovey Tracey Beds crop out over only ten miles, they are more than 640 feet thick, the full depth of the deposit being unknown. The basin was originally a hill-girt lake, gradually silted up by the detritus from the Dartmoor upland. At the upper, north-western, end sands were dumped, together with masses of wood, and lake gravels are found around the margins; the finer detritus, from the decomposed feldspars of the granite, was carried to the lower end, where the clays form seams among the sands and have been used for pottery. The vegetation of the hillside slopes and lake margins is represented by seams of lignite as well as by the fossils found in the clays and sands. The dominant form is the redwood genus, *Sequoia*, but there were also thick growths of the palm *Calamus*, the fern *Osmunda* and water plants such as *Stratiotes*. A smaller outcrop of similar sands, clays and lignites is found in north Devon, near Petrockstow, and may have a similar derivation.

Unfortunately there is little detailed evidence for the age of the Bovey beds, since the plants are long-range forms. They are usually considered to be Oligocene, with a slight bias in favour of Middle Oligocene. Much of the interest of these deposits, however, lies in their rarity. During periods of emergence and erosion the hilly country of any age must have contained many such silted up basins, rather in the same way that large and small lakes among the hills of northern England and Scotland are silting up at the present day, and others of late glacial age are wholly filled. These lakes contain clays, sands and pebble beaches and the plant-remains are now found as peat. Such deposits, however, tend to be geologically transient, and all too often are swept away by a later phase of rejuvenation and erosion. It may be that much of the Bovey Tracey Beds has already gone, and that we owe the remainder to the exceptionally deep basin, perhaps down-faulted, in which they accumulated.

The Mochras borehole (p. 311) penetrated more than 1000 feet of Tertiary sediments, mainly clays.

THE TERTIARY VOLCANIC PROVINCE

A detailed study of igneous activity in the British Isles is no part of this book, which principally seeks to trace the surface history—of the seas, and as far as may be, of the lands. As such it is largely concerned with sedimentary rocks and the conditions under which they were formed. Tertiary volcanic history is not intimately bound up with Tertiary sedimentation, as were the two aspects, for instance, of Caledonian history. Consequently it is treated summarily here, except for certain topics that relate to the lavas, their age and associated sediments.

Geographically the British volcanic province comprises the Inner Hebrides (Fig. 8), Arran, Antrim (Fig. 53) and other parts of north-east Ireland. Outside Britain similar lavas and plutonic centres have a much greater extent around the North Atlantic: in the Faroe Islands, Greenland, and in Iceland where there is still some active vulcanicity.

In Britain the igneous history can be divided, broadly, into three stages, but they are not sharply divided, and in one region or another there was much overlap between them. The first was an extrusive phase of very widespread eruption of basaltic rocks, including tholeiitic and particularly olivine basalts. This was followed by the emplacement of many large intrusions, with long and complex histories. In Scotland these include the plutonic centres of Skye, Mull, Rhum, Ardnamurchan and Arran; in north-east Ireland there are the Mourne Mountains, Slieve Gullion and Carlingford, and the age of 50 to 55 million years determined for the Lundy Granite establishes it as a southern outpost of Tertiary vulcanicity. Both lavas and central intrusions are cut by dyke swarms (chiefly doleritic), whose dominant direction is north-west. Those centred on Mull are widespread in south and west Scotland, and include the Acklington Dyke of Northumberland and the Cleveland Dyke of north-east Yorkshire. From their petrological similarity to the Scottish rocks it is probable that certain dolerite dykes in North Wales and the Midlands belong to this phase of intrusion. A much greater area must have originally been covered by the Tertiary lavas than the eroded remnants now remaining.

In their early phases the lavas welled up gently, probably from shield volcanoes of Hawaiian type, and spread out on to a Tertiary land surface, overwhelming forests and lakes. Layers of sediment are occasionally found below the lavas, and more rarely between successive flows. Most of these are only a few feet thick, but the Interbasaltic Bed of Antrim is a more massive example. Only in Mull is there much evidence of extrusion below water level, and this was in a large crater lake, where there are now many layers of pillow lavas.

In Scotland there is a considerable variety of sediments associated with the early lavas—sandstones, mudstones, lignites and conglomerates. A thin basal mudstone underlies the basalts in Mull, Ardnamurchan and Morvern and is itself underlain by a coal seam in Ardnamurchan. Comparable sediments in Skye include sands, lignites and tuffs, although in general tuffs are rare. The cliffs of Canna expose over 400 feet of lavas, sedimentary conglomerates and volcanic agglomerates, suggesting a basaltic plateau traversed by a great river, or by shifting streams.

Among these several outcrops some plants are found, such as leaves of oak, plane, hazel, ginkgo and magnolia in the leaf-beds of Mull and Ardnamurchan; the remains of a large, upright charred tree trunk is exposed in the lower basalts on the cliffs of western Mull and several more are known from Antrim. In Antrim lignite-bearing beds are found beneath the lavas, but no comparable plant beds. On the other hand the Interbasaltic Bed, which separates the Upper and Lower Basalts, is a substantial zone of weathered lava and decomposition products. During this interval the surface was colonized by a considerable range of plants —hazel, alder and maple and conifers, with monocotyledons and ferns. A lignite-bearing channel cuts down 20 feet into the lava at Portrush, where the lignite has been worked for fuel.

No Tertiary sediments survive above the lavas of Scotland, but the much larger outcrop on Antrim is partly overlain by the Lough Neagh Clays. These are presumably the products of an earlier and more extensive Lough Neagh; they appear at the surface on the south-west of that lake and overlap farther to the west on to Cretaceous and Triassic rocks. Including the portion still submerged they cover some 200 square miles and have been penetrated by boreholes to over 1,100 feet—dimensions which place them among the more substantial Tertiary formations of the British Isles.

Lithologically these sediments are chiefly pale-grey, silty clays with beds of sand and pebbles in the upper and lower parts; there are also many lignite seams, some accompanied by ironstone nodules. In the lower levels plants are abundant, and there are some freshwater shells. The plants and pollen have been compared to upper Oligocene floras of Germany, but fragments of *Sphagnum* suggest a later date. The clays have also been affected by the major faults that cut the basalts beneath. In all this there is little evidence for the age of the clays, nor have they been much help in determining the age of the Antrim lavas.

Until recently the principal clues to the age of Tertiary lavas (and hence the age of this vulcanicity in Britain as a whole) were the leaves and pollen in the associated sediments—and very varying results were obtained. The Mull leaf-beds were considered for many years to be probably Eocene, but later it was thought that the pollen from them was not older than Upper Oligocene and that much of the igneous activity might be Miocene. Then if the sparse and poorly preserved pollen of the Antrim Intrabasaltic Bed was that of *Sequoia* an Oligocene age was likely, but if *Metasequoia* then it might be late Miocene or early Pliocene. Part of the difficulty was that the leaves of Tertiary plants are commonly not a good guide to specific identification.

These divergences are now largely of historical interest, because a number of isotopic dates are available from the minerals of the granitic intrusions in Ireland, Arran and the Hebrides, and no doubt more will be forthcoming. Although there is some variation, all appear to fall into the Palaeocene or Eocene periods. It thus seems probable that there was no very long unrecorded time-gap in late Cretaceous and early Tertiary history in this province. The chalk of the *mucronata* zone was deposited, and some of the later zones in Ireland and possibly elsewhere; the region was then uplifted, faulted and eroded, and transformed into wooded fertile country, which was in turn overwhelmed by the vast flows of basaltic lavas. From time to time these ceased long enough for the flow surfaces to weather, when lateritic 'boles', soil and vegetation were established, and then the eruptions began again. The great central intrusions, mainly of acidic magma, were accompanied by explosive extrusive activity. After the last stages of dyke intrusion there was a long period, in later Tertiary times, during which much of the basalt plateaux was eroded and the plutonic masses deeply dissected.

REFERENCES

Arkell, 1947*b*; Chandler, 1957; Charlesworth, 1963, chh. 15 and 16; Curry, 1958, 1960*a*, *b*, 1962, 1965; Curry and Wisden, 1960; Davies, 1939; Davis and Elliott, 1958; Hallam, 1965; Pitcher *et al.* 1960; Reid and Chandler, 1933; Stewart, 1965; Wilson and Robbie, 1966; Wrigley and Davis, 1937; Wooldridge, 1926.

British Regional Geology. Hampshire Basin, London District, south-west England, Tertiary Volcanic Districts.

12

PLIOCENE AND PLEISTOCENE DEPOSITS AND HISTORY

The outstanding feature of the Pleistocene period was widespread glaciation, so that it has also been called the Great Ice Age or the Quaternary Ice Age, and is the best known by far of the various ice ages that have left their traces on the earth's surface. Most of the British Isles was covered by ice during this time and much of it more than once. In contradistinction, while a gradual cooling took place during Tertiary times, there are no known traces of land ice in Pliocene deposits, which are in that sense pre-glacial. But the appearance of the major ice-sheets, producing spreads of boulder clay, or till, was not synchronous over wide areas. The ice-caps started to grow on high ground and from there spread on to the lowlands, and moreover waxed and waned several times during the Pleistocene as a whole. In a country unaffected by the ice-sheets some or all of the glacial periods were represented by less dramatic evidence of a cold or arctic climate.

For many years it was customary to equate, roughly, Pleistocene history in the British Isles with recognizable relics of glacial, interglacial or periglacial deposits. Older than these, and locally below them, lie the 'crags' of East Anglia; these are dominantly marine, shelly sands and clays, and being pre-glacial were accordingly considered to be Pliocene. Nevertheless, a detailed examination of the faunas and microfloras shows that they contain many 'cold' species, and the general history of glaciation in Europe confirms that most of the crags were laid down near the margin of an early North Sea, when ice-sheets had already developed in other parts of Europe. Accordingly nearly all the crags are now considered to be early, or Lower, Pleistocene, and only the earliest, Coralline Crag, is retained in the Pliocene. In other parts of Britain Pliocene relics are very few, and some of the small outliers which have commonly borne this title have either, by comparison with East Anglia, been transferred to the Pleistocene, or contain a fauna of uncertain age, or no fauna at all.

The term 'Holocene' (denoting that all the species are still living) is not used here. It is approximately equivalent to post-glacial, but the ice-sheets disappeared gradually from Britain, as from elsewhere—in places not more than 10,000 years ago—and the post-glacial phase fits in more logically as the last Pleistocene stage. Indeed the earlier interglacial periods were probably much longer than this, and one at least had a warmer climate than that of the present time.

PLIOCENE RELICS

The crags extend through the eastern parts of East Anglia, from the borders of Essex on the south to Cromer on the north coast of Norfolk. The Coralline Crag forms only a small part of this outcrop, near Orford

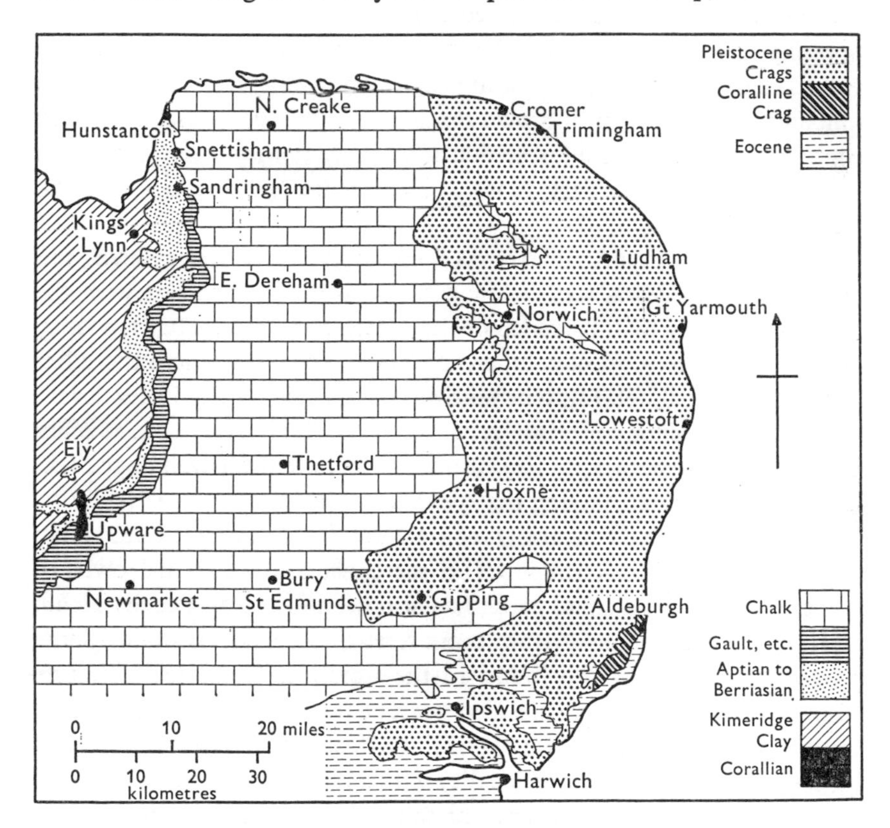

Fig. 71. Geological sketch map of East Anglia showing outcrops of Jurassic and Cretaceous rocks, Coralline Crag and Pleistocene Crags; glacial and interglacial deposits not shown but some important Pleistocene localities are given.

and Aldeburgh in the south (Fig. 71). It consists of white and yellow, current-bedded sands, with great numbers of molluscan shells, at some levels whole, at others broken into shell debris. Lesser components are echinoids, terebratulids and barnacles, and the term 'coralline' refers not to the occasional coral but to the abundant polyzoa. The whole, which is not more than 60 feet thick, was laid down as shelly sand-banks in a shallow sea, swept by currents. That the climate was still relatively warm is shown by the Mediterranean affinities of many species and the absence of the cold or boreal forms that dominate much of the later North Sea faunas. The nearest comparison is with the Astian, or Upper Pliocene, beds of the type area in Italy.

At the base of this and other crags in the south of the outcrop lies the Suffolk Bone Bed or 'Nodule Bed'. This is a remanié deposit, incorporating worn and often phosphatized fossils from the nearby Jurassic and Cretaceous outcrops, including bones and teeth; there is also much flint and eroded pebbles of brown sandstone called 'boxstones'. The last contain the casts of molluscs, doubtful in age; they may have been derived from some Tertiary deposit (? Oligocene or Miocene) which no longer survives.

At present the East Anglian crags do not occur far above sea-level, and their base in places is well below O.D., a position consonant with the general tendency to subsidence in the North Sea basin. That tendency does not apply to most of southern England, however, where there has been a substantial, intermittent uplift (or else a eustatic fall in sea-level), although with minor periods of reversal. Accordingly any Pliocene relics in south-east or south-west England must be looked for well above sea-level, and it is not surprising that very few remain.

The Lenham Beds probably come into this category. Their main occurrence is in solution hollows in the chalk of the North Downs of Kent (Fig. 66), but a small patch has also been found on Beachy Head. Most of the marine faunas from these ferruginous sands contain a large proportion of species also found in the Coralline Crag, and although Miocene and living forms are also present there seems little reason to dispute the accepted age for most of these deposits. There are, however, blocks of sandstone from one locality with a rather different assemblage, nearer to faunas referred at present to the Upper Miocene of Belgium (previously Lower Diestian = Lower Pliocene); if that position is maintained—the Miocene/Pliocene boundary is under discussion—it is possible that we have here the only known remnant of Miocene beds in this country.

While the formal age is perhaps a rather artificial point, the implications of the Lenham Beds and their fauna are far reaching. Lying as they do, over 600 feet up on the North Downs, they entail a marine invasion of most of the London basin during Pliocene times, and uplift of that amount since then, whereas the Coralline Crag lies almost at sea-level. Lenham Beds are not found farther west; the fauna of the Netley Heath Beds of the Surrey Downs shows them to be of Red Crag age and hence Lower Pleistocene.

Further problems arise in the ages ascribed to several residual patches of sands and gravels in south-west England. Typically these lie on uplifted erosion surfaces and are unrelated to the more recent valley systems. Such examples include the Crousa gravels of the Lizard, on a surface at 360 feet O.D., and a smaller patch between 300 and 400 feet south of Bideford, the Orleigh Court gravels. The latter resemble the beds capping the Haldon Hills (p. 358) in that they contain flint casts of Upper Chalk fossils and they have sometimes been similarly ascribed to the Eocene. At a comparable height (350–420 O.D.) there is a belt of sands around St Agnes Beacon on the north Cornish coast. Little direct deduction can be made from such outcrops unless they are clearly related to other, and better dated, deposits.

The only fossiliferous beds of this type known from Cornish peninsula are those at St Erth, at about 100 feet, on the neck of land between St Ives Bay and Mounts Bay. Although the shelly clays here have not been accessible for many years, they were seen, interbedded with sands, in the late nineteenth century, and over 250 species of molluscs collected from them. In this assemblage littoral genera were conspicuously absent, and the Mediterranean affinities pronounced. The clays were deduced to have been laid down in quiet waters at a depth of at least 25 metres. It was natural that the shells did not closely resemble the crag faunas of the North Sea, as there was no seaway through the Straits of Dover at this time. They were usually ascribed to the Pliocene in rather general terms, but more recently a Pleistocene age has been put forward. The St Erth Beds are thus of some stratigraphical importance in their own right as an indication of the form of Cornwall, which was then not a peninsula but an archipelago; they also bear indirectly on the age of the marked erosion surfaces in this part of the country.

LOWER PLEISTOCENE MARINE BEDS

To continue the stratigraphical history we must return to East Anglia, where the Red Crag is taken as the lowest Pleistocene formation. Its fauna shows a marked increase in cold or boreal species compared with that of the Coralline Crag, and is closest to the Calabrian stage of south

Italy, the type area for the marine Lower Pleistocene. The Red Crag also contains the first remains, in this country, of the true horses and elephants (*Equus* and *Elephas*) which characterize the Villafranchian, or equivalent continental stage. The boundary is also a convenient one in East Anglia in that a period of erosion elapsed between the accumulation of Red and Coralline crags so that the junction is marked by a minor unconformity.

The classic descriptions of the East Anglian formations, dating from the turn of the century, were largely derived from cliff and inland exposures, particularly the former. More recent information from boreholes has somewhat revised the classification, which is given below as a general guide. The faunas found in the surface outcrops were chiefly the mollusca and rarer vertebrate remains; those in the boreholes include especially foraminifera.

Weybourne Crag
Norwich Crag
Red Crag

The stratigraphical distinctions between these three are not very clear, partly because they are rarely visible in simple vertical sequence, but tend to crop out in different regions. Moreover, there was originally thought to be a gradual refrigeration of the climate, marked by a corresponding regular increase in boreal species, and the stratigraphical succession was partly based on this assumption. It is more probable, however, from a study of the foraminifera and pollen, that there were fluctuations in the climate—perhaps three cold periods, one in the Red Crag, one in the lower part of the Norwich Crag and one in the Weybourne Crag and upper part of the Norwich Crag.

Where exposed at the surface the Red Crag owes its name to the oxidation of the iron compounds, and consists of shelly, current-bedded sands with seams of clay and pebbles towards the top; the thickness is about 40 feet and the beds lie on the irregular, slightly eroded Coralline Crag or on the London Clay. In the Ludham borehole (Norfolk) the lithology is similar but the sands and clays are grey.

The later crags are more extensive and in surface exposures they replace the Red Crag northwards; where the Norwich Crag rests on chalk there is a basal flint conglomerate which also includes the bones of elephant, mastodon and deer. The Chillesford Clay is a local facies, probably of Norwich Crag age, which overlies the Red Crag in Suffolk, and suggests quieter conditions than the shelly sands beneath, for in addition to a marine fauna it has yielded the articulated skeleton of a whale.

The Weybourne Crag of north Norfolk contains cold-water shells and an impoverished assemblage of foraminifera; it marks the last stage in this part of Pleistocene history, before the Scandinavian ice-sheets approached the English coasts. The vegetation of the nearby land is reflected in the pollen that drifted out to sea; the two cold periods are marked by a sparse heath vegetation, not unlike that of north Norway at the present day, and a warmer phase between them by temperate spruce forests. Presumably the climatic oscillations correspond to those in the early history of the Alpine glaciations (? Günz), but correlation of all the early Pleistocene glacial stages is difficult, because their evidence tends to be obliterated by or confused with that of later glaciations and the geographical surroundings of the two regions are very different. It is probable that there was also an early Pleistocene development of the Scandinavian ice-sheet, but nothing is known, or precisely dated, of its history at this stage. On the other hand, the climatic succession deduced in East Anglia fits well with that already known from the periglacial deposits of Holland.

Nearly all the deposits that remain from the Lower Pleistocene sea are confined to the East Anglian outcrops, but that it did invade the London Basin is shown by remnants in Surrey and Hertfordshire, both belonging to the Red Crag. The Netley Heath Beds contain blocks of ferruginous sandstone with a Red Crag fauna and lie at about 600 feet on the North Downs east of Guildford. The beds do not show fluviatile or marine characters at present and are associated with flint pebbles, clays and sands, so that they may have been disturbed or somewhat redistributed by periglacial action. The Red Crag at Rothamstead in Hertfordshire is at a lower level (under 400 feet) but reinforces the evidence for this marine invasion.

EROSION SURFACES

The succession of the late Tertiary erosion surfaces can best be traced in south-east England, because there they lie closest to contemporary sediments or other forms of chronological evidence. First we must hark back a little to the preceding chapter. At some period after the deposition of the Oligocene beds of the Hampshire basin, southern England was strongly affected by the Alpine earth movements; these produced the pronounced folding, in places accompanied by thrusting, of the Isle of

Wight and Dorset, the doming of the Wealden anticlinorium and the minor structures of the Hampshire basin and Salisbury Plain. These effects are sometimes called Miocene, though perhaps the less definite 'Mid-Tertiary' is a better term.

Thereafter the following history is deduced in the south-east. The folding of the Weald, London basin and Chilterns was succeeded by a considerable period of denudation, in which much of the land surface was reduced to something approaching a peneplain; the remnants of this surface form the summit plane of the chalk around the London basin, now at about 800 feet. It is sometimes called the Mio-Pliocene surface, but there is little firm evidence for its age except that it is younger than the earth movements and older than the next lowest surface—that at 600 feet, which in places forms the surface of the lower parts of the North Downs, but more characteristically is a bench cut into their northern dip slope and into the southern dip slope of the Chilterns. Here we meet a difficulty. The Netley Heath Beds (Lower Pleistocene) lie on this bench and it is usually considered to be cut by marine erosion of about the same period. Was there then an earlier, Pliocene, invasion to about the same level, which left traces in the Lenham Beds, or was the London basin continuously a marine area through later Pliocene until early Pleistocene times? To this question there is no satisfactory answer at present. It is, however, probable that the '600-feet sea' entered the Hampshire basin (though no known sediments survive) for the level of

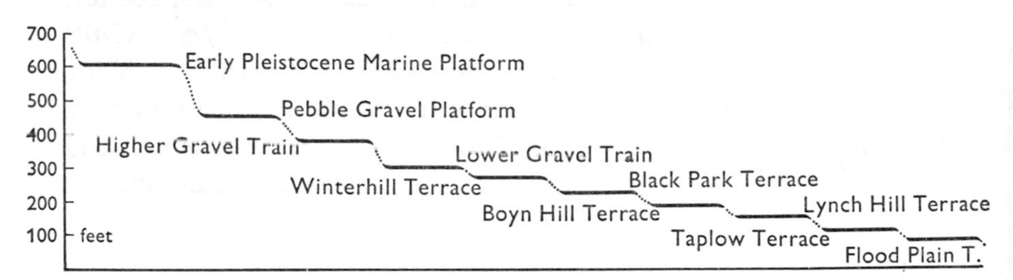

Fig. 72. The terrace sequence in the Middle Thames Valley (redrawn from Wooldridge, 1960, p. 121).

the chalk falls well below 600 feet in the intervening area around Basingstoke.

The next stage is represented by many remnants of Pebble Gravels capping hills in the higher parts of the London Basin, at 400 feet or somewhat above, but whether these were originally the shingle of a retreating Pleistocene sea or fluviatile gravels is uncertain. From here onwards there is a descending sequence of Thames terraces (Fig. 72), from about 360 feet downwards, which can be related to the various Middle and Upper Pleistocene glacial and interglacial stages, with the Boyn Hill, or '100-foot', terrace as a particularly important datum level.

The primary cause of this impressive descending flight of erosion surfaces is problematic—was it due to a eustatic fall in sea-level or to a tectonic uplift of the land? If the latter then the lack of warping (particularly marked in the 600-foot bench) is remarkable: southern England had reacted very differently to the Mid-Tertiary movements. A eustatic effect entails rather fewer difficulties. It demands a complementary tectonic down warping of East Anglia where the Red Crag now lies near sea-level, and still more in Holland where it is far below; but a subsiding basin in the southern part of the North Sea is an accepted part of Plio-Pleistocene history.

A sequence of uplifted erosion surfaces comparable to those of the London basin has been described from the seaward slope of the South Downs and some also from the Dorset coast. In Cornwall erosion plateaux are particularly well developed; they are cut in harder rocks than in the south-east, conspicuously planing off their complex, Hercynian structures. They border the present coast and the best of them bear convincing signs of marine erosion. That at 430 feet is conspicuous in many parts of the peninsula. It shelves slightly towards the sea and is bounded inland by a marked rise or degraded cliff-line; some of the harder rock masses, such as the granite of Lands End, stand up as residual 'islands'. Above there are less pronounced plateaux at 750 feet and 1,000 feet. The 430-foot level is commonly called Pliocene, and sometimes related to the St Erth Beds. On such a view the shelly clays were laid on the sea floor some 300 feet below sea-level while the marine platform was being cut farther inshore. The higher, earlier levels would then belong to some previous phase of erosion. This ascription, however, involves two assumptions—that the St Erth Beds *are* Pliocene, which is not universally accepted, and that they are genetically related to the 430-foot platform. The results are interesting, but the evidence not wholly complete.

The coastal surfaces of South Wales are almost as impressive as those of Cornwall, especially the lowest which lies at about 200 feet in Glamorgan and falls a little westwards into Pembrokeshire, where it planes off very varied rocks and structures. In the Vale of Glamorgan a 400-foot surface is recognizable, and a 600-foot inland in Pembrokeshire and Cardiganshire. The upper parts of certain South Wales rivers are graded to one or more of these levels. No Tertiary rocks are available for comparison, but the 200-foot surface cuts across, and is thus later than, the gentle folds of the Glamorganshire Lias. These are almost certainly due to the same Mid-Tertiary movements that affected southern England; hence this planation has been called Pliocene, but it could well be early Pleistocene.

In between Cornwall and South Wales a number of high-level plateaux have been described in north Devon, including the 430-foot on which lie the gravels of Orleigh Court. Some of them have also been called Pliocene, and the high surfaces of Exmoor, Miocene. Some lower ones, from their relations to periglacial deposits, erratics and boulder clay in the Bideford-Barnstaple region are more securely attributed to the Pleistocene.

Similar features have been found farther north, for instance in Anglesey, North Wales, the Lake District and southern Scotland, though they are rarely so well preserved. Erosion surfaces have been described from various parts of Ireland and Scotland, and there are also higher and more debated features, such as the High Plateau of Wales; this has been interpreted as an inclined surface, falling from 2,100 feet in North Wales to 1,400 feet in Mid-Wales, or as a series of much dissected but unwarped levels.

It is not intended to discuss these northerly or westerly surfaces. They have much geomorphological interest and significance, not least in their relation to the history and adjustment of the accompanying drainage patterns. But the farther such features are from known Tertiary deposits, or from structures reasonably attributable to Mid-Tertiary earth movements, the more difficult it is to determine their age. Only in the broadest way can they be used to fill in the history of the British Isles between the Lower Tertiary deposition of the south-east or the vulcanicity of the north-west, and the arrival of the ice-sheets. What is clear, however, is that, with the exception of the North Sea basin, this was generally a period of uplift; it is also probable that the broad outline of the British Isles, except in the south-east where it was still connected with the continent, was established in this interval.

GLACIAL AND INTERGLACIAL DEPOSITS

At the present time there are two major ice-sheets in the world, of Greenland and Antarctica; although the first is probably nearer in its régime to the Pleistocene ice-sheets than is the second, neither forms a very good analogue. In particular both abut on a coast shelving steeply into deep water, so that the ice moves out as a floating mass or calves off into icebergs. Presumably this happened to much of the northern and western ice from Scandinavia and Britain at the height of the glacial periods, but Pleistocene history is deduced most successfully where the ice-sheets advanced on to lower land (in northern Europe to the south or south-east), and there declined through melting or ablation.

In Europe there was one great ice-sheet, the Scandinavian, and lesser ones in the British Isles and Alps. By Middle Pleistocene times the uplands of Britain had developed their own ice-caps, particularly in the west and north, where precipitation was highest, and the Scandinavian ice, which at an early period approached our eastern coasts, was forced away from them by the pressure of native ice.

The early attempts at a European correlation of glacial and interglacial periods were based on the famous fourfold stages of the Alps—Günz, Mindel, Riss and Würm, and the three major glaciations of Britain were compared (as they sometimes still are) with the last three. But the Alpine glaciations do not really form a good standard sequence, for in these warmer latitudes the ice-sheets, even at their maximum, did not extend far on to the neighbouring lowlands, and in the mountains each one tended to destroy the evidence for its predecessors. Moreover most of the interglacial deposits lack palaeontological data. A better international sequence can be deduced from the glacial and interglacial deposits of the northern European plain, in Holland, Denmark, North Germany and Poland; East Anglia is in some sense a westerly continuation of this plain and provides one of the best composite successions, though one in which many queries still remain.

Once the ice-sheets were established in the British Isles a new régime developed. In the place of a few fragments of marine deposits, largely confined to south-east England, we have to hold in view all the processes of glacial erosion and deposition through which a complex series of terrestrial deposits was formed, and often removed. For such deposits a greater range of stratigraphical methods have been employed than those needed in more orthodox systems. Nevertheless, the aim is the same: the establishment of a standard, dated sequence, with which less perfect exposures and sections may be compared. The difference between 'orthodox' and Pleistocene (in the sense of glacial) stratigraphy is chiefly that in the latter the sections showing uncontrovertible evidence of conformable, dated, formations are few and short.

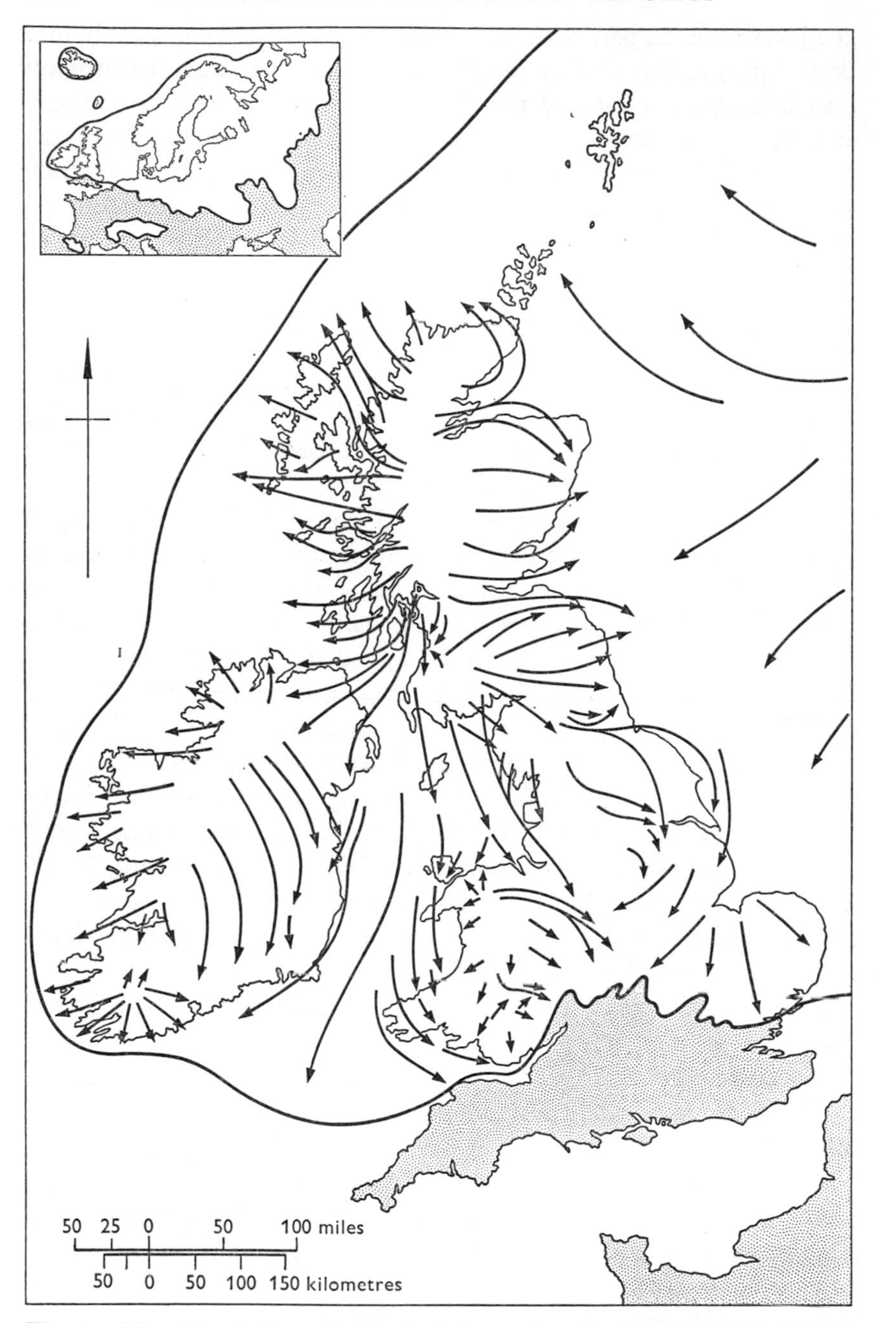

Fig. 73. The extent of maximum glaciation and generalized directions of ice-flow. Inset: the ice-sheets of Europe. (From several published sources.)

At no time during the glacial periods was the whole of the British Isles covered with ice-sheets; even at their greatest extent southern England, south of a line approximately from the Severn to the Thames, was free of permanent ice and periglacial conditions obtained. Within the glaciated areas it has long been observed that Pleistocene history is least easy to follow on the uplands, because here erosion was dominant rather than deposition. In some areas different suites of erratics may provide evidence of ice movement from more than one direction, but a succession of superposed tills is very rare and substantial glacial deposits are usually referable only to the last glaciation. Interglacial sands and gravels are present, but hardly ever contain a flora or fauna of determinable age. In places it has been found relatively easy to distinguish the last glaciation, with its fresher deposits and land forms, from one or more older ones. Hence there arose the terms Older and Newer Drift—Newer Drift being, in a more sophisticated terminology, Würm or Weichsel of the continent, or the Last Glaciation. In Scotland in particular late Würm-Weichsel history was one of oscillating ice-margins; each minor re-advance covering a lesser area, and separated not by true interglacial periods but by lesser climatic ameliorations called interstadials. Such oscillations presumably accompanied the growth and decline of earlier glaciations as well, but the record is most legible in the last.

Although till was undoubtedly formed within the area of the ice-sheet its exact nature and origin are not very clear. The term is better than 'boulder clay' because though the deposit is unsorted neither boulders nor clay is an essential ingredient. Little of it appears to correspond to a true ground moraine, which under modern ice-sheets is a much thinner formation; to a large extent till was probably carried in the lower layers of the ice, and so was chiefly englacial rather than sub-glacial. Some may have been dumped in sheltered depressions, such as transverse valleys, in the path of the ice advance, but most thick lowland till was probably deposited near the ice-margin where there was little active movement.

Where the tills spread out on the lowlands they can be recognized and correlated to some extent by their lithology and erratic content, and hence such terms as Chalky Boulder Clay or Scandinavian Drift. The system has its drawbacks, however, if only because more than one ice-sheet in eastern England passed over chalk lands and incorporated debris from them. More recently it has been shown that the erratic pebbles tend to be aligned with their long axes parallel to the movement of the ice. Since the ice movement of successive glaciations was usually slightly different in a given area, the 'stone orientation' of the tills may be significantly different, and serve to distinguish them.

Till is thus an unusual deposit, in the same way that major glaciations are unusual geological events. But it is much more, and more interesting, than just a thin veneer, a minor nuisance to the geologist who is in search of solid rocks. 320 feet of glacial and interglacial deposits have been recorded at Crewe, up to 300 feet at Cromer and 470 feet in a borehole in the Stour Valley of Suffolk. The thicker drift records often relate to buried hollows and channels in the surface of the solid rocks below, and they tend to be filled with sands and gravels rather than with abnormal amounts of till. It is also well known that the sub-drift surface is below sea-level in many parts of eastern England and in some of the lowlands of Lancashire. Glacial deposits in some regions are thus thicker than the combined total of the East Anglian crags, far more than any English remnant of Pliocene beds, and not far short of a major Eocene formation such as the London Clay.

Part of Pleistocene correlation depends on tracing the margins of the ice-sheets at each period of glacial maximum. These are best defined by terminal or end-moraines; in many places, however, no such moraine was formed (perhaps because the ice was stagnant, or only maintained the final position for a short period) and there is only a feather-edge to the till. Immediately beyond the ice-front various outwash deposits were laid down, in which the glacial debris was re-sorted by the streams issuing from the ice. Since the latter sometimes blocked the normal drainage lines water was occasionally held up between the ice-edge and higher ground, forming pro-glacial lakes, which have left their traces in bedded deposits and temporary or permanent deflexions of the drainage. These various fluviatile and lacustrine deposits may help to locate the position of the ice-front, but not all water-laid sands and gravels were necessarily formed beyond it. In the lower, melting, parts of an ice-sheet heavily laden streams are known to run on, in, and under the ice, and some of their deposits have survived, and thus lie within the area of the contemporary glaciation.

In regions more remote from the ice-sheets glacial and interglacial history is chiefly traced in the river terraces, such as those of the Thames, Lower Severn and Warwickshire Avon in this country, or the Somme, Elbe and Nile abroad. In England the land surface was subject to permafrost, when the ground is permanently frozen at depth and the surface layers, lacking drainage, are liable to solifluxion; in these conditions the bedding becomes disturbed, or unsorted masses of rock debris and mud slide down the hill slopes. Such deposits, as 'head' or 'coombe rock' (as it is called in chalk country), were formed during cold periods, and can occasionally be related to a specific glaciation. Solifluxion tends to affect the wetter periglacial regions, whereas loess is characteristic of the drier; it is important in central and eastern Europe, where successive loess formations and their intervening soil horizons are a means of correlation, but is very rare in the British Isles.

Methods of dating and correlation

Isotopic methods, that is by measurements of radiocarbon (p. 5), are an admirable tool for dating the deposits of the last 30,000 years or so. This takes us back well into the Last Glaciation. Special techniques are needed at present for estimating radiocarbon from earlier deposits and results here are few. Nevertheless, we may look in the future to a great increase in the radiocarbon dates, which, with pollen analysis, will make late Pleistocene history much clearer.

Changing levels due to eustatic causes appear at first sight to be a very satisfactory method because the effects should be world wide. The volume of water locked up in the ice-caps during glacial maxima must have lowered sea-level by some hundreds of feet; the amount is much debated, because of the difficulty of estimating the thickness of the ice. Nevertheless, glacial periods should coincide with depressed sea-levels and interglacial with elevated ones. In practice correlation is not so simple as this, because of other, more local, factors. Many parts of the world have recently been affected by tectonic depression or elevation, especially related to Alpine orogeny, and the Alps themselves and the Apennines have risen tectonically during Pleistocene times. Then the weight of the ice-caps depressed, isostatically, the landmasses that were thickly covered, and after the ice melted they gradually rose, as Scandinavia is still doing. Southern Britain does not seem to have been much affected by either complication, but the raised beaches of Scotland show the effects of late and post-glacial isostatic rise that in a miniature way resemble those in Scandinavia. A succession of Pleistocene beaches has been recognized in the Mediterranean, and some correlations made elsewhere, but in the absence of palaeontological confirmation the results have to be interpreted cautiously.

Within a limited region a good deal can be done by tracing river terraces, especially where they occur in a regular order and can be dated by isolated palaeontological finds or their relationship to known tills. In southern England there is a succession of Pleistocene terraces, in general from the older, upper remnants to younger, lower ones. Even so the lower reaches of a river tend to be dominantly affected by changes in base-level, so that the high, interglacial, sea-levels are periods of aggradation and the glacial periods of those down-cutting; conversely in the upper reaches the greater load of detritus produced by the neighbouring ice-sheets tends to increase aggradation during glaciations. The interaction of such factors complicates the river régime, and the history of the Pleistocene Thames, for instance, is far from simple.

Palaeontological methods are mainly based on three groups: land mammals, Pleistocene man and his artefacts, and pollen. Macroscopic plants and freshwater and marine shells are useful occasionally. The zonation, or sequences, so established very rarely correspond to the ideal (and perhaps somewhat mythical) zone; few species are known to have appeared and become extinct the world over during the Pleistocene period, the time being too short. The effective sequences are essentially climatic, which is one reason why Pleistocene correlation is so troublesome over wide ranges of latitude.

The most important mammals that roamed over northern Europe during interglacial periods, or outside the ice-sheets, were elephants, hippopotamus, rhinoceros, deer, horses and oxen; they include southern forms, such as the hippopotamus, and northern, such as reindeer, but some groups now restricted to warmer climates had a greater range in Pleistocene times. The elephants are particularly useful, and are one of the few groups in which species were sufficiently short-lived to characterize part of the period:

Elephas primigenius	Upper Pleistocene
Elephas antiquus	Middle Pleistocene
Elephas meridionalis	Lower Pleistocene

These stratigraphical ranges are only approximate and the first two, for instance, overlapped in the Last Interglacial period. Complete elephant skeletons are rare, but fortunately the teeth are more common and species are recognizable from them. Mammalian remains are found in terrace gravels, interglacial beds and some types also in caves.

The hominids as a group are rare as fossils, and in spite of all the searches of anthropologists, only a few examples of Pleistocene man are thoroughly known; of these only one is British—Swanscombe 'Man' (in fact probably a female)—an incomplete skull from the Boyn Hill Terrace, or later part of the Hoxnian interglacial period (p. 380). Despite their great evolutionary interest, skeletal remains are usually dated by the beds in which they are found rather than vice versa. Artefacts, on the other hand, are locally abundant. In Britain they were commonly fashioned out of flint, admirable material for flaking and easily to hand in the south and east of England. Today an expert flint-knapper can produce a hand-axe in about five minutes, and Pleistocene man probably made such tools when they were needed, discarded them when blunted and made some more. It has been estimated that when Swanscombe Man flourished the human population of Britain was very small, perhaps 200–2,000, and probably nearer the lower figure in glacial periods. Each individual made and discarded many tools, and it is thus not surprising that skeletons are rare, and artefacts on some sites plentiful.

A general succession of human cultures during 'the ice-age' was established many years ago, and three different types distinguished—the hand-axe (or core-tool or biface), better considered as an all-purpose cutting and skinning implement; the flake cultures, and the blade cultures. The first two continued side by side during much of the Middle Pleistocene and lower part of the Upper Pleistocene, and the last appeared after the onset of the Last Glaciation. The earlier types in particular were very widespread and show a rough equivalence in Europe and Africa (North America apparently had no human population till much later), and although there are some anomalies, as tribes did or did not migrate into certain regions, the general succession stands.

The analysis of pollen from interglacial deposits is more recent and has been much advanced in the past decade. Pollen grains are very resistant to decay and tree pollen in particular may be carried long distances by the wind. Thus the proportion of grains from different plants, as they are preserved in peat or lake-clays, reflects the main constituents of the vegetation up to some 50 miles around. The flora of each interglacial period passed from a cold, sparse, tundra type of vegetation to temperate forests and back again, tree pollen being dominant in the central, warmer section, and 'non-tree pollen' before and after. If this pattern were all that could be deduced it would not serve to distinguish one interglacial period from another; but fortunately the detailed proportions of the dominant forest trees (chiefly birch, pine, spruce, fir, elm, oak, hornbeam, lime, alder and yew) show patterns of change that are somewhat different in each. The late glacial and post-glacial floras and climates can be traced in the same way.

From detailed pollen analyses it appears the vegetational changes in northern Europe were sufficiently widespread for correlation of pollen zones to be possible from East Anglia to northern Germany, and in Britain it is one of the most successful methods. Even so the critical

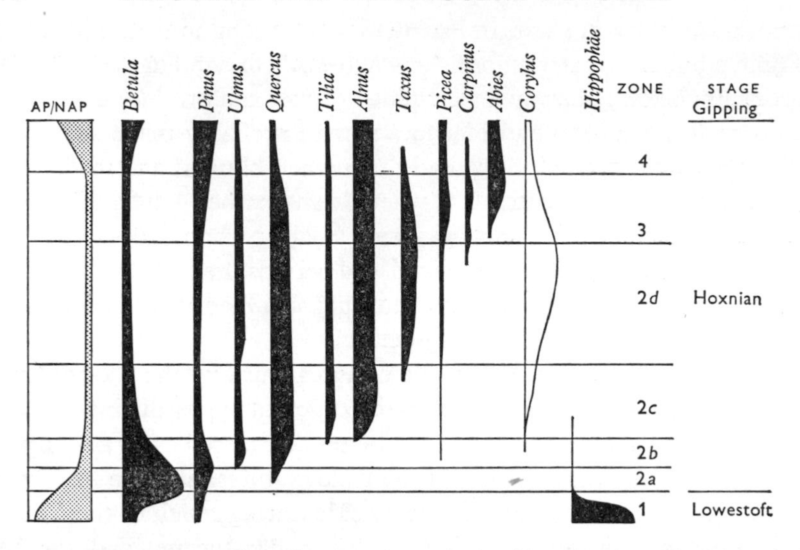

Fig. 74. Composite pollen-diagram through the Hoxnian interglacial period, based on a number of sites. The thicknesses of the black lines denote percentages of total tree pollen. The panel on the left shows the relative proportions of tree pollen and non-tree pollen (AP/NAP). (Redrawn from West, 1963, p. 175.)

sections, in which pollen-bearing beds are clearly related to an identifiable glacial deposit, are not many, and the piecing together of short sequences, which sometimes has to be done in other stratigraphical successions, is a particularly necessary and delicate part of Pleistocene studies.

Glacial and Interglacial history, Cromerian to Ipswichian

The stages and typical deposits of glacial and post-glacial history in the British Isles are shown in Table 60, which thus summarizes much of Middle and Upper Pleistocene stratigraphy, but it must be emphasized that many of the correlations are provisional and the age of certain older formations is particularly debatable. The Newer Drift (Weichsel or Würm) has the advantage of radiocarbon dates in the later horizons, and the disagreements here are on rather a smaller scale—the equivalence of re-advances and interstadials rather than of full glacial and interglacial deposits.

The Cromerian stage refers to the Cromer Forest Bed and associated sediments of north Norfolk. They consist of a middle 'estuarine' group of water-laid sands and clays with uprooted tree stumps, mammalian bones and marine shells, flanked below and above by 'freshwater' beds, the upper of which is based on a rootlet horizon. Pollen from the upper bed, and some of the mammals, indicate a fairly warm climate, from which the Cromerian interglacial stage takes its name.

During the early part of the next cold phase (Lowestoft), Scandinavian ice approached the East Anglian coast and produced boulder clay with Scandinavian erratics in Aberdeenshire and probably also on the Durham coast. At its maximum the British ice moved into East Anglia from the north-west, and overrode the Chiltern scarp, with lobes extending into the Oxford region and the northern side of the Thames drainage basin. According to the views embodied in the table there was a relatively minor recession during this glaciation, when the Corton Sands, near Lowestoft, were laid down between the North Sea drift and Lowestoft till. (On another view they are a true interglacial deposit separating two glaciations, and the correlations of the table would not hold.) In the west there are few deposits definitely attributable to the Lowestoft glaciation, but the lowest, Bubbenhall, clay of the Midlands has pollen-dated support; to this stage also belong the higher terraces of the Thames. The Lowestoft stage lacks end-moraines and its limits are only deduced from the presence or absence of recognizable tills. Even its separation from the Gipping till above is so far based largely on lithology, erratics and their orientation, and the interglacial deposits of Hoxne. There is as yet no section available showing unequivocal superposition of the two distinct tills.

The succeeding, Hoxnian, interval is sometimes called the Great Interglacial, by comparison with that so-named (Mindel-Riss, or Elster-Saale) on the continent. The type section at Hoxne in Suffolk, where carbonaceous clays and silts lie in a small lake basin on the surface of the Lowestoft till, contains pollen representing a full sequence from cold conditions to temperate and back to cold; the Hoxne beds are overlain not by the Gipping till, but by periglacial deposits containing erratics typical of it. A temperate climate is also suggested by a few other contemporary interglacial deposits, including those at Nechells (Birmingham) and Clacton (Essex); the last also includes marine beds with the warm-water shell *Corbicula*, *Elephas antiquus*, *Hippopotamus* and Clactonian artefacts. There is widespread evidence for a high sea-level during this interglacial phase—at Clacton itself, in the Nar Valley of west Norfolk (where the land was submerged to a depth of 50 feet and clays laid down containing *Corbicula*, *Turritella* and *Littorina*) and at Kirmington at 80 feet O.D. in Lincolnshire, where there are estuarine and marine clays. Raised beaches in Sussex and South Wales are also

Table 60. *Representative Pleistocene successions in England and Scotland* (*based on West*, 1963, *p.* 166)

Stage	East Anglia	Southern England, Thames, etc.	Midlands	Lincolnshire and East Yorkshire	Northern England	Scotland
FLANDRIAN	Peat, alluvium, raised beaches, estuarine clays					
WEICHSEL OR WÜRM	Hunstanton Till	Buried Channels Coombe Rock Arctic peats of Lower Thames	Terraces Severn and Avon Main Irish Sea Glaciation Chelford Peat	Holderness Tills including Hessle Till	Scottish Re-advance Main Dales, Main Irish Sea Glaciation (Newer Drift)	Loch Lomond Perth Aberdeen (Re-advances) Strathmore Glaciation
IPSWICHIAN	March Gravels Interglacial deposits at Cambridge, Ipswich and Sutton	Upper Floodplain and Ilford Terraces Interglacial deposits at Trafalgar Square	Terraces of Severn and Avon	Interglacial deposits at Sewerby		
GIPPING	Gipping Till	Taplow Terrace Main Coombe Rock Solifluxion deposits	Oldest terraces of Severn and Avon Older Irish Sea Glaciation	Chalky Till of Wolds	Older Drift here or earlier	Greater Highland Glaciation
HOXNIAN	Interglacial deposits at Clacton, Hoxne and Nar Valley	Boyn Hill Terrace Swanscombe Gravels	Interglacial deposit at Nechells, Birmingham	Interglacial deposit at Kirmington, Lincs.		
LOWESTOFT	Lowestoft Till Cromer Till and Norwich Brick-earth	Gravel Trains Winter Hill and Black Park Terraces	Bubbenhall Clay		? Till with Scandinavian erratics (Durham)	
CROMERIAN	Cromer Forest Bed Series					
(EARLY PLEISTOCENE)	Weybourne Crag Norwich Crag Red Crag	Pebble Gravels Crag at Netley Heath and Rothamstead				

attributed to this interval, as is the Boyn Hill or 100-foot terrace of the Thames.

The Gipping till has a similar distribution to the Lowestoft (in so far as the latter is known) and also lacks end-moraines over much of its limits. To this period probably belong the Chalky Boulder Clay of the Midlands, much of the 'Older Drift' of northern England, and one or more, older, tills in Wales, the areas around the Irish Sea and in Scotland. There is a good deal of evidence for some advances and recessions of the ice-margin in Wales and the west, and also in East Anglia. The one small deposit of genuine till known in south-west England, at Fremington, near Barnstaple, has been tentatively ascribed to this glaciation, and presumably reflects a temporary extension of Welsh ice across the Bristol Channel.

In the west Midlands the lower Severn drainage was blocked by western ice, and eastern ice blocked the previous course of the Avon which ran north-eastwards into the Trent; there was thus produced Lake Harrison, one of the largest of English pro-glacial lakes, which left its traces in bedded lake deposits. On the retreat of the Gipping ice the Severn drainage was restored, but the Avon took a new course and became a tributary of it.

By the Ipswichian interglacial period most of the modern river valleys were established and the interglacial deposits are commonly related to them. In the type section, at Bobbitshole, near Ipswich, the pollen-bearing clays lie in a channel cut in the Gipping till. Comparable sections are known from Stutton, Suffolk, and in the high terrace of the Cam near Cambridge, the former being rich in southern mollusca. Pollen from the Upper Flood Plain terrace of the Thames, at Trafalgar Square, is Ipswichian in composition, and to this period also belong certain terraces of the Severn and Avon, formed after the disappearance of the Gipping ice-sheets. The sea-level does not seem to have been so generally high as in the Hoxnian interglacial period, and the East Anglian deposits are not far from modern sea-level, but a submergence of 25 feet on the shores of the English Channel is suggested from interglacial deposits of the Sussex coast.

The Irish Sea existed in some form in early Pleistocene times, for marine shells of Red Crag age have been brought up by later ice and incorporated in the coastal drift of Co. Wicklow and the Isle of Man. Less is known about the glacial and interglacial periods in Ireland than in England; Palaeolithic artefacts have not been found and interglacial deposits are few. Hoxnian and Ipswichian pollen sequences are known from Waterford, and the former also from Galway, and an older series of drifts can be distinguished from a younger series. The Lowestoft glaciation may be chiefly represented by a number of erratics stranded on the coasts by floating ice at a time of low sea-level, but Gipping ice covered almost all the country, and three glacial phases of this stage are recognized in south-east Ireland. The main ice centre was the Dalradian uplands of the north-west, but Irish Sea ice, with Scottish erratics, was powerful on the eastern side and the Cork and Kerry mountains supported an independent ice-cap.

The Last Glaciation (Weichsel-Würm) and post-glacial history

During the last major glaciation the ice-sheets did not reach so far south as in the earlier ones. The glacial topography is also much clearer, including the great sheets of drumlins and eskers of the Irish plain, and the deposits are less weathered. They are also found in the modern valleys, in contradistinction to much of the Older Drift, which, in the Pennines, for instance, forms weathered and dissected patches on the interfluves. End-moraines are relatively common, such as the York and Escrick moraines of the Vale of York or the Tipperary moraine, taking a sinuous line across south central Ireland. The irregularity of the Last Glaciation ice-front (Fig. 75) is caused especially by the lobes that extended southwards down the lowlands (York, Cheshire plain and Irish Sea) and the failure of the ice to override some of the uplands, like the Cleveland Hills of Yorkshire, the southern Pennines, Leinster mountains and others in southern Ireland.

In spite of this greater topographic clarity, however, the line is not wholly agreed, and it may be that different parts represent different Weichselian phases rather than the ice-front position at any one time. Part of the difficulty is the virtual absence of fossiliferous interglacial deposits under the Newer Drift—those described in the foregoing section all lying outside its limits.

There are two early radiocarbon dates related to Weichsel deposits. Peat from the Middle Sands of Cheshire, below the till of the Main Irish Sea glaciation is 57,000 years old and organic remains from the Main Severn Terrace, outwash from this ice, 41,900 years old. Outside the ice-sheets the history of this stage is reflected in the lowest and latest terraces of the Severn, Avon and Thames, and a late phase of low, glacial sea-level that resulted in the cutting of the steep, narrow, buried channels below the present Thames and Severn estuaries.

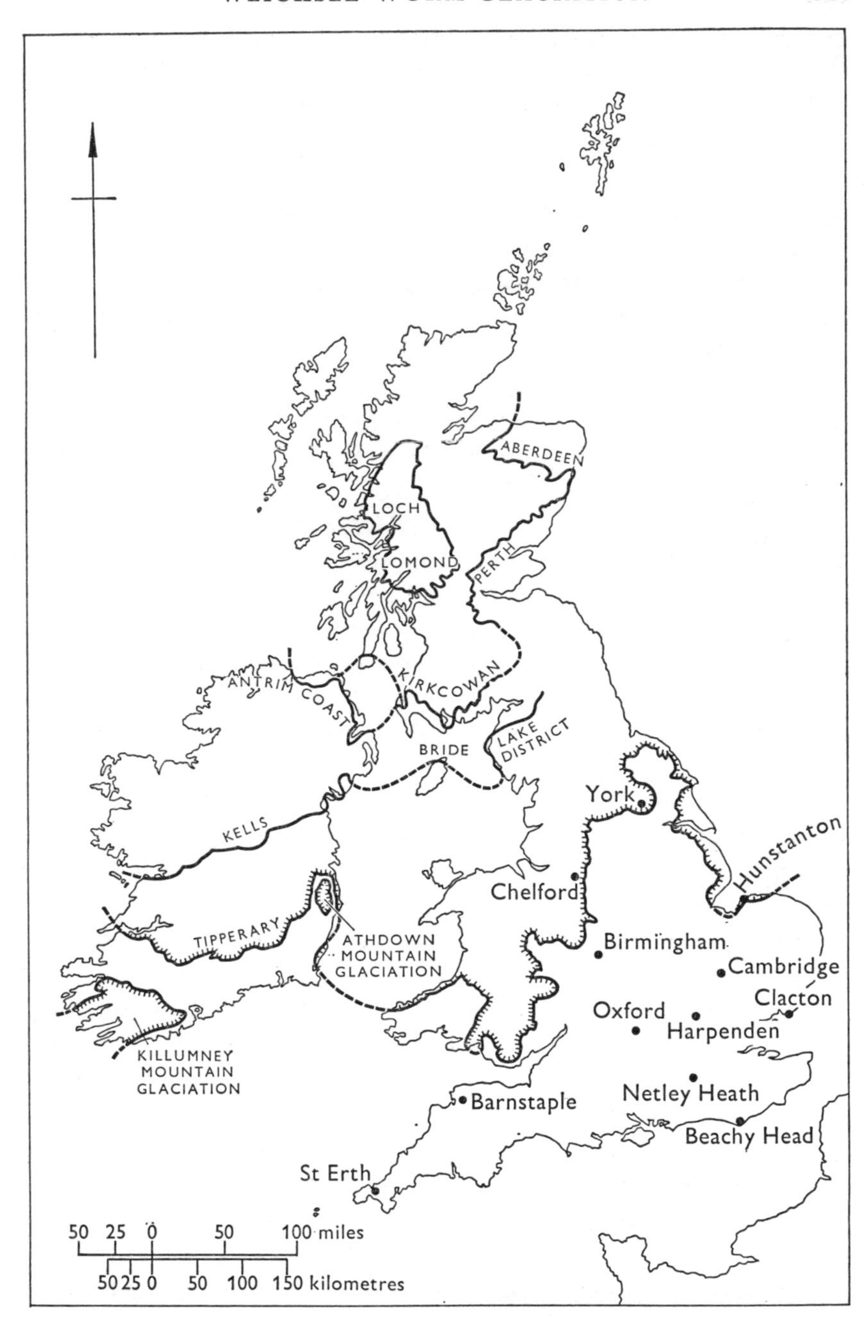

Fig. 75. Map showing the limit of the Weichsel-Würm (Last) Glaciation and the lines of three successive re-advances, including an alternative interpretation between Antrim and south-west Scotland. (Adapted from Charlesworth (1957), Penny (1964) and other sources.)

L. I. H. E.
THE MARKLAND LIBRARY
STAND PARK RD., LIVERPOOL, L16 9JD

During the recession of the ice-front water was again held up between the ice and higher ground. Pro-glacial lakes were formed around the Leinster mountains and in the Cleveland Hills. The eastward drainage from the Vale of Pickering to the North Sea was permanently blocked by drift and the Yorkshire Derwent now flows south-westwards into the Ouse. One of the most famous glacial diversions belongs to this stage, when the previous Severn drainage, northwards to the Irish Sea, was blocked by ice from that direction. The waters of the upper Severn cut the Iron Bridge gorge and still take this devious way into the lower Severn valley.

Subsidiary fluctuations in climate during the later part of the Weichsel glaciation produced periods of standstill or active readvances of the ice. Traces of these are left in moraines or other ice-marginal features, such as those of the Scottish Re-advance north of the Lake District and in northern England, which succeeded the Main Glaciation (earlier Weichsel) of those regions. In Scotland two further re-advances covered diminishing areas, namely the Perth–Aberdeen and Loch Lomond. The central and most persistent ice area was in the western highlands, because, as at present, there was greater precipitation on that side.

Late glacial raised beaches were formed when ice still covered parts of Scotland. They are highest in central Scotland, east of Stirling, not far from the centre of isostatic uplift which was probably in the south-west Grampian Highlands; from Stirling they fall in height outwards and are absent, for instance, in the Orkneys and much or all of the Outer Hebrides. In the coastal regions and Scottish lowlands marine clays and silts were laid down while ice still covered the higher ground. Their peri-glacial climate is shown by cold species as well as by boulders dropped by ice floes.

Post-glacial, or Flandrian, time in the British Isles is presumed to begin when the last major ice-sheets melted from Scotland, some 10,000 years ago. During early post-glacial time there was a relative emergence over much of Scotland and peat was formed which has been dated at 8500–9500 B.P.[1] At about this time the open park-like vegetation over lowland Britain gave place to denser forests, dominated at first by pine and birch and later by a succession of various trees. This changing vege-

[1] In the more recent radiocarbon determinations the length of the Christian era is a significant fraction of the total, and dates may be given as 'B.C.' or 'B.P.'. The latter, which is the more modern method, is abbreviated from 'before present', although since some fixed date is necessary, 'present' in fact is A.D. 1950.

tation is reflected in the late and post-glacial pollen zones which also show that a climatic optimum, warmer than the present day, took place about 5,000 years ago. Some 7,000 years ago peat, or 'moor log', was forming on the site of the Dogger Bank, at that time a lowland bordering a reduced North Sea. Since then this region has been submerged in the Flandrian, eustatic rise in sea-level, which culminated 5000–5500 B.P. It is reflected in the highest and oldest post-glacial shore-lines of central Scotland, where subsequently through a combination of isostatic uplift and minor eustatic oscillations lower beaches were formed. There were other effects in southern England. During the glacial periods, when ice stretched from Britain to Scandinavia, the northern drainage of Europe, from the Elbe to the Rhine, was blocked, and the waters were probably deflected into a westerly course down a broad valley that lay on the site of the upper English Channel. At the height of the Flandrian transgression, in the Neolithic period, this valley was drowned and the Straits of Dover were formed.

Pleistocene geology, in all its diversity, does not lend itself to simple or generalized conclusions but one aspect may be taken as an epilogue. There is no significant division between the geological past and the present. It is not so much that, in a uniformitarian manner, we may interpret the past by references to the present—a generalization that will not bear scrutiny in all its aspects, as witness the difficulty of interpreting till, or the absence of modern oolitic iron ores: it is rather that Pleistocene processes, sediments, faunas and floras pass smoothly into those of modern times, and studies of both past and present gain from the interpretations of the other.

Palaeolithic cultures give place to Mesolithic, Neolithic, Bronze Age and Romano-British, and the historian takes over from the archaeologist. The distribution of modern animals, and still more plants, has been strongly influenced by the fluctuations of ice-sheets, climates and sea-level, opening some routes of migration and closing others. The present land surface in glacial and periglacial regions has been affected by the same factors, and weathering and denudation are already altering the composition, texture and relief of the latest drifts in the same way that they altered the older ones.

Post-glacial sediments are separated only arbitrarily, and for convenience, from late glacial on one hand and modern on the other, and the three combine to record a continuous history of sedimentation in the Fenlands and the Wash. Possibly the term 'modern' itself suggests an unreal boundary. As soon as a sediment comes to rest it is subject to physical and chemical changes—the expulsion of water from the pore spaces, the disturbance of bedding structures by organisms, and the change in chemical conditions (especially towards reduction) as soon as a millimetre or two of covering layers have accumulated. A few centimetres down in the bottom sediments of the Humber, the Wash or the Firth of Clyde, the muds or silts are already in the first stages of diagenesis, which in time will transform them into lithified rocks, comparable to the shales and siltstones of the Jurassic or the Coal Measures. Given yet deeper burial, enhanced temperature and pressure, they may become slates and schists such as those of the Dalradian Series.

REFERENCES

Pliocene and non-glacial Pleistocene. Charlesworth, 1957; Funnell, 1961; Hollingworth, 1938; Jones, 1956; King, 1963; King and Oakley, 1936; Linton, 1951; West, 1963; Wooldridge and Linton, 1955; Wooldridge, 1960.

Glacial and post-glacial deposits. Charlesworth, 1957; Flint, 1957; Godwin, 1956; Oakley, 1953, 1961; Penny, 1964; Shotton, 1953, 1962; Sissons, 1965; West, 1961, 1963; West and Donner, 1956; Zeuner, 1959.

General. Oakley and Baden Powell, 1963.

British Regional Geology. East Anglia (together with Pleistocene or Glacial Geology in several handbooks).

13

EPILOGUE

THE DATA OF PALAEOGEOGRAPHY

From one aspect the stratigraphy reviewed in the preceding chapters could be described as a series of changing geographies. Palaeogeographical maps illustrate this aspect and the history of the British Isles has been amply covered by those of Wills (1951); in an earlier book (1929) this author also described the evolution of Britain in physiographical terms. The data used in compiling palaeogeographical maps are naturally very incomplete compared with those of contemporary maps and the aim is somewhat different. The most striking feature of a contemporary map is the coast-line, and only very rarely can this be drawn at all precisely for past periods.

Over the range of geological time shore-lines are transitory, even when associated with pronounced features such as cliffs. Emergent shore-lines tend to become blurred and finally obliterated by subsequent subaerial erosion and their uplifted relics survive only here and there, from relatively recent periods (p. 369). If the shore is gently shelving—for instance, a sandy intertidal zone backed by sand dunes—small changes in the relative levels of land and sea will cause wide lateral shifts in the shore-line, and the sediments originally deposited in one environment may be reworked in another. In such conditions it may well be very difficult to identify with certainty deposits formed above and below high tide level unless they contain an indigenous, diagnostic, flora or fauna.

The shore-line most likely to survive is the subsequent one where the younger sediments lap round and over older rocks of some relief. Two such examples were quoted in chapter 4, near Builth (Jones and Pugh, 1949) and south of the Longmynd (Whittard, 1932). These are small in extent, however, and outstanding exceptions among the Lower Palaeozoic outcrops in this country.

Time is another awkward factor. Obviously a map can only be an approximation for even part of a system, and much of its value lies not so much in attempting to delimit, precisely, land and sea, as in separating the principal areas of uplift and erosion from those of subsidence and deposition. To the latter can be added the main facies or environments, which are not always marine. Thus a land-and-sea map of the British

Isles during the deposition of the Lower and Middle Coal Measures is, strictly, an impossibility. During the inundations of the major marine bands much of the British Isles was sea, albeit probably one with restricted access to oceanic waters; during the periods of peat accumulation (the coal seams) it was land. What can be shown on a palaeogeographical map are main areas of deposition, in either environment, as in plates IX and X of Wills (1951), and the separation of the south-western coal basin from that of the Midlands and north. Somewhat similarly the London Platform was an important positive element in the Mesozoic history of south-east England (*op. cit.* pls. XIII–XVII), but it did not form a precisely definable landmass or island throughout that time.

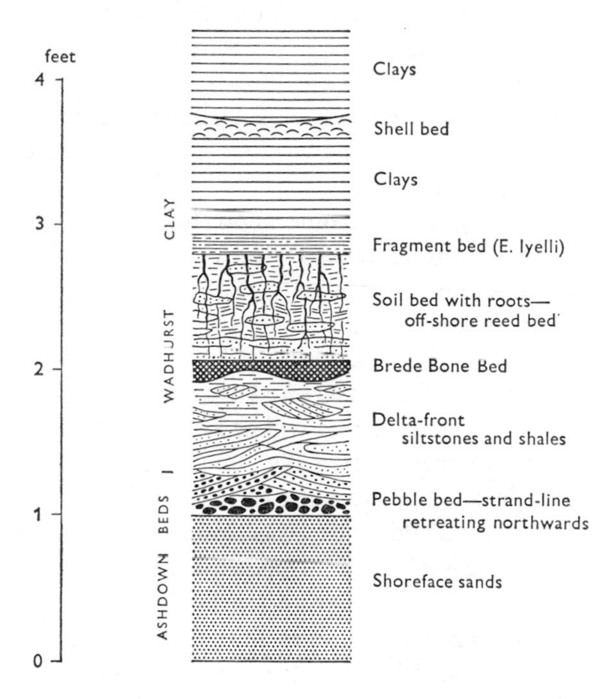

Fig. 76. Diagrammatic section of the junction between the Ashdown Sands and Wadhurst Clay. (After Allen, 1949, p. 267; 1959, p. 287.)

A large proportion of the data of palaeogeography concerns environments—their sedimentology, palaeontology, and the attempts to interpret them by comparison with modern analogues. In the descriptive chapters of this book the stratigraphical successions and the environments they represent were described in broad terms so that the picture remained an appreciable whole, not overwhelmed by detail. The detail is important, however, and to balance the generalizations it is worth looking at a few examples more closely to see how stratigraphy is built up.

The bulk of the cyclic Wealden sediments summarized in chapter 10 consists of the deltaic facies, dominantly sandy, and the clays and silts. Their interpretation, set out by P. Allen (1959), is grounded on the detailed succession at the junctions of these two facies, of which that between the Ashdown Sands and overlying Wadhurst Clay may be taken as an example. Typically the transition comprises only a few feet of strata which are remarkably constant over all the Wealden outcrops, particularly in their two most important members—the Top Ashdown Pebble Bed and Brede Soil Bed, with the remains of *Equisetum lyelli*. The section at Brede, Sussex, is set in detail by Allen (1949, p. 267) and the interpretation put on this and succeeding transitions is reproduced in Fig. 76. This author has also analysed the detrital components of uppermost Ashdown strata, taking samples from the 4-inch pebble bed itself and from a 2-inch band in the sandstone below. From very large assemblages, collected from 700 square miles and treated statistically, Allen has shown that the detritus was brought in by rivers from the north-east, north-west and south-west, each contributing its characteristic suite of pebbles and mineral grains (1949, 1954, 1959).

It is from this scrupulous attention to detail and quantitative data that the most thorough knowledge of past environments is built up. The Wealden beds are favourably placed in that the extensive outcrops allow an areal assessment to be made and the tectonic complications are small. It was also possible to compare the sediments with those of modern analogues, such as the Rhone and Mississippi deltas.

Few other formations have been subjected to such a rigorous examination as the Wealden or on a comparable scale, but examples of local studies which lead to wider estimates of environment can be taken from Devonian, Carboniferous and Permian beds. Six cycles in the Lower Old Red Sandstone have been described by J. R. L. Allen (1964) as part of a systematic study of the Anglo-Welsh province. An example from Mitcheldean in Gloucestershire is shown in Fig. 77, together with the interpretation put on it by Allen; the age is known to be Breconian since this is one of the very few quarries that have yielded vertebrates of that stage (*Rhinopteraspis dunensis*). As in the previous example the sediments, their structures and deduced environments have been carefully compared with modern régimes, and Allen concludes (1964, p. 190) that 'This cyclothem is probably fluviatile in origin. In general succession it

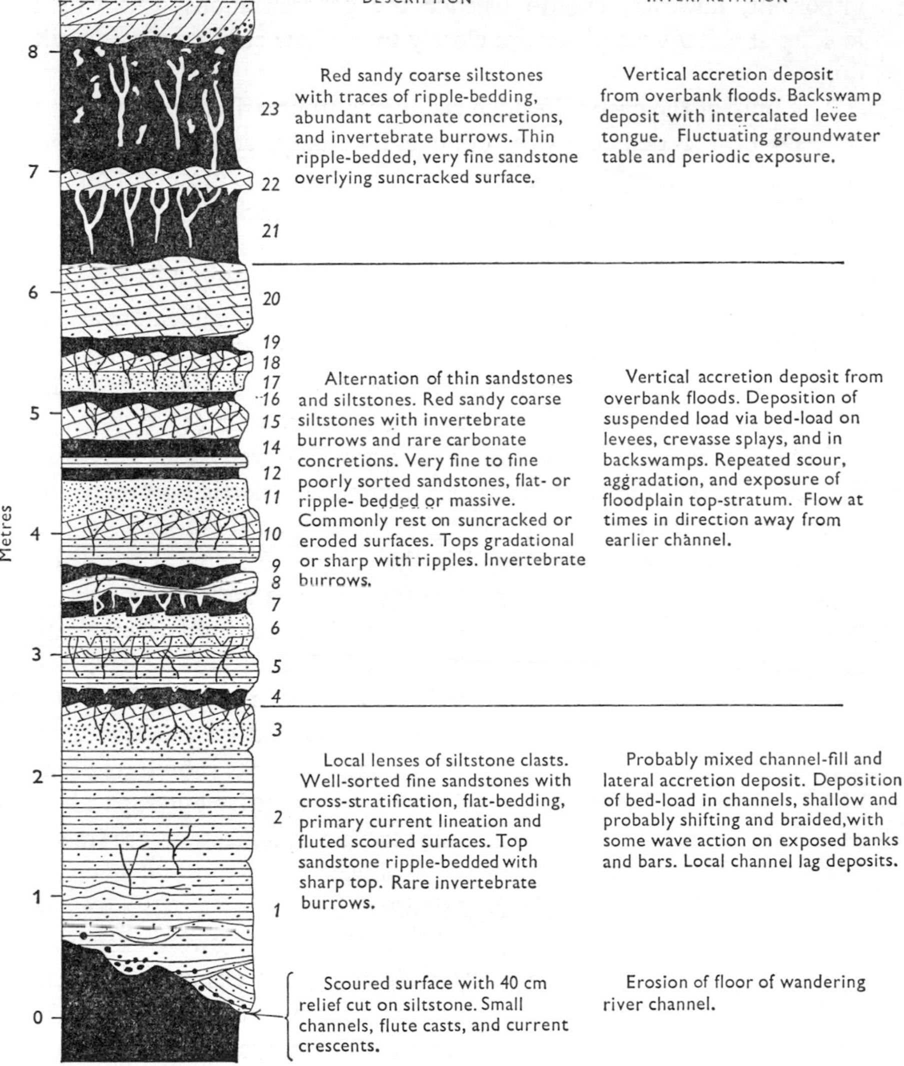

Fig. 77. Generalized succession and interpretation of a Breconian cyclothem at Mitcheldean, Gloucestershire. (Redrawn from Allen, 1964, p. 188.)

closely resembles a modern alluvial sequence, in that uniform coarse beds underlie coarse and fine deposits, followed by uniform fine sediments.'

The exceptional economic value of the Coal Measures has produced a corresponding amount of stratigraphical data, though rather on the thickness, extent and correlation of the coal seams that on environmental studies such as those we are concerned with. Nevertheless, three aspects may be briefly considered. Cyclic sedimentation is a key feature in the productive coal measures, but so also is variation in the type of cycle. That mentioned in chapter 7 (cf. Fig. 78), from the Middle Coal Measures of Nottinghamshire, was originally studied because the marine

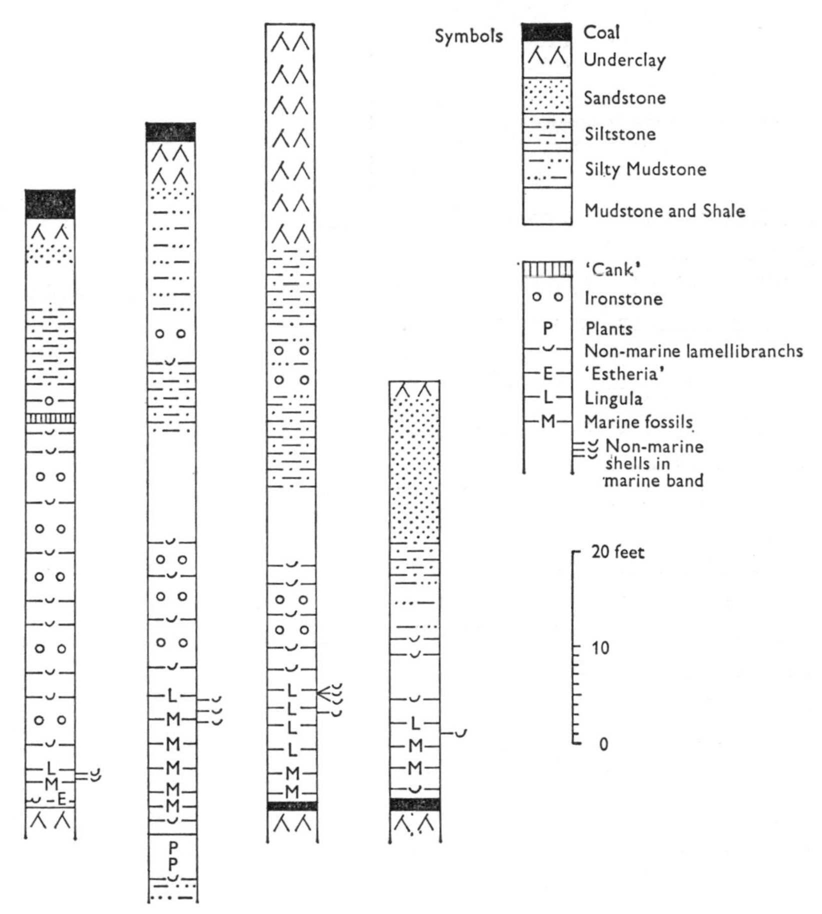

Fig. 78. Details of the Clay Cross cyclothem (beginning with the Clay Cross marine band) from borehole sections in Nottinghamshire and Derbyshire. (Redrawn from Edwards and Stubblefield, p. 215.)

bands at the base of the cyclothems were exceptionally well developed (Edwards and Stubblefield, 1948; Dunham, 1948). A more recent, statistical, treatment of all cyclothems in the East Pennine Coalfield (Duff and Walton, 1962) shows that although this is the fullest ('composite') type it is not the commonest (or 'modal'), which is more restricted and lacks the marine strata. In this paper Duff and Walton analyse 1,200 cycles, from borehole sections, and find that the marine type compares well with the sediments forming near delta distributaries in the Gulf of Mexico:

Marsh deposits
Delta front silts and sands
Pro-delta silty clays
Off-shore clays } with marine
Marginal deposits } shells

The commoner, non-marine, type similarly agrees fairly well with the regions further removed from marine influence, and perhaps represents the filling up of lakes on the delta surface, the process culminating in growth of vegetation.

The marine bands themselves are not uniform and the thicker ones show a considerable range of fossils, which probably reflects minor variations in salinity. Such changes have been described from the Bullion Mine marine band of the Burnley Coalfield (Lower Coal Measures in north Lancashire, Earp *et al.* 1961). Some 10 feet of shale can be divided into thin layers characterized by very varied faunas: fish debris, ostracods, lamellibranchs and goniatites; there are also layers with *Lingula* and plants. Of these the thick-shelled goniatites (*Gastrioceras listeri* at this horizon) represent the most fully marine conditions. Similar small-scale sequences have been recorded from marine bands elsewhere—such as those in the Coal Measures of South Wales (Woodland and Evans, 1964, p. 18), and the Millstone Grit of Derbyshire (Ramsbottom *et al.* 1962), where an exceptionally full comparison was made of lithology and faunal phases in a group of boreholes. Some Westphalian marine bands maintain the same detailed variations in sequence over very wide areas, from Britain to Germany, and must reflect an impressive degree of uniformity during these periods of inundation.

The coal seams similarly have some microscopic variations, not only in their petrology but also in their spore content. The changes in spore assemblages observed through certain seams in the Yorkshire Coalfield reflect the changes in the dominant vegetation during the formation of the peat (Smith, 1962). They are analogous to the pollen sequences in the interglacial peats (chapter 12, p. 377), and indeed the latter could be taken as an excellent example of palaeontological minutiae which nevertheless have substantial environmental and stratigraphical value.

The relation of fauna and flora to the sediments containing them is part of palaeoecology—the environment of fossil organisms. A classical palaeoecological study of Carboniferous faunas has been made by Craig

(1954). The Top Hosie Shale of Scotland underlies the basal Namurian Top Hosie Limestone. From the top 2 feet of this shale Craig collected some 2,000 macrofossils and 3,000 microfossils. Two ecological communities were distinguished. The earlier, in a dark pyritous mudstone, was characterized by the lamellibranch *Posidonia* (but very few other macrofossils) and a varied microfauna; it was considered to live on the mud surface, the lower layers being anaerobic. With slightly stronger currents, accompanied by a slight increase in scouring and grain size of the sediment, the area was invaded by a community of burrowers, especially *Lingula* and lamellibranchs, and a much more varied macrofauna. Both communities were considered to represent a shallow, subtidal environment.

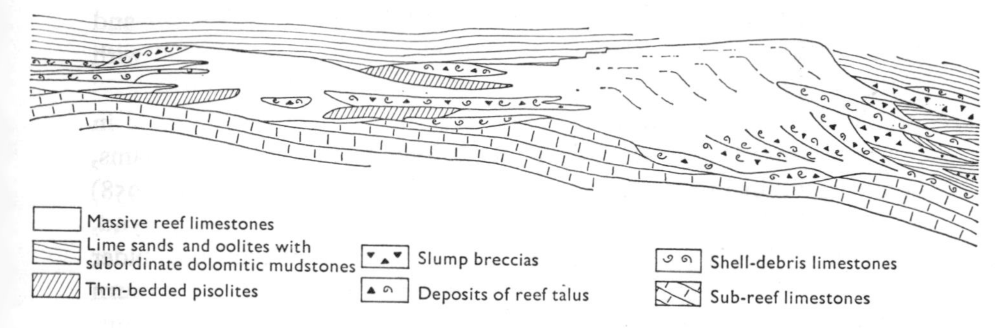

Fig. 79. Diagrammatic section showing the principal components of the Permian (Zechstein) reef of County Durham. (Redrawn from Smith, 1958, p. 78.)

The Permian reef of Durham (p. 243) has long been known through the work of Trechmann (e.g. 1931) and detail has been added by Smith (1958). The latter demonstrates how the upward impoverishment of the reef fauna (crinoids, polyzoa, molluscs and brachiopods), probably owing to increasing salinity, was accompanied by an increase in algal rocks. The higher parts of the reef in particular, as exposed in several quarries, are characterized by dome-shaped bodies and sheets of limestone, up to about 30 feet in height. Smith considers that these structures were principally formed by algal colonies, which acted as reef-binders, and that they tended to grow upward and outward so that the reef gradually shifted eastward towards the Zechstein basin. Although recrystallization has severely affected these rocks, enough remains of cellular and filamentous structures to support the evidence of the macroscopic forms, which are strongly characteristic of algal growth.

The detailed mapping of these rocks and other components of the reef allows a cross-section to be reconstructed (Fig. 79) in which there is a lateral transition from the basin, through the steep outer slope and major (outer) reef, to the inner edge and back-reef lagoon. The interpretation is assisted not only by comparison with modern reefs but with the great Guadaloupe reef complex of Texas and New Mexico; this is also Permian in age and despite the difference in scale the faunal associations are remarkably similar and comparable fore-reef, reef and back-reef associations are recognized.

These half dozen examples are among the most convincing detailed environmental studies in British stratigraphy and it will be seen that they relate to shallow marine, lacustrine, fluviatile or terrestrial conditions. The modern analogues thus have the advantage of being reasonably accessible. For various reasons the earlier systems are less productive. By and large the rock sequences are much thicker and tectonic and metamorphic effects greater, and research has commonly dealt with problems on a larger scale.

Certain marginal or exceptional conditions have been interpreted in some detail, for instance the erosion of the Ballantrae massif (Williams, 1962) or the ignimbrites of Snowdonia (Rast, Beavon and Fitch, 1958) and a good deal is known of the south-eastern shelf of the Longmynd, Wenlock Edge, Ludlow, etc. Among Dalradian sediments the boulder beds of Portaskaig and elsewhere (p. 68) have been described by Kilburn *et al.* (1965) who assemble evidence for its glacial origin. The most important geosynclinal sediments, however, are the turbidite-greywackes. Some sequences have been described in detail such as those of the Cambrian at St Tudwals (Basset and Walton, 1960), the Ordovician in Galloway (Kelling, 1961) and the lower Silurian Aberystwyth Grits (Wood and Smith, 1959); there are also voluminous writings on turbidites in general. Nevertheless, although modern turbidites probably bulk large among the deposits of the deeper seas, many aspects of their interpretation are still wide open to conjecture and the fossil representatives, and greywackes in general, can rarely be used as detailed or precise palaeogeographical data.

Palaeoclimates

The principal components of climate are radiation, temperature, and precipitation; at any given place they are influenced by a number of factors such as latitude, altitude, the distribution of land and sea and the relation to wind belts and ocean currents. Although on a global basis the first is the most important, a glance at climatic maps will show how far the actual climatological average diverges from the latitudinal average.

Palaeoclimates—those of past ages—are part of the total environment but with few exceptions, usually of climatic extremes, they are difficult to assess with certainty. The difficulties are aggravated by the abnormality of the present time. Owing to the Pleistocene glaciation and Tertiary orogenies both climatic belts and topographic relief are unusually pronounced and the existing ice-caps still retain sufficient oceanic water for the sea-level to be abnormally low. In periods of little relief and widespread inundation, such as the late Cretaceous, there was a greater uniformity. We should not expect, therefore, that latitude and climate should be associated in quite the same pattern in the past as they are at present. Many aspects of palaeoclimates have recently been summarized by Hollingworth (1962) and in two books edited by Nairn (1961, 1964); they also have an important bearing on problems of continental drift, but this aspect is reserved for the last section of this chapter.

Climatic indicators may be biological, physical or chemical. Deductions from the first are based largely on the assumption that plant and animal groups of the past had the same environmental restrictions as their modern relatives. Clearly this argument is more convincing for fairly recent floras and faunas where the degree of relationship is close and the time for adaptation less. It has been applied with considerable success to the early Tertiary floras; among these the London Clay is outstanding, with a high percentage of genera whose relatives live in tropics and subtropics at the present time. The succeeding floras and faunas fit well into a picture of gradual cooling during Tertiary time and show how the evidence from one group reinforces that from another (Reid and Chandler, 1933).

The more closely a living group is restricted by its climatic requirements, the more helpful its fossil relatives are likely to be. The reptiles lack a dermal insulating layer, are cold blooded, and the larger forms cannot flourish except in a warm climate (the smaller ones survive severer conditions better, by burrowing); it is extremely probable that the large reptiles and amphibia were similarly restricted, and they are among the best fossil climatic indicators. Corals are frequently quoted as warm-water inhabitants, at least when abundant or reef-forming, since reefs are not built at present in waters colder than about 18° C. But while Palaeozoic corals may have had similar limitations, they are too distant in relationship and time for this to be a secure assumption.

Fortunately here, as elsewhere in the assessment of climates, it is rarely necessary to base much on one line of evidence alone.

Physical and chemical indicators include evidence of widespread land ice and certain components of evaporite and carbonate deposition. Many marine organisms secrete a skeleton of calcium carbonate and among some, such as molluscs and foraminifera, the relative proportions of the two oxygen isotopes, O^{16} and O^{18}, depends on the temperature of the water; it is, however, also affected by the salinity, through small variations in the isotopic composition of sea water, and the original skeleton must be clearly distinguished from any later, diagenetic calcium carbonate. If these and other problems are overcome, however, the isotope ratio can be a sensitive indicator and has been used in estimating fluctuations in Cretaceous and Pleistocene climates.

Major ice-sheets, with their glaciated pavements, boulders and all the accessory characters mentioned in the last chapter, are the most impressive, least questionable evidence of palaeoclimates. As we saw there, however, the distribution of land-ice is by no means governed simply by temperature and still less by latitude. Boulders dropped from icebergs are even more wayward in their distribution. At the present time bergs from Greenland carry their debris as far as 40° N., south of Newfoundland; at the glacial maxima the drift ranged farther into the Atlantic and erratics have been dredged up off the Azores. A boulder bed formed in this way, such as that in the Dalradian and other late Pre-Cambrian successions is presumed to be, tells us very little of either the immediate palaeoclimate or palaeogeography; it is only indicative of an ice-sheet somewhere at that time.

The best known pre-Pleistocene glaciation is the late Palaeozoic of the southern hemisphere, affecting parts of South America, South Africa, peninsular India, Antarctica and Australia. Stratigraphical and palaeontological evidence from the associated sediments suggests that it spanned a long period from Carboniferous to early Permian, and that there were fluctuations comparable to the Pleistocene glacial and interglacial phases. These presumably caused comparable fluctuations in sea-level and it has been argued that the latter were a possible cause of the Carboniferous cyclic sequences of the northern hemisphere. In Britain the cycles are best seen in the Yoredale facies but similar examples are widespread in North America. On this view the southern ice-sheets would have exercised a kind of remote control over sedimentation many thousands of miles away.

The most satisfactory single indicator of aridity and high temperature is probably the deposition of thick beds of evaporites. This is supported not only by their distribution today, especially that of halite, but is also firmly based on aspects of geochemistry. The precipitation of the more soluble salts from the brines can only take place at relatively high temperatures; even so the processes are far from simple, the problem of distinguishing primary from secondary (diagenetic) minerals being particularly complex in evaporites.

Two other sedimentary features are often associated with evaporites —red beds and desert sands. The redness itself (the oxidation of existing iron-compounds or the introduction of red iron oxide from elsewhere) is not reliable evidence alone, since it can be produced in several ways, both primary and secondary. Lateritic soils, leached of all major components except iron oxides and alumina, are characteristic of warm, humid climates, and may be a source of red detritus which is then carried into a neighbouring area of deposition. On the other hand most modern desert deposits are grey, yellow or brown, and much hematite in red beds is a secondary, post-depositional product.

In a similar fashion hot deserts are not the only conditions producing aeolian erosion, deposition and sand dunes, but they are the most widespread. The Permian Dune Sandstone and similar beds, such as the Penrith Sandstone (p. 253), embody many features of desert sands—the aeolian bedding, the red colour, the surface textures and rounding of the grains. The wind directions deduced from the foreset dips of the barchan dunes are remarkably constant in the various outcrops and were probably those of the north-east trade winds of Permian time (Shotton, 1956). The Permian climate of Britain is in fact deduced, with unusual security, from a number of sources; of these the Zechstein evaporites are probably the most important, but there are impressive reinforcements from other lines of evidence. The hot arid belt certainly stretched across the North Sea into Germany where a similar succession occurs, and further important deposits of halite and other evaporites are found in the Caspian–Ural region, Texas, New Mexico and neighbouring states.

Warm waters also tend to promote thick carbonate deposits, especially of dolomite, whether primary or early diagenetic; but again there are other contributing factors, such as suitable regions of shallow water immune from heavy influx of detritus. At present the relatively high topographic relief, and hence drainage from the continents, has increased the amount of detritus entering the seas and suitable regions for carbonate deposition are correspondingly decreased.

We may conclude this section by considering the evidence bearing on the climate of the Upper Carboniferous Coal Measures of north-western Europe—not because it is a clear-cut issue, but is rather a balance of probabilities, and as such a better example of the problems of palaeoclimates than the Permian period. In the first place it must be admitted that the plants themselves tell us little about temperature, being too distant from their few surviving relatives. They do, however, suggest a well-forested region that was decidedly humid. The horsetails live in wet conditions today (and did so also in Jurassic and Cretaceous times) and the lycopod roots show adaptations to a waterlogged substratum. Humidity also is the governing factor in peat accumulation; thick peats are known from some tropical and near tropical regions, such as Java and Sumatra, but they are much more widespread in high latitudes. There are, however, some indications of warm, as well as wet conditions from the sediments associated with the coal seams and those that succeeded them. The contemporary soils (the kaolin-bearing 'tonsteins') are best matched in either the conditions of the monsoon belt or tropical and subtropical rain forests; the reddened measures towards the top are attributable to the weathering and transport of lateritic soils of the forest uplands in the later Carboniferous or early Permian times; by late Permian times the evaporites were being precipitated in the Zechstein sea.

From a variety of evidence, some direct and some indirect, we can safely assume a humid swamp forest, which was most probably warm, perhaps subtropical. It should be noted, however, that this does not necessarily apply to all coal-bearing formations and each has to be examined individually. The interglacial peats of the Pleistocene (if they should ever survive to become coals) would clearly belong to quite a different régime, as do the Coal Measures of South Africa and New South Wales which succeed the tillite horizons of the late Palaeozoic glaciation.

THE BRITISH SEAS

This book is specifically concerned with the British Isles, but stratigraphy does not stop short at national boundaries or the sea coast, and some brief comparisons have already been made with structure and history of the neighbouring parts of Europe. Something also is known of the sea floors surrounding these islands, which form part of the continental shelf. This is the submerged edge of the continental mass, similar in composition, structure and history to the adjacent lands which, at this fleeting moment of geological time, stand up above sea-level.

Although the sea floor is partly mantled by recent or reworked

Pleistocene deposits, a good deal can be gained from borings, echo-sounding, seismic and other geophysical techniques, which bring up samples of the solid rocks or produce information on underwater topography and on the strata and structures at depth. The geology of the British seas is summarized by Donovan (1963). Most has been published on the English Channel and western approaches (e.g. King, 1954; Curry, 1962; Hill and King, 1953; Day *et al.* 1956; Whittard, 1962; Curry *et al.* 1962). Many submarine outcrops here are bare or covered by only a few inches of sand or gravel; the scour of tides and currents keeps the river-borne detritus on the move, ultimately to slip down the continental slope beyond the western approaches. In the eastern part there are folds (?Mid-Tertiary) like those of the Weald and the Isle of Wight; chalk is the dominant rock with Lower Cretaceous or Jurassic in anticlinal areas.

The western section, and western approaches, are possibly older. Chalk again forms an important outcrop and in the centre is overlain by an extensive outlier of Palaeogene and Pliocene. To the south granite rocks like those of Brittany and the Channel Islands form an underwater fringe to the French coast (including the Cotentin Peninsula), while to the north the metamorphosed Devonian rocks similarly extend a few miles offshore from Devon and Cornwall. Between these and the chalk outcrop, however, there is a wide and deep belt of New Red Sandstone, which comes ashore in south-east Devon, at Teignmouth. It is probable that the Bristol Channel occupies a comparable downwarp, on the line of a Mesozoic depression; it contains some New Red Sandstone and Jurassic outcrops. One important outcome of this south-westerly marine exploration is that the folded Hercynian strata of Brittany, Cornwall and Ireland show no indication of continuing beyond the continental shelf, and this truncation of pre-Cretaceous structures has been observed elsewhere on Atlantic coasts. Little has been published on the sea floor to the north and west, though (as might be expected) there is a large area of basaltic rocks off northern Scotland, which links up with the Tertiary lavas of the Inner Hebrides, the Faroes and Iceland.

Until recently the geology of the North Sea area was virtually unknown, but in the south at least the position is now revolutionized by the results obtained during the explorations in the gas fields. Although only an outline is as yet available (Kent, 1967), it appears that the region was a sedimentary basin for much of Mesozoic and Tertiary times, and that the geology both resembles and differs from that of eastern England. Relatively little is known of pre-Carboniferous rocks, but the Coal

Measures probably extend right across from England to Holland and Germany north of the London-Brabant Massif (cf. inset in Fig. 39). The Permian and Trias are much more like the German outcrops than the English. The Lower Permian is well represented and the Zechstein evaporites are thick, especially the rock salt. This salt has moved under the weight of the overlying strata, and salt-dome tectonics dominate the southern North Sea basin. Bunter, Muschelkalk and Keuper are all well developed. The first is marly and contains thin beds of evaporites, and the Keuper salt is up to 1,500 feet thick. The Muschelkalk consists of alternating layers of mudstone, dolomite, salt and gypsum, and is 800 feet thick 40 miles east of the Humber.

Rhaetic, Jurassic and Lower Cretaceous beds tend to be thin or absent, and thus partly resemble those of north-eastern England. In many places Red Chalk, or a thin Lower Cretaceous remnant, rests on Trias; the Chalk, however, extends almost throughout. A particularly striking feature is the eastward thickening cover of Tertiary and Quaternary deposits, which average 4–5,000 feet (locally 8,000) in a north–south belt, east of the mid-line and related to the outflow of the German and Dutch estuaries. Between Scotland and Norway there may be as much as 10,000 feet. So far much less information has emerged on this northern part of the basin, but the 'sedimentary fill' overlying the Caledonian folded belt is known to be enormously thick.

CONTINENTAL DRIFT

Once our investigations have carried us beyond the continental shelf to consider the relation of British stratigraphy to that of other countries around the North Atlantic, all interpretations are governed by the view taken of continental drift. This book is not the place to examine so large, and still controversial, a topic in detail; recent reviews or compilations will be found in Bullard (1964), Runcorn (1962), King (1962), Holmes (1965, p. 1193) and Blackett, Bullard, Runcorn *et al.* (1965). It is desirable, however, to summarize some of its basic elements.

The postulate of drift has two sources. Certain rocks, as they are formed (lavas cooled or sediments deposited), acquire a stable, or 'remanent', magnetism that reflects the latitude at the time of formation. In the older systems this latitude is markedly different from that of the present day. It is thus deduced that continents have moved in relation to the poles, or that there has been 'pole wandering'. But the position of the poles estimated for, say, the Permian or Carboniferous system, is not the same in the different continents; consequently there has also been independent drifting of the continents and their spatial relations have altered.

It may be said that to the geologist working well within the confines of a single continent the effects of drift are not immediate, although the assessment of palaeoclimates might well create problems. The British Isles, however, are conspicuously on the edge of a continent mass. Their position in relation to oceans, seas and continental margins of the past is constantly before us and it becomes incumbent on the stratigrapher to examine, briefly, the evidence of his subject in relation to the hypothesis of drift. The history of the North Atlantic is especially important in this context with emphasis on any outstanding structural and stratigraphical links across that ocean and the implications of palaeoclimates.

In restricting this discussion to one part of the earth's surface we are not dealing with the most famous interrelation of palaeoclimates and inferred drift. This concerns South Africa, South America, Australia, Antarctica and peninsular India, parts of which were affected by the late Palaeozoic glaciation. For this and other reasons they are considered to have formed a single landmass at that time around the south pole (Gondwanaland), and to have since drifted apart. The similarities in structure and outline of the opposing coasts of South Africa and South America have long been quoted as one of the most striking results of this rifting. (Cf. Holmes, 1965, p. 1217; Bullard *et al.* 1965.)

In British stratigraphy much of the climatic evidence, already summarized, is suggestive rather than conclusive, but taken altogether and correlated with parallel climatic changes elsewhere on the globe, there is considerable support for changes in latitude. The significant correspondences across the North Atlantic belong to the Palaeozoic systems and are lithological, faunal or structural. Some similarities are not necessarily significant, if the conditions responsible for them could have developed independently on continents as far apart as Europe and America at the present day. A good example is the considerable resemblance between the Permian evaporite sequences of the Zechstein basin and Texas. Both were formed under comparable (not identical) conditions and governed by the same geochemical laws, but they are not otherwise related and their implications are climatic, not geographical.

To some extent the same argument applies to the lithological and environmental similarities of the Upper Carboniferous coal-bearing strata in both continents. If, however, there were many similarities in the flora and freshwater fauna of the two regions, then these would constitute a modest piece of evidence against the existence of a major ocean between them. In fact both the floras and non-marine faunas do show some affinities, such as those between the lamellibranchs of Britain and of the Appalachian and Acadian regions (cf. Trueman, 1946, p. lxvi).

The most impressive stratigraphical evidence is the marked resemblance, already briefly mentioned, in the Cambro-Ordovician, orthoquartzite-carbonate facies of several countries around the North Atlantic, and its accompanying Lower Cambrian (Olenellid) and Lower Ordovician faunas. This facies only occupies a narrow strip in Scotland, but is more extensive in Spitsbergen, East Greenland and much of North America, including southern Quebec and north-western Newfoundland. In south-eastern Newfoundland, New Brunswick, and fragmentary outcrops of the New England states the Lower Palaeozoic rocks and faunas belong to the 'Atlantic' (i.e. on the European side the 'Welsh-Scandinavian') province, where *Holmia*, *Callavia*, *Paradoxides*, *Olenus*, etc., are the characteristic Cambrian trilobites.

In Britain these faunas are sharply distinct, but an occasional mixing of species is found in Newfoundland and at a few other places in eastern North America. The Cambro-Ordovician carbonates of Greenland are distinctly similar to those of Scotland; they are overlain with marked unconformity by Middle Old Red Sandstone and are clearly part of the Caledonian mobile belt, though not 'geosynclinal' in facies, as are the Dalradian or Welsh rocks. In Spitsbergen there is no marked break between the Lower Cambrian carbonates and the tillite-bearing, late Pre-Cambrian below and the latter is commonly correlated with the boulder beds of the Dalradian Series. There are thus both resemblances and differences in the various outcrops of this facies, and the arguments for a much greater approximation of the several landmasses, and absence of a Lower Palaeozoic ocean, rest chiefly on two aspects. First, although trilobites are wholly extinct their structure is complex and mode of life fairly well established; with a few possible exceptions they were probably bottom-living forms, to which 3,000 miles of deep ocean would be a formidable barrier. Second, there is much in the limestone-dolomite facies to suggest shallow waters and possibly even occasional emergence.

The remaining Lower Palaeozoic sequences have less to offer, though there are some transatlantic affinities in the later Ordovician and Silurian faunas of this country. The distribution of the Old Red Sandstone or continental Devonian, however, is another important parallel, for it is found in many of the same regions bordering the North Atlantic. It is well developed in Spitsbergen, East Greenland and maritime Canada and red detritus was carried into the Appalachian geosyncline. Though the vertebrate faunas have their local characters there are enough genera in common to make a broad correlation possible. Some of the fishes inhabited fresh waters; other early armoured types (such as the pteraspids) probably originated as coastwise travellers, later penetrating the continent along its rivers. Typical vertebrates of the Old Red Sandstone are very rare in fully marine facies and it is difficult to avoid the conclusion that a single major landmass, resulting from the Caledonian movements, shed its debris into all these basins, though each had its local variants.

The Caledonian landmass continued to influence sedimentation in the British Isles for much of late Palaeozoic time. After the invasion of the Dinantian seas it was still a source of detritus, notably in Donegal, and throughout the Carboniferous period marine facies tend to diminish towards the north. In Permian and Triassic times the position is much more uncertain, partly because outcrops are so meagre in the north and west, but there is little to suggest a nearby margin to the continent on those directions.

On the structural side there are three items that are suggestive, though not conclusive. The Hercynian and Caledonian folded belts of Europe disappear westwards and all the available evidence is against their continuation on the ocean floor. There are, however, plausible continuations in, respectively, the Appalachian and Taconic-Acadian belts of North America. In Pembrokeshire and Ireland the two converge; in North America they cross, as Bailey (1929) pointed out many years ago. The orogenic phases and characters are not identical but sufficiently alike. As in the stratigraphical resemblances the fit is not perfect but fairly good.

The third feature is the Great Glen Fault of Scotland, which, with a lateral displacement of over 60 miles, is a global rather than a local structure. It has been suggested that there is a possible continuation in the Cabot Fault, which is similarly a sinistral transcurrent fracture, extending south-westwards from Newfoundland to the Bay of Fundy in eastern Canada. In addition the Great Glen Fault has an historic significance. It provided one of the first demonstrations (by Kennedy in 1946), through routine geological mapping and the matching of evidence on either side, that sections of the earth's crust really had moved laterally by amounts of this order.

From Mesozoic times onward British stratigraphy has little to contribute to the problem of drift, nor is there any local indication of when

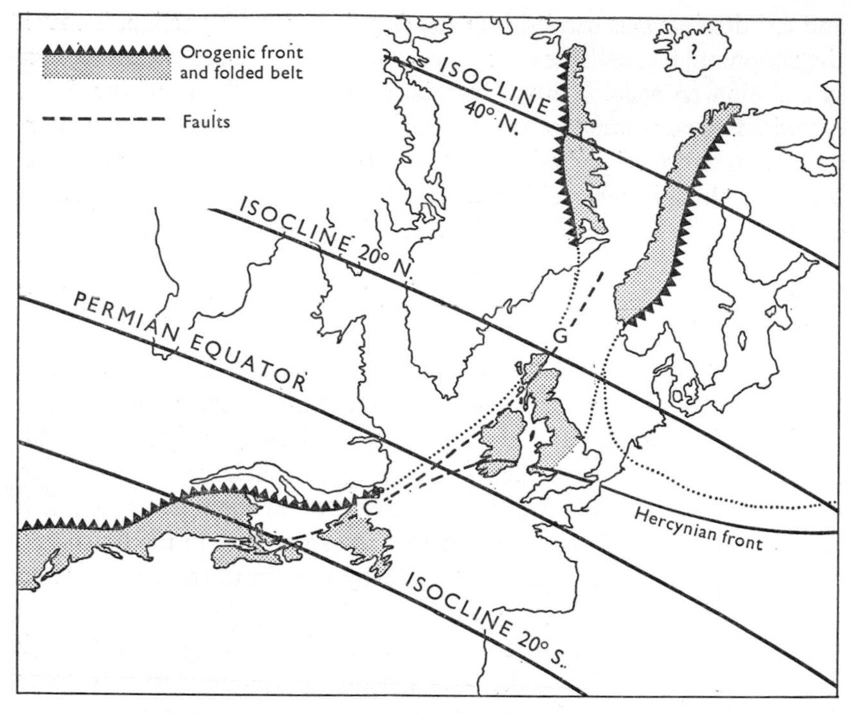

Fig. 80. Tentative reassembly of lands on either side of the North Atlantic, based on the mean position deduced for the Permian poles; the isoclines come into line as shown. Regions of Caledonian folding in Europe and Greenland, and Taconic-Acadian in North America, stippled; the fronts of these folded belts are also indicated. A possible continuation of the Caledonian front is shown through the North Sea and into the Netherlands. *G*, Great Glen Fault; *C*, Cabot Fault. (Redrawn from Holmes, 1965, p. 1231; cf. Bullard *et. al.* 1965.)

the North Atlantic 'rift' might have taken place. The pole positions deduced for Europe and North America do not help much as the necessary data are not available between the Triassic and later Tertiary periods. There is little evidence in Jurassic sediments of other than shallow, epicontinental seas, but late Cretaceous marine strata are widespread in several countries on either side of the North Atlantic. A rift widening gradually, from south to north, is sometimes postulated, beginning perhaps in Cretaceous, or late Cretaceous, times; in the succeeding early Tertiary the great upwelling of basaltic magma is similarly construed as linked with a state of regional crustal tension. The resulting lavas are seen today from Ireland and north-west Scotland to the Faroe Islands, Spitsbergen and East Greenland (the Brito-Arctic province). Remnants of this vulcanism still persist in Iceland, which lies on the Mid-

Atlantic ridge; this is considered to mark the latest position of the rift and Iceland appears to be still expanding, with the development of fissures in the central trough that runs through the island.

It is convenient here to summarize the postulated drift of Europe as it is deduced from palaeomagnetic evidence, bearing in mind that this can only provide data on latitude and on the direction of the contemporary poles, not on longitude; and that there are as yet rather few determinations for the pre-Carboniferous positions. Briefly it is considered that in late Pre-Cambrian and Cambrian times Europe lay south of the equator, since when there has been gradual and possibly rather irregular drift northward, crossing the equator in the late Palaeozoic. In Carboniferous and Permian times the British Isles were within 20 degrees of the equator and were in low northern latitudes during the Trias. Northerly movement continued during the late Mesozoic to about 40° N. in the Eocene period and thence to the present position.

This history can only be an approximation; experimental errors amount to as much as 15° and obviously no one position can really represent, for instance, the whole 65 million years of the Carboniferous period, or even a major part of it. Moreover, there may well have been some distortion of the continental masses themselves, particularly in regions of recent orogeny. Nevertheless, the most pronounced climatic phases, the warm Coal Measures, arid New Red Sandstone and gradual cooling through the Tertiary, agree well with this pattern of drift.

Conclusion

As far as the relation of the British Isles with their westerly neighbours is concerned much of the foregoing hypothesis is illustrated in Fig. 80 which shows the presumed latitude of Europe during the Permian period and the position of the North Atlantic countries before rifting took place. It will be seen that the fit is not ideal nor would one expect it to be. The true margin is the edge of the continental shelf and not the present coast-line; Scottish and Irish structures will not be found neatly matched in Greenland or Newfoundland.

A few more generalizations can be added, though the reader must accept them as somewhat tentative and personal. Much of the known geological history of north-western Europe is suggestive of a landmass or shallow seas to the west of the British Isles. In the past such stratigraphical demands have met in some degree by the concept of a North Atlantic continent ('North Atlantis') or some fairly substantial land bridge. Neither is any longer acceptable. The geophysical investigations of the last twenty years have shown with increasing force that the continents and the ocean floor are fundamentally different from each other, especially in chemical composition and density, and that they exhibit different structures. They are not, so to speak, interchangeable; the ocean floor has not been uplifted to form major continental masses, nor the continents submerged to lie beneath the deep oceans. 'North Atlantis' thus disappears into the limbo of other legendary lands. In its place, however, there arises the probability (or possibility according to one's vicw) of a single landmass which suffered Cretaceous rifting. When this pattern of continents and oceans is compared with that of the globe as a whole, and with the more reliable data on a palaeoclimates and latitude, it seems to the writer that a reasonable framework emerges. It is neither complete, nor perfect, but is one that accommodates logically the stratigraphy described in this book. The concluding words lie with the geophysicist:

'Past experience suggests that it is too much to expect a hypothesis concerning a major geological problem to be wholly correct. It is sufficient if it is not clearly absurd either geologically or physically and if it suggests relations and other hypotheses that can be tested. The idea of continental drift is exceptionally fruitful in this way; it suggests large programmes of investigation on land, at sea, and in the laboratory. If these are pursued vigorously for ten or twenty years it is probable that general agreement will be reached. From what we know at present it seems likely that the decision will be that the continents have moved' (Bullard, 1964, p. 24).

APPENDIX

STRATIGRAPHICAL DIVISIONS AND ZONAL TABLES

Zonal tables are given for those systems where a standard, fossiliferous sequence is available in the British Isles; thus the New Red Sandstone, for instance, is not included. In many cases the zones are applicable in other parts of Western Europe and occasionally farther afield. In matters of palaeontological and stratigraphical terminology I have consulted in particular two recent, authoritative groups of publications—the *Lexique Stratigraphique International*[1] and the three handbooks of the British Museum (Natural History) on Palaeozoic, Mesozoic and Cainozoic Fossils. The tables given below have been taken largely from the last three and also are very similar to those in Morley Davies and Stubblefield (1961); other authorities are quoted in the notes.

Finally the reader will probably be aware that there is nothing sacrosanct about such tables. Zones are the tools of the stratigraphical palaeontologist and these lists will with time be revised or emended as one species (or assemblage) is found to be more useful than another, or as the species themselves are more accurately named.

With the exception of the Eocene, Tertiary successions are poorly represented in the British Isles and a full standard sequence of zones, comparable to those of earlier systems, cannot be applied. Certain international stages or zones are given by Morley Davies and Stubblefield (1961, p. 290) and by Funnell (in Harland *et al.* 1964, p. 179).

[1] The published sections on the British Isles will be found in the List of References under the following authors: Anderson (1959), Stubblefield (1959), Whittard (1960, 1961), Simpson (1959*b*), Trotter (1960), Donovan and Hemingway (1963), Curry (1958), Oakley and Baden Powell (1963) and Gill (1956).

CRETACEOUS SYSTEM

Stages	Zones	English formations, etc.
	UPPER CRETACEOUS	
LOWER MAASTRICHTIAN	*Liostrea lunata*	UPPER CHALK
SENONIAN	*Belemnitella mucronata* *Gonioteuthis quadrata* *Offaster pilula* *Marsupites testudinarius* *Uintacrinus socialis* *Micraster coranguinum* *Micraster cortestudinarium*	
TURONIAN	*Holaster planus*	
	Terebratulina lata *Inoceramus labiatus*	MIDDLE CHALK
CENOMANIAN	*Holaster subglobosus* *Schloenbachia varians*	LOWER CHALK
	LOWER CRETACEOUS	
ALBIAN	*Stoliczkaia dispar* *Mortoniceras inflatum* *Euhoplites lautus* *Hoplites dentatus*	GAULT
	Douvilleiceras mammillatum *Leymeriella tardefurcata*	LOWER GREENSAND [of southern England]
APTIAN	*Hypacanthoplites jacobi* *Parahoplites nutfieldensis* *Cheloniceras martinoides* *Tropaeum bowerbanki* *Deshayesites deshayesi* *Deshayesites forbesi* *Prodeshayesites fissicostatus*	
BARREMIAN	*Costidiscus recticostatus* *Heteroceras asterianum* *Crioceratites emericianus*	[Stages present in Yorkshire and Lincolnshire, see text.]
HAUTERIVIAN	*Pseudothurmannia angulicosta* *Subsaynella sayni* *Crioceratites duvali* *Acanthodiscus radiatus*	
VALANGINIAN	*Kilianella roubaudiana* [unnamed gap]	
BERRIASIAN	*Subthurmannia boissieri*	

Largely from *Mesozoic Fossils* (B.M.N.H.) and *Treatise on Invertebrate Paleontology*, vol. L, p. 128. The zonal indices of the Upper Cretaceous are drawn from several phyla; those of the Lower Cretaceous are all ammonites. The zones below the Aptian are not much used in Britain, being those of the standard southern (Tethyan) sequence. The sparse ammonite faunas of north and east England belong to the boreal province; the stage Ryazanian has been used here to replace Berriasian; it contains two zones—*Riasanites rjasanensis* below and *Surites stenomphalus* above.

JURASSIC SYSTEM

Stages	Zones	English formations, etc.
PORTLANDIAN	*Cypridea setina*	PURBECK BEDS
	Cypridea granulosa	
	'Cypris' purbeckensis	
	Titanites giganteus	PORTLAND BEDS
	Glaucolithites gorei	
	Zaraiskites albani	
KIMERIDGIAN	*Pavlovia pallasioides*	KIMERIDGE CLAY
	Pavlovia rotunda	
	Pectinatites pectinatus	
	Subplanites wheatleyensis	
	Subplanites spp.	
	Gravesia gigas	
	Gravesia gravesiana	
	Aulacostephanus pseudomutabilis	
	Aulacostephanoides mutabilis	
	Rasenia cymodoce	
	Pictonia baylei	
OXFORDIAN	*Ringsteadia anglica*	CORALLIAN BEDS, etc.
	Decipia decipiens	
	Perisphinctes cautisnigrae	
	Perisphinctes plicatilis	
	Cardioceras cordatum	OXFORD CLAY, etc.
	Quenstedtoceras mariae	
CALLOVIAN	*Quenstedtoceras lamberti*	
	Peltoceras athleta	
	Erymnoceras coronatum	
	Kosmoceras jason	
	Sigaloceras calloviense	KELLAWAYS BEDS
	Macrocephalites macrocephalus	UPPER CORNBRASH
BATHONIAN	*Clydoniceras discus*	LOWER CORNBRASH
	Clydoniceras hollandi	GREAT OOLITE
	Oppelia aspidoides	
	Tulites subcontractus	
	Gracilisphinctes progracilis	
	Oppelia fallax	
	Zigzagiceras zigzag	INFERIOR OOLITE
BAJOCIAN	*Parkinsonia parkinsoni*	
	Garantiana garantiana	
	Strenoceras subfurcatum	
	Stephanoceras humphriesianum	
	Sonninia sowerbyi	
	Graphoceras concavum	
	Ludwigia murchisonae	
	Tmetoceras scissum	
	Leioceras opalinum	

JURASSIC SYSTEM (*cont.*)

Stages		Zones	English formations, etc.
TOARCIAN	YEOVILIAN	*Dumortieria levesquei*	UPPER LIAS
		Grammoceras thouarsense	
		Haugia variabilis	
	WHITBIAN	*Hildoceras bifrons*	
		Harpoceras falciferum	
		Dactylioceras tenuicostatum	
UPPER PLIENSBACHIAN (DOMERIAN)		*Pleuroceras spinatum*	MIDDLE LIAS
		Amaltheus margaritatus	
LOWER PLIENSBACHIAN (CARIXIAN)		*Prodactylioceras davoei*	LOWER LIAS
		Tragophylloceras ibex	
		Uptonia jamesoni	
SINEMURIAN		*Echioceras raricostatum*	
		Oxynoticeras oxynotum	
		Asteroceras obtusum	
		Caenisites turneri	
		Arnioceras semicostatum	
		Arietites bucklandi	
HETTANGIAN		*Schlotheimia angulata*	
		Psiloceras planorbis	
		(Pre-*planorbis* beds, no ammonites)	

All the zonal indices are ammonites except those of the Purbeck Beds which are ostracods. From *Mesozoic Fossils* (B.M.N.H.).

PERMIAN AND TRIASSIC SYSTEMS

Macrofossils are rare or absent in these systems in the British Isles, and although certain genera and species are indicative of the various divisions of the Magnesian Limestone, no standard sequence of zones can be applied. The Permian standard succession and stages are derived from the Russian platform and Ural mountains and those of the Trias chiefly from the eastern Alps. (See for instance Morley Davies and Stubblefield, 1961, pp. 295–6; Smith, in Harland *et al.* 1964, p. 212.) Microfossils have recently been applied in Britain to the correlation of continental Triassic sediments with those of Germany.

Carboniferous system

Compiled from many sources, especially *Palaeozoic Fossils* (B.M.N.H.), Crookall (1955, p. 2) and Butterworth and Millott (1960).

CARBONIFEROUS SYSTEM

Series	Stage	Substage	Zone	Floras and spores	Lamellibranchs	Goniatites
COAL MEASURES UPPER	? STEPHANIAN		I	*Verrucososporites obscurus*	*Anthraconaia prolifera*	
COAL MEASURES UPPER	MORGANIAN	WESTPHALIAN D	H		*Anthraconauta tenuis*	
COAL MEASURES UPPER	MORGANIAN	WESTPHALIAN C	G	*Torispora securis*	*Anthraconauta phillipsii*	
COAL MEASURES MIDDLE	AMMANIAN	WESTPHALIAN C	F	*Novisporites magnus*	*Anthracosia similis* and *Anthraconaia pulchra* — Upper	*Anthracoceras* A
COAL MEASURES MIDDLE	AMMANIAN	WESTPHALIAN B	E	*Dictyotriletes bireticulatus*	*Anthracosia similis* and *Anthraconaia pulchra* — Lower	
COAL MEASURES MIDDLE	AMMANIAN	WESTPHALIAN B			*Anthraconaia modiolaris*	
COAL MEASURES LOWER	AMMANIAN	WESTPHALIAN A	D	*Cirratriradites aligerens*	*Carbonicola communis*	
COAL MEASURES LOWER	AMMANIAN	WESTPHALIAN A	C	*Schulzospora ovata*	*Anthraconaia lenisulcata*	*Gastrioceras* G_2
NAMURIAN		C	C	*Schulzospora ovata*		*Gastrioceras* G_1
NAMURIAN		B	B			*Reticuloceras* R_2 R_1
NAMURIAN		B/A	B			*Homoceras* H_2 H_1
NAMURIAN		A	A	*Cincturasporites carnosus*	**Corals and brachiopods**	*Eumorphoceras* E_2 E_1
DINANTIAN	VISÉAN			*Rotaspora knoxi*	*Dibunophyllum* D_2 D_1	*Goniatites* P_2 P_1
DINANTIAN	VISÉAN			*Camptotriletes verrucosus*	*Seminula* S_2	*Beyrichoceras* B_2 B_1
DINANTIAN	VISÉAN				*Caninia* C_2 S_1 C_1	*Pericyclus*
DINANTIAN	TOURNAISIAN				*Zaphrentis* Z_2 Z_1	*Gattendorfia*
DINANTIAN	TOURNAISIAN				*Cleistopora* K_2 K_1	

DEVONIAN SYSTEM

Stages (marine)		Ammonoid zones	Vertebrate zones	Anglo-Welsh stages	
UPPER DEVONIAN	FAMENNIAN	*Wocklumeria* *Clymenia* *Platyclymenia* *Cheiloceras*		FARLOVIAN ? probably not represented	UPPER OLD RED SANDSTONE
	FRASNIAN	*Manticoceras*			
MIDDLE DEVONIAN	GIVETIAN	*Maenioceras*		?	MIDDLE OLD RED SANDSTONE
	EIFELIAN	*Anarcestes*			
LOWER DEVONIAN	EMSIAN	*Mimosphinctes*	*Rhinopteraspis dunensis*	BRECONIAN	LOWER OLD RED SANDSTONE
	SIEGENIAN		*Cymripteraspis leachi* *Belgicaspis crouchi* *Simopteraspis leathensis*	DITTONIAN	
	GEDINNIAN	No ammonoid zones	*Traquairaspis symondsi* *Traquairaspis pococki* (no zone fossil) *Hemicyclaspis* *Cyathaspis*	DOWNTONIAN	

Largely from *Palaeozoic Fossils* (B.M.N.H.). The correlation between continental and marine stages is only approximate. Vertebrates are used as stratigraphical guides in certain parts of the Middle and Upper Old Red Sandstone of Scotland (cf. Waterston, 1965, pp. 291, 300).

SILURIAN SYSTEM

Series		Graptolite zones
LUDLOW		[No graptolites known in Britain]
		Monograptus leintwardinensis
		Monograptus tumescens
		Monograptus nilssoni and *Monograptus scanicus*
		Monograptus vulgaris
WENLOCK		*Cyrtograptus lundgreni*
		Cyrtograptus rigidus
		Cyrtograptus linnarssoni
		Cyrtograptus symmetricus
		Monograptus riccartonensis
		Cyrtograptus murchisoni
LLANDOVERY	Upper	*Monograptus crenulatus*
		Monograptus griestonensis
		Monograptus crispus
		Monograptus turriculatus
		Rastrites maximus
		Monograptus halli
		Monograptus sedgwicki
	Middle	*Cephalograptus cometa*
		Monograptus convolutus
		Monograptus leptotheca
		Diplograptus magnus
	Lower	*Monograptus triangulatus*
		Monograptus cyphus
		Monograptus acinaces
		Monograptus atavus
		Akidograptus acuminatus
		Glyptograptus persculptus

From *Palaeozoic Fossils* (B.M.N.H.). The Ludlow beds of the Welsh Borders are divisible on the basis of their shelly faunas; see for instance Strachan (in Harland *et al.* 1964, p. 239), and Chapter 4, p. 102.

ORDOVICIAN SYSTEM

Series	Graptolite zones	Trilobite-brachiopod stages
ASHGILL	*Dicellograptus anceps*	
	Dicellograptus complanatus	
CARADOC	*Pleurograptus linearis*	Pusgillian
		Onnian
	Dicranograptus clingani	Actonian
		Marshbrookian
		Longvillian
	Climacograptus wilsoni	Soudleyan
	Climacograptus peltifer	Harnagian
	Nemagraptus gracilis	Costonian
LLANDEILO	*Glyptograptus teretiusculus*	
LLANVIRN	*Didymograptus murchisoni*	
	Didymograptus bifidus	
ARENIG	*Didymograptus hirundo*	
	Didymograptus extensus	
TREMADOC	*Shumardia pusilla* (trilobite)	
	Clonograptus tenellus	
	Dictyonema flabelliforme	

In North Wales and the Welsh Borders the *wilsoni* and *peltifer* zones are represented by the single zone of *Diplograptus multidens*; the trilobite-brachiopod stages are based primarily on the shelly faunas of Shropshire. In Britain the Tremadoc Series is sometimes included in the Cambrian System. From *Palaeozoic Fossils* (B.M.N.H.), Dean (1958, 1959); a slightly different version may be found in Whittington and Williams (in Harland *et al.* 1964).

CAMBRIAN SYSTEM

	Series or groups	Zones
UPPER CAMBRIAN	DOLGELLY	*Acerocare* *Ctenopyge* and *Peltura* *Eurycare* and *Leptoplastus* *Parabolina spinulosa*
	FFESTINIOG	
	MAENTWROG	*Olenus* *Agnostus pisiformis*
MIDDLE CAMBRIAN	MENEVIAN	*Lejopyge laevigata* *Paradoxides forchhammeri* *Paradoxides davidis* *Paradoxides hicksi*
	SOLVA	*Paradoxides salopiensis* *Paradoxides oelandicus* *Paradoxides harknessi*
LOWER CAMBRIAN	No major groups definable in Britain	*Lapworthella* *Protolenus* *Strenuella* *Eodiscus bellimarginatus* *Callavia* *Holmia* *Acrothele* *Obolella*

The table is based on the Anglo–Welsh–Scandinavian faunas and does not apply to north-west Scotland. All the zonal indices are trilobites except *Acrothele* and *Obolella* (brachiopods) and *Lapworthella* (? annelid). From *Palaeozoic Fossils* (B.M.N.H.).

REFERENCES

ADAMS, F. D. (1938). *The Birth and Development of the Geological Sciences*, v + 506 pp. Baltimore: Williams and Wilkins.

ALLEN, J. R. L. (1960). Cornstone. *Geol. Mag.* **97**, 43–8.

ALLEN, J. R. L. (1962). Intraformational conglomerates and scoured surfaces in the Lower Old Red Sandstone of the Anglo-Welsh cuvette. *L'pool Manchr Geol. J.* **3**, 1–20.

ALLEN, J. R. L. (1963). Depositional features of Dittonian rocks: Pembrokeshire compared with the Welsh Borderland. *Geol. Mag.* **100**, 385–400.

ALLEN, J. R. L. (1964). Studies in fluviatile sedimentation: six cyclothems from the Lower Old Red Sandstone, Anglo-Welsh basin. *Sedimentology*, **3**, 163–98.

ALLEN, J. R. L. (1965). The sedimentation and palaeogeography of the Old Red Sandstone of Anglesey, North Wales. *Proc. Yorks. Geol. Soc.* **35**, 139–85.

ALLEN, J. R. L. and TARLO, L. B. (1963). The Downtonian and Dittonian facies of the Welsh Borderland. *Geol. Mag.* **100**, 129–55.

ALLEN, P. (1949). Wealden petrology: The Top Ashdown Pebble Bed and the Top Ashdown Sandstone. *Q. Jl Geol. Soc. Lond.* **104**, 257–321.

ALLEN, P. (1954). Geography and geology of the London–North Sea uplands in Wealden times. *Geol. Mag.* **91**, 498–508.

ALLEN, P. (1955). Age of the Wealden in north-western Europe. *Geol. Mag.* **92**, 265–81.

ALLEN, P. (1959). The Wealden environment: Anglo-Paris basin. *Phil. Trans. R. Soc.* B, **242**, 283–346.

ALLEN, P. (1965). L'âge du Purbecko-Wealdien d'Angleterre. Pp. 321–6 in Colloque sur le Crétacé inférieur (Lyon, septembre 1963). *Mém. Bur. Rech. Géol. Min.* no. 34, Paris.

ANDERSON, J. G. C. (1947). The geology of the Highland Border; Stonehaven to Arran. *Trans. R. Soc. Edinb.* **61**, 479–515.

ANDERSON, J. G. C. (1953). The stratigraphical succession of the late Pre-Cambrian and Cambrian of Scotland and Ireland. *Int. Geol. Congr.* (19th Algeria), fasc. 1, 9–19.

ANDERSON, J. G. C. (1959). Pre-Cambrian. In *Lexique Stratigraphique International*, 1 (3aII), 87 pp. Paris.

ARKELL, W. J. (1933). *The Jurassic System in Great Britain*, xii + 681 pp. Oxford.

ARKELL, W. J. (1947*a*). *The Geology of Oxford*, vi + 267 pp. Oxford.

ARKELL, W. J. (1947*b*). The geology of the country around Weymouth, Swanage, Corfe and Lulworth. *Mem. Geol. Surv. U.K.* xii + 386 pp. London: H.M.S.O.

ARKELL, W. J. (1956). *The Jurassic Geology of the World*, xv + 806 pp. Edinburgh: Oliver and Boyd.

Atlas of Britain (1963), xii + 200 pp. Oxford. [Geological systems and selected bore-holes, pp. 9–16; regional geological maps, pp. 55–67.]

BAILEY, E. B. (1929). The palaeozoic mountain systems of Europe and America. *Rep. Br. Ass. Advmt Sci.* (for 1928), pp. 57–76.

BAILEY, E. B. and HOLTEDAHL, O. (1938). Northwestern Europe, Caledonides. *Regionale Geologie der Erde*, **2**(2), 1–84.

BAKER, J. W. (1955). Pre-Cambrian rocks in Co. Wexford. *Geol. Mag.* **92**, 63–8.

BALL, H. W., DINELEY, D. L. and WHITE, E. I. (1961). The Old Red Sandstone of Brown Clee hill and the adjacent area. *Bull. Br. Mus. Nat. Hist. Geol.* **5**, no. 7, 175–310.

BASSETT, D. A. (1963). The Welsh palaeozoic geosyncline: a review of recent work on stratigraphy and sedimentation. Pp. 35–69, in *The British Caledonides* (Johnson and Stewart, 1963).

BASSETT, D. A. and WALTON, E. K. (1960). The Hell's Mouth Grits: Cambrian greywackes in St Tudwal's peninsula, North Wales. *Q. Jl Geol. Soc. Lond.* **116**, 85–110.

BEAVON, R. V., FITCH, F. J. and RAST, N. (1961). Nomenclature and diagnostic characters of ignimbrites with reference to Snowdonia. *L'pool Manchr Geol. J.* **2**, 600–9.

BLACK, M. (1934). The Middle Jurassic rocks. Pp. 261–74 in Wilson, Black and Hemingway (1934).

BLACK, M. (1953). [Lecture on the constitution of the Chalk.] *Proc. Geol. Soc. Lond.* no. 1499, lxxxi–lxxxvi.

BLACK, M. (1957). Sedimentation in relation to the Caledonian movements in Britain. *Int. Geol. Congr.* (20th, Mexico) sect. 5, 139–52.

BLACK, G. P. and WELSH, W. (1961). The Torridonian succession of the island of Rhum. *Geol. Mag.* **98**, 265–76.

BLACKETT, P. M. S. (1961). Comparison of ancient climates with ancient latitudes deduced from rock magnetic measurements. *Proc. R. Soc.* A, **263**, 1–30.

BLACKETT, P. M. S., BULLARD, E. C. and RUNCORN, S. K. *et al.* (1965). A symposium on continental drift. *Phil. Trans. R. Soc.* A, **258**, x + 1–323.

BOTT, M. H. P. (1964). Gravity measurements in the north-eastern part of the Irish Sea. *Q. Jl Geol. Soc. Lond.* **120**, 369–96.

BRITISH MUSEUM (NATURAL HISTORY) (1962). *British Mesozoic Fossils*, v + 205 pp. London.

BRITISH MUSEUM (NATURAL HISTORY) (1963). *British Caenozoic Fossils*, v + 132 pp. London.

BRITISH MUSEUM (NATURAL HISTORY) (1964). *British Palaeozoic Fossils*, vi + 208 pp. London.

British Regional Geology [Geol. Surv.] (also listed under separate authors):
Bristol and Gloucester District, Kellaway and Welch (1948).
Central England, Edmunds and Oakley (1947).
East Anglia, Chatwin (1961).
East Yorkshire and Lincolnshire, Wilson (1948).
The Hampshire Basin, Chatwin (1960).
London and Thames Valley, Sherlock (1960).
Northern England, Eastwood (1953).
North Wales, George (1961).
Pennines and adjacent areas, Edwards and Trotter (1954).
South Wales, Pringle and George (1948).
South-West England, Dewey (1948).
The Wealden District, Gallois (1965).
The Welsh Borderland, Pocock and Whitehead (1948).
The Grampian Highlands, Macgregor (1948).
The Midland Valley of Scotland, Macgregor and Macgregor (1948).
The Northern Highlands, Phemister (1960).
The South of Scotland, Pringle (1948).
The Tertiary Volcanic Districts, Macgregor and Anderson (1961).

BROWN, P. E., MILLER, J. A., SOPER, N. J. and YORK, D. (1965). Potassium-argon age pattern of the British Caledonides. *Proc. Yorks. Geol. Soc.* **35**, 103–38.

BULLARD, E. C. (1964). Continental drift. *Q. Jl Geol. Soc. Lond.* **120**, 1–33.

BULLARD, E. C., EVERETT, J. E. and GILBERT SMITH, A. (1965). The fit of the continents around the Atlantic. *Phil. Trans. R. Soc.* A, **258**, 41–51.

BUTTERWORTH, M. A. and MILLOTT, J. O'N. (1960). Microspore distribution in the coalfields of Britain. *Proc. Int. Comm. Coal Petrol.* no. 3, 157–63.

CASEY, R. (1961). The stratigraphical palaeontology of the Lower Greensand. *Palaeontology*, **3**, 487–621.

CASEY, R. (1963). The dawn of the Cretaceous period in Britain. *Bull. S.-East Un. Scient. Socs*, no. 117, 15 pp.

CHANDLER, M. E. J. (1957). The Oligocene flora of the Bovey Tracey lake basin, Devonshire. *Bull. Brit. Mus. Nat. Hist. Geol.* **3**, no. 3, 71–123.

CHARLESWORTH, J. K. (1957). *The Quaternary Era*, 2 vols., xlvii + 591, xxi + 592–1700 pp. London: Arnold.

CHARLESWORTH, J. K. (1963). *Historical Geology of Ireland*, xxiii + 565 pp. Edinburgh: Oliver and Boyd.

CHARLESWORTH, J. K. *et al.* (1960). The geology of north-east Ireland. *Proc. Geol. Ass.* **71**, 429–60.

CHATWIN, C. P. (1960). *British Regional Geology:* The Hampshire Basin and adjoining areas, 3rd edn., iv + 99 pp. London: H.M.S.O.

CHATWIN, C. P. (1961). *British Regional Geology:* East Anglia and adjoining areas, 4th edn., vi + 100 pp. London: H.M.S.O.

COE, K. (Ed.) (1962). *Some Aspects of the Variscan Fold Belt*, vii + 163 pp. Manchester: The University Press.

COX, A. H. (1925). The geology of the Cader Idris range. *Q. Jl Geol. Soc. Lond.* **81**, 539–94.

COX, A. H. and WELLS, A. K. (1927). The geology of the Dolgelley district, Merionethshire. *Proc. Geol. Ass.* **38**, 265–318.

COX, A. H. *et al.* (1930). The geology of the St David's district, Pembrokeshire. *Proc. Geol. Ass.* **41**, 241–73.

CRAIG, G. Y. (1954). The palaeoecology of the Top Hosie Shale (Lower Carboniferous) at a locality near Kilsyth. *Q. Jl Geol. Soc. Lond.* **110**, 103–19.

CRAIG, G. Y. (Ed.) (1965*a*). *The Geology of Scotland*, xv + 556 pp. Edinburgh: Oliver and Boyd.

CRAIG, G. Y. (1965*b*). Permian and Triassic. Pp. 383–400 in Craig (1965*a*).

CRAIG, G. Y. and WALTON, E. K. (1959). Sequence and structure in the Silurian rocks of Kirkcudbrightshire. *Geol. Mag.* **96**, 209–20.

CRAMPTON, C. B. and CARRUTHERS, R. G. (1914). The geology of Caithness. *Mem. Geol. Surv. U.K.* viii + 194 pp. Edinburgh: H.M.S.O.

CROOKALL, R. (1955). Fossil plants of the Carboniferous rocks of Great Britain, second section. *Mem. Geol. Surv. U.K. Palaeont.* **4**, pt. 1, 84 pp.

CROOKALL, R. (1959). Fossil plants of the Carboniferous rocks of Great Britain, second section. *Mem. Geol. Surv. U.K. Palaeont.* **4**, pt. 2, 85–216.

CUMMINS, W. A. (1957). The Denbigh Grits; Wenlock greywackes in Wales. *Geol. Mag.* **94**, 433–51.

CUMMINS, W. A. (1962). The greywacke problem. *L'pool Manchr Geol. J.* **3**, 51–72.

CURRY, D. (1958). Palaeogene. In *Lexique Stratigraphique International*, **1** (3*a* XII), 82 pp. Paris.

CURRY, D. (1960*a*). Eocene limestones to the west of Jersey. *Geol. Mag.* **97**, 289–98.

CURRY, D. (1960*b*). The Tertiary. Pp. 11–19 *in* The Isle of Wight. *Geol. Ass. Guides*, no. 25, 19 pp.

CURRY, D. (1962). A Lower Tertiary outlier in the central English Channel, with notes on the beds surrounding it. *Q. Jl Geol. Soc. Lond.* **118**, 177–205.

CURRY, D. (1965). The Palaeogene beds of south-east England. *Proc. Geol. Ass.* **76**, 151–73.

CURRY, D., MARTINI, E., SMITH, S. J. and WHITTARD, W. F. (1962). The geology of the western approaches of the English Channel. I. Chalky rocks from the upper reaches of the continental slope. *Phil. Trans. R. Soc.* B, **245**, 267–90.

CURRY, D. and WISDEN, D. E. (1960). Geology of the Southampton area. *Geol. Ass. Guides*, no. 14, 16 pp.

DAVIES, G. M. (1939). *Geology of London and South-east England*, iv + 198 pp. London: Murby.

DAVIES, G. M. (1956). *The Dorset Coast*, 2nd edn. vii + 128 pp. London: A. and C. Black.

DAVIS, A. G. and ELLIOTT, G. F. (1958). The Palaeogeography of the London Clay sea. *Proc. Geol. Ass.* **68**, 255–77.

DAY, A. A., HILL, M. N., LAUGHTON, A. S. and SWALLOW, J. C. (1956). Seismic prospecting in the western approaches of the English Channel. *Q. Jl Geol. Soc. Lond.* **112**, 15–44.

DEAN, W. T. (1958). The faunal succession of the Caradoc Series of south Shropshire. *Bull. Br. Mus. Nat. Hist. Geol.* **3**, 191–231.

DEAN, W. T. (1959). The stratigraphy of the Caradoc Series in the Cross Fell inlier. *Proc. Yorks. Geol. Soc.* **32**, 185–227.

DEAN, W. T. (1964). The geology of the Ordovician and adjacent strata in the southern Caradoc district of Shropshire. *Bull. Br. Mus. Nat. Hist. Geol.* **9**, 257–69.

DEAN, W. T., DONOVAN, D. T. and HOWARTH, M. K. (1961). The Liassic ammonite zones and subzones of the north-west European province. *Bull. Br. Mus. Nat. Hist. Geol.* **4**, no. 10, 435–505.

DEARNLEY, R. (1962). An outline of the Lewisian Complex of the Outer Hebrides in relation to that of the Scottish mainland. *Q. Jl Geol. Soc. Lond.* **118**, 143–76.

DEWEY, H. (1948). *British Regional Geology:* South-West England, 2nd edn., vi + 73 pp. London: H.M.S.O.

DEWEY, J. F. (1963). The Lower Palaeozoic stratigraphy of central Murrisk, County Mayo, Ireland, and the evolution of the south Mayo trough. *Q. Jl Geol. Soc. Lond.* **119**, 313–44.

DONOVAN, D. T. (1963). The geology of British seas. *Univ. Hull Publ.* 24 pp.

DONOVAN, D. T. (1966). *Stratigraphy*, 199 pp. London: Murby.

DONOVAN, D. T. and HEMINGWAY, J. E. (1963). Jurassic. In *Lexique Stratigraphique International*, 1 (3*a*x), 294 pp. Paris.

DUFF, P. MCL. D. and WALTON, E. K. (1962). Statistical basis for cyclothems: a quantitative study of the sedimentary succession in the East Pennine coalfield. *Sedimentology*, **1**, 235–55.

DUNHAM, K. C. (1948). Petrographical descriptions of two cycles of sedimentation from the Middle Coal Measures, Woodborough Borehole. Appendix, pp. 249–53, to Edwards and Stubblefield (1948).

DUNHAM, K. C. (1960). Syngenetic and diagenetic mineralization in Yorkshire. *Proc. Yorks. Geol. Soc.* **32**, 229–84.

DUNHAM, K. C., HEMINGWAY, J. E., VERSEY, H. C. and WILCOCKSON, W. H. (1953). A guide to the geology of the district round Ingleborough. *Proc. Yorks. Geol. Soc.* **29**, 77–115.

DUNHAM, K. C. and ROSE, W. C. C. (1949). Permo-Triassic geology of south Cumberland and Furness. *Proc. Geol. Ass.* **60**, 11–40.

EARP, J. R., MAGRAW, D., POOLE, E. G., LAND, D. H. and WHITEMAN, A. J. (1961). Geology of the country around Clitheroe and Nelson. *Mem. Geol. Surv. U.K.* ix + 346 pp. London: H.M.S.O.

EASTWOOD, T. (1953). *British Regional Geology:* Northern England, 3rd edn., iv + 72 pp. London: H.M.S.O.

EDEN, R. A., ORME, G. R., MITCHELL, M. and SHIRLEY, J. (1964). A study of part of the margin of the Carboniferous Limestone 'massif' in the Pindale area of Derbyshire. *Bull. Geol. Surv. Gt. Br.* no. 21, 73–118.

EDMUNDS, F. H. and OAKLEY, K. P. (1947). *British Regional Geology:* The Central England District, 2nd edn., iv + 80 pp. London: H.M.S.O.

EDWARDS, W. (1951). The concealed coalfield of Yorkshire and Nottinghamshire, 3rd edn. *Mem. Geol. Surv. U.K.*, x + 285 pp. London: H.M.S.O.

EDWARDS, W. and STUBBLEFIELD, C. J. (1948). Marine bands and other faunal marker-horizons in relation to the sedimentary cycles of the Middle Coal Measures of Nottinghamshire and Derbyshire. *Q. Jl Geol. Soc. Lond.* **103**, 209–60.

EDWARDS, W. and TROTTER, F. M. (1954). *British Regional Geology:* The Pennines and adjacent areas, 3rd edn., v + 86 pp. London: H.M.S.O.

ELLIOTT, R. E. (1961). The stratigraphy of the Keuper series in southern Nottinghamshire. *Proc. Yorks. Geol. Soc.* **33**, 197–234.

EVANS, J. W. and STUBBLEFIELD, C. J. (Eds.) (1929). *Handbook of the Geology of Great Britain*, xii + 556 pp. London: Murby.

FALCON, N. L. and KENT, P. E. (1960). Geological results of petroleum exploration in Britain 1945–1957. *Mem. Geol. Soc. Lond.* no. 2, 56 pp.

FENTON, C. L. and FENTON, M. A. (1945). *The Story of the Great Geologists*, xvi + 301 pp. New York: Doubleday, Doran and Co.

FLINN, D. (1959). On certain geological similarities between north-east Shetland and the Jotunheim area of Norway. *Geol. Mag.* **96**, 473–81.

FLINN, D. (1961). Continuation of the Great Glen Fault beyond the Moray Firth. *Nature, Lond.* **191**, 589–91.

FLINT, R. F. (1957). *Glacial and Pleistocene Geology*, xiii + 553 pp. New York: Wiley.

FORD, T. D. (1954). The Upper Carboniferous rocks of the Ingleton Coalfield. *Q. Jl Geol. Soc. Lond.* **110**, 231–65.

FORD, T. D. (1958). Pre-Cambrian fossils from Charnwood Forest. *Proc. Yorks. Geol. Soc.* **31**, 211–17.

FOWLER, A. and ROBBIE, J. A. (1961). Geology of the country around Dungannon. *Mem. Geol. Surv. N. Ireland*, xii + 274 pp. Belfast: H.M.S.O.

FRANCIS, E. H. (1959). The Rhaetic of the Bridgend district, Glamorganshire. *Proc. Geol. Ass.* **70**, 158–78.

FRANCIS, E. H. (1965*a*). Carboniferous. Pp. 309–57 in *The Geology of Scotland* (Craig, 1965*a*).

FRANCIS, E. H. (1965*b*). Carboniferous-Permian igneous rocks. Pp. 359–82 in *The Geology of Scotland* (Craig, 1965*a*).

FUNNELL, B. M. (1961). The Palaeogene and early Pleistocene of Norfolk. Pp. 340–64 in Larwood and Funnell (1961).

GALLOIS, R. W. (1965). *British Regional Geology:* The Wealden District, 4th edn., xii + 101 pp. London: H.M.S.O.

GEIKIE, A. (1905). *The Founders of Geology*, 2nd edn., xi + 486 pp. London: Macmillan.

GEORGE, T. N. (1958*a*). The geology and geomorphology of the Glasgow district. Pp. 17–61 in *The Glasgow Region* (Eds. Miller and Tivy), xix + 325 pp. Glasgow.

GEORGE, T. N. (1958*b*). Lower Carboniferous palaeogeography of the British Isles. *Proc. Yorks. Geol. Soc.* **31**, 227–318.

GEORGE, T. N. (1960*a*). The stratigraphical evolution of the Midland Valley. *Trans. Geol. Soc. Glasg.* **24**, 32–107.

GEORGE, T. N. (1960*b*). Lower Carboniferous rocks in County Wexford. *Q. Jl Geol. Soc. Lond.* **116**, 349–64.

GEORGE, T. N. (1961). *British Regional Geology:* North Wales, 3rd edn., vi + 96 pp. London: H.M.S.O.

GEORGE, T. N. (1962*a*). Devonian and Carboniferous foundations of the Variscides in north-west Europe. Pp. 19–47 in *Some Aspects of the Variscan Fold Belt* (Coe, 1962).

GEORGE, T. N. (1962*b*). Tectonics and palaeogeography in southern England. *Sci. Prog., Lond.* **50**, 192–217.

GEORGE, T. N. (1963*a*). Tectonics and palaeogeography in northern England. *Sci. Prog., Lond.* **51**, 32–59.

GEORGE, T. N. (1963*b*). Palaeozoic growth of the British Caledonides. Pp. 1–33 in *The British Caledonides* (Johnson and Stewart, 1963).

GEORGE, T. N. (1965). The geological growth of Scotland. Pp. 1–48 in *The Geology of Scotland* (Craig, 1965*a*).

GEORGE, T. N. and OSWALD, D. H. (1957). The Carboniferous rocks of the Donegal syncline. *Q. Jl Geol. Soc. Lond.* **113**, 137–83.

GILETTI, B. J., MOORBATH, S. and LAMBERT, R. ST J. (1961). A geochronological study of the metamorphic complexes of the Scottish Highlands. *Q. Jl Geol. Soc. Lond.* **117**, 233–72.

GILL, W. D. (1956). Ireland. In *Lexique Stratigraphique International,* **1** (3*b*), 98 pp. Paris.

GILL, W. D. (1962). The Variscan fold belt in Ireland. Pp. 49–64 in *Some Aspects of the Variscan Fold Belt* (Coe, 1962).

GODWIN, H. (1956). *The History of the British Flora,* viii + 384 pp. Cambridge University Press.

GOLDRING, R. (1962). The bathyal lull: Upper Devonian and Lower Carboniferous sedimentation in the Variscan geosyncline. Pp. 75–91 in *Some Aspects of the Variscan Fold Belt* (Coe, 1962).

GOODLET, G. A. (1957). Lithological variation in the Lower Limestone group of the Midland Valley of Scotland. *Bull. Geol. Surv. Gt Br.* no. 12, 52–65.

GRABAU, A. W. (1913). *Principles of Stratigraphy,* xxxiv + 1185 pp. New York: Seiler. [Reprinted Dover Publications, 2 vols. 1960.]

GREEN, G. W. and WELCH, F. B. A. (1965). Geology of the country around Wells and Cheddar. *Mem. Geol. Surv. U.K.*, x + 225 pp. London: H.M.S.O.

GREENSMITH, J. T., HATCH, F. H. and RASTALL, R. H. (1965). *Petrology of the Sedimentary Rocks,* 4th edn., 408 pp. London: Murby.

HALLAM, A. (1958). The concept of Jurassic axes of uplift. *Sci. Prog., Lond.* **46**, 441–9.

HALLAM, A. (1960*a*). A sedimentary and faunal study of the Blue Lias of Dorset and Glamorgan. *Phil. Trans. R. Soc.* B, **243**, 1–94.

HALLAM, A. (1960*b*). The White Lias of the Devon coast. *Proc. Geol. Ass.* **71**, 47–60.

HALLAM, A. (1965). Jurassic, Cretaceous and Tertiary sediments. Pp. 401–16 in *The Geology of Scotland* (Craig, 1965*a*).

HAMILTON, E. I. (1965). *Applied Geochronology*, xv+267 pp. London: Academic Press.

HANCOCK, J. M. (1961). The Cretaceous system in Northern Ireland. *Q. Jl Geol. Soc. Lond.* **117**, 11–36.

HARLAND, W. B., GILBERT SMITH, A. and WILCOCK, B. (Eds.) (1964). The Phanerozoic time-scale. *Q. Jl Geol. Soc. Lond.* **120 S**, viii+458 pp.

HARPER, J. C. (1948). The Ordovician and Silurian rocks of Ireland. *Proc. L'pool Geol. Soc.* **20**, 48–67.

HARRIS, A. L. (1962). Dalradian geology of the Highland border near Callander. *Bull. Geol. Surv. Gt Br.* no. 19, 1–15.

HEMINGWAY, J. E. (1951). Cyclic sedimentation and the deposition of ironstone in the Yorkshire Lias. *Proc. Yorks. Geol. Soc.* **28**, 67–74.

HENDRIKS, E. M. L. (1937). Rock succession and structure in south Cornwall; a revision, with notes on the central European facies and Variscan folding there present. *Q. Jl Geol. Soc. Lond.* **93**, 322–67.

HENDRIKS, E. M. L. (1959). A summary of present views on the structure of Cornwall and Devon. *Geol. Mag.* **96**, 253–7.

HICKLING, H. G. A. (1908). The Old Red Sandstone of Forfarshire. *Geol. Mag.* (5), **5**, 396–408.

HILL, M. N. and KING, W. B. R. (1953). Seismic prospecting in the English Channel and its geological interpretation. *Q. Jl Geol. Soc. Lond.* **109**, 1–19.

HODSON, F. (1954). The beds above the Carboniferous Limestone in north-west county Clare, Eire. *Q. Jl Geol. Soc. Lond.* **109**, 259–83.

HODSON, F. (1957). Marker horizons in the Namurian of Britain, Ireland, Belgium and Western Germany. *Ass. Étude Paléont. Stratigr. Houill.* Publ. no. 24, 25 pp.

HODSON, F. (1959). The palaeogeography of Homoceras times in western Europe. *Bull. Soc. Belge Géol.* **68**, 134–50.

HODSON, F. and LEWARNE, G. C. (1961). A Mid-Carboniferous (Namurian) basin in parts of the counties of Limerick and Clare, Ireland. *Q. Jl Geol. Soc. Lond.* **117**, 307–33.

HOLLAND, C. H. (1959). The Ludlovian and Downtonian rocks of the Knighton district, Radnorshire. *Q. Jl Geol. Soc. Lond.* **114**, 449–82.

HOLLAND, C. H. and LAWSON, J. D. (1963). Facies patterns in the Ludlovian of Wales and the Welsh Borderland. *L'pool Manchr Geol. J.* **3**, 269–88.

HOLLINGWORTH, S. E. (1938). The recognition and correlation of high-level erosion surfaces in Britain. *Q. Jl Geol. Soc. Lond.* **94**, 55–84.

HOLLINGWORTH, S. E. (1942). The correlation of gypsum-anhydrite deposits and the associated strata in the north of England. *Proc. Geol. Ass.* **53**, 141–51.

HOLLINGWORTH, S. E. (1962). The climatic factor in the geological record. *Q. Jl Geol. Soc. Lond.* **118**, 1–21.

HOLLINGWORTH, S. E. and TAYLOR, J. H. (1951). The Northampton Sand Ironstone. Stratigraphy structure and reserves. *Mem. Geol. Surv. U.K.* viii+211 pp. London: H.M.S.O.

HOLMES, A. (1965). *Principles of Physical Geology*, 2nd edn., xv+1288 pp. London: Nelson.

HOLTEDAHL, O. (1952). The structural history of Norway and its relation to Great Britain. *Q. Jl Geol. Soc. Lond.* **108**, 65–98.

HOUSE, M. R. (1963). Devonian ammonoid successions and facies in Devon and Cornwall. *Q. Jl Geol. Soc. Lond.* **119**, 1–27.

HOUSE, M. R. and SELWOOD, E. B. (1966). Palaeozoic palaeontology in Devon and

Cornwall. Pp. 45–86 in *Present views of some aspects of the geology of Cornwall and Devon* (ed. Hosking, F. K. G. and Shrimpton, G. J.). Truro: O. Blackford.

Howitt, F. (1964). Stratigraphy and structure of the Purbeck inliers of Sussex (England). *Q. Jl Geol. Soc. Lond.* **120**, 77–113.

Hudson, J. D. (1964). The petrology of the sandstones of the Great Estuarine Series and the Jurassic palaeogeography of Scotland. *Proc. Geol. Ass.* **75**, 499–527.

Hughes, N. F. (1958). Palaeontological evidence for the age of the English Wealden. *Geol. Mag.* **95**, 41–9.

Humphries, D. W. (1964). The stratigraphy of the Lower Greensand of the south-west Weald. *Proc. Geol. Ass.* **75**, 39–59.

Hutchins, P. F. (1963). The Lower New Red Sandstone of the Crediton Valley *Geol. Mag.* **100**, 107–28.

James, J. H. (1956). The structure and stratigraphy of part of the Pre-Cambrian outcrop between Church Stretton and Linley, Shropshire. *Q. Jl Geol. Soc. Lond.* **112**, 315–37.

Johnson, G. A. L. (1959). The Carboniferous stratigraphy of the Roman Wall district in western Northumberland. *Proc. Yorks. Geol. Soc.* **32**, 83–130.

Johnson, M. R. W. (1965*a*). Torridonian and Moinian. Pp. 79–113 in *The Geology of Scotland* (Craig, 1965*a*).

Johnson, M. R. W. (1965*b*). Dalradian. Pp. 115–60 in *The Geology of Scotland* (Craig, 1965*a*).

Johnson, M. R. W. and Stewart, F. H. (Eds.) (1963). *The British Caledonides*, ix + 280 pp. Edinburgh: Oliver and Boyd.

Jones, O. T. (1938). On the evolution of a geosyncline. *Q. Jl Geol. Soc. Lond.* **94**, lx–cx.

Jones, O. T. (1956). The geological evolution of Wales and the adjacent regions. *Q. Jl Geol. Soc. Lond.* **111**, 323–51.

Jones, O. T. and Pugh, W. J. (1949). An early Ordovician shore-line in Radnorshire, near Builth Wells. *Q. Jl Geol. Soc. Lond.* **105**, 65–99.

Kellaway, G. A. and Welch, F. B. A. (1948). *British Regional Geology:* Bristol and Gloucester district, 2nd edn., iv + 99 pp. London: H.M.S.O.

Kelling, G. (1961). The stratigraphy and structure of the Ordovician rocks of the Rhinns of Galloway. *Q. Jl Geol. Soc. Lond.* **117**, 37–75.

Kennedy, W. Q. (1946). The Great Glen Fault. *Q. Jl Geol. Soc. Lond.* **102**, 41–76.

Kennedy, W. Q. (1955). The tectonics of the Morar anticline and the problem of the north-west Caledonian front. *Q. Jl Geol. Soc. Lond.* **110**, 357–90.

Kennedy, W. Q. (1958). The tectonic evolution of the Midland Valley of Scotland. *Trans. Geol. Soc. Glasg.* **23**, 106–33.

Kent, P. E. (1949). A structure contour map of the surface of the buried pre-Permian rocks of England and Wales. *Proc. Geol. Ass.* **60**, 87–104.

Kent, P. E. (1955). The Market Weighton structure. *Proc. Yorks. Geol. Soc.* **30**, 197–227.

Kent, P. E. (1966). The structure of the concealed Carboniferous rocks of north-eastern England. *Proc. Yorks. Geol. Soc.* **35**, 323–52.

Kent, P. E. (1967). Outline geology of the southern North Sea Basin. *Proc. Yorks. Geol. Soc.* **36**, 1–22.

Kilburn, C., Pitcher, W. S. and Shackleton, R. M. (1965). The stratigraphy and origin of the Portaskaig Boulder Bed Series (Dalradian). *Geol. J.* **4**, 343–60.

King, C. A. M. (1963). Some problems concerning marine planation and the formation of erosion surfaces. *Trans. Pap. Inst. Br. Geogr.* no. 33, 29–43.

KING, L. C. (1962). *The Morphology of the Earth*, xii + 699 pp. New York: Hafner.
KING, W. B. R. (1954). The geological history of the English Channel *Q. Jl Geol. Soc. Lond.* **110**, 77–101.
KING, W. B. R. and OAKLEY, K. P. (1936). The Pleistocene succession in the lower part of the Thames valley. *Proc. Prehist. Soc.* (N.S.), **2**, 52–76.
KIRKALDY, J. F. (1939). The history of the Lower Cretaceous period in England. *Proc. Geol. Ass.* **50**, 379–417.
KNILL, J. L. (1963). A sedimentary history of the Dalradian Series. Pp. 99–121 in *The British Caledonides* (Johnson and Stewart, 1963).
KRUMBEIN, W. C. and SLOSS, L. L. (1963). *Stratigraphy and Sedimentation*, 2nd edn., xvi + 660 pp. San Francisco: W. H. Freeman.
LARWOOD, G. P. (1961). The Lower Cretaceous deposits of Norfolk. Pp. 280–92 in Larwood and Funnell (1961).
LARWOOD, G. P. and FUNNELL, B. M. (Eds.) (1961). The geology of Norfolk. *Trans. Norfolk Norwich Nat. Soc.* **19**, 269–375.
LAWSON, J. D. (1955). The geology of the May Hill inlier. *Q. Jl Geol. Soc. Lond.* **111**, 85–116.
LEE, G. W. and PRINGLE, J. (1932). A synopsis of the Mesozoic rocks of Scotland. *Trans. Geol. Soc. Glasg.* **19**, 158–224.
LEGGO, P. J., COMPSTON, W. and LEAKE, B. E. (1966). The geochronology of the Connemara granites and its bearing on the antiquity of the Dalradian Series. *Q. Jl Geol. Soc. Lond.* **122**, 91–118.
LINTON, D. L. (1951). Midland drainage: some considerations on its origin. *Advmt Sci.* **7**, 449–56.
LLOYD, W. (1933). The geology of the country around Torquay. *Mem. Geol. Surv. U.K.* ix + 169 pp. London: H.M.S.O.
MACGREGOR, A. G. (1948). *British Regional Geology:* The Grampian Highlands, 2nd edn. viii + 83 pp. Edinburgh: H.M.S.O.
MACGREGOR, A. G. and ANDERSON, F. W. (1961). *British Regional Geology:* Scotland: The Tertiary Volcanic Districts. 3rd edn., viii + 120 pp. Edinburgh: H.M.S.O.
MACGREGOR, M. and MACGREGOR, A. G. (1948). *British Regional Geology:* The Midland Valley of Scotland, 2nd edn., vi + 95 pp. Edinburgh: H.M.S.O.
MAGRAW, D., CLARKE, A. M. and SMITH, D. B. (1963). The stratigraphy and structure of part of the south-east Durham coalfield. *Proc. Yorks. Geol. Soc.* **34**, 153–208.
MENEISY, M. Y. and MILLER, J. A. (1963). A geochronological study of the crystalline rocks of Charnwood Forest, England. *Geol. Mag.* **100**, 507–23.
MERCY, E. L. P. (1965). Caledonian igneous activity. Pp. 229–67 in *The Geology of Scotland* (Craig, 1965*a*).
MITCHELL, G. H. (1956). The geological history of the Lake District. *Proc. Yorks. Geol. Soc.* **30**, 407–463.
MITCHELL, G. H. and MYKURA, W. (1962). The geology of the neighbourhood of Edinburgh. *Mem. Geol. Surv. U.K.* xii + 159 pp. Edinburgh: H.M.S.O.
MOORE, D. (1958). The Yoredale Series of Upper Wensleydale and adjacent parts of north-west Yorkshire. *Proc. Yorks. Geol. Soc.* **31**, 91–148.
MOORE, D. (1959). Role of deltas in the formation of some British Lower Carboniferous cyclothems. *J. Geol.* **67**, 522–39.
MORLEY DAVIES, A. and STUBBLEFIELD, C. J. (1961). *An Introduction to Palaeontology*, 3rd edn., ix + 322 pp. London: Murby.
NAIRN, A. E. M. (Ed.) (1961). *Descriptive Palaeoclimatology*, xi + 380 pp. New York: Interscience.

NAIRN, A. E. M. (Ed.) (1964). *Problems in Palaeoclimatology*, xiii + 705 pp. London: Interscience.
NEVILL, W. E. (1961). The Westphalian of Ireland. *C.R. Int. Congr. Carb. Stratigr. Palaeont. (Heerlen)*, **2**, 453–60.
OAKLEY, K. P. (1953). Swanscombe man. *Proc. Geol. Ass.* **63**, 271–300.
OAKLEY, K. P. (1961). *Man the Toolmaker*, 5th edn., vi + 98 pp. London: British Museum (Natural History).
OAKLEY, K. P. and BADEN POWELL, D. F. W. (1963). Neogene and Pleistocene. In *Lexique Stratigraphique International*, I (3a XIII), 166 pp. Paris.
PEAKE, N. B. and HANCOCK, J. M. (1961). The Upper Cretaceous of Norfolk. Pp. 293–339 in Larwood and Funnell (1961).
PEACH, B. N. and HORNE, J. (1907). The geological structure of the North-West Highlands of Scotland. *Mem. Geol. Surv. U.K.* xviii + 668 pp. Glasgow: H.M.S.O.
PENNY, L. F. (1964). A review of the Last Glaciation in Great Britain. *Proc. Yorks. Geol. Soc.* **34**, 387–411.
PETTIJOHN, F. J. (1957). *Sedimentary Rocks*, 2nd edn., xvi + 718 pp. New York: Harper.
PHEMISTER, J. (1960). *British Regional Geology:* Scotland: the Northern Highlands, 3rd edn., vii + 104 pp. Edinburgh: H.M.S.O.
PITCHER, W. S. and CHEESMAN, R. L. (1954). Summer field meeting in north west Ireland with an introductory note on the geology. *Proc. Geol. Ass.* **65**, 345–71.
PITCHER, W. S. and SHACKLETON, R. M. (1966). On the correlation of certain Lower Dalradian successions in northwest Donegal. *Geol. J.* **5**, 149–56.
PITCHER, W. S. *et al.* (1960). The London Region. *Geol. Ass. Guides.* no. 30, 41 pp.
PITCHER, W. S., ELWELL, R. W. D., TOZER, C. F. and CAMBRAY, F. W. (1964). The Leannan fault. *Q. Jl Geol. Soc. Lond.* **120**, 241–73.
POCOCK, R. W. and WHITEHEAD, T. H. (1948). *British Regional Geology:* The Welsh Borderland, 2nd edn., vi + 82 pp. London: H.M.S.O.
POCOCK, R. W., WHITEHEAD, T. H., WEDD, C. B., and ROBERTSON, T. (1938). Shrewsbury district. *Mem. Geol. Surv. U.K.* xxi + 297 pp.
PRENTICE, J. E. (1960*a*). Dinantian, Namurian and Westphalian rocks of the district south-west of Barnstaple, north Devon. *Q. Jl Geol. Soc. Lond.* **115**, 261–89.
PRENTICE, J. E. (1960*b*). The stratigraphy of the Upper Carboniferous rocks of the Bideford region, north Devon. *Q. Jl Geol. Soc. Lond.* **116**, 397–408.
PRENTICE, J. E. (1962). The sedimentation history of the Carboniferous in Devon. Pp. 93–108 in *Some Aspects of the Variscan Fold Belt* (Coe, 1962).
PRINGLE, J. (1948). *British Regional Geology:* The south of Scotland, 2nd edn., vi + 87 pp. Edinburgh: H.M.S.O.
PRINGLE, J. and GEORGE, T. N. (1948). *British Regional Geology:* South Wales, 2nd edn., vi + 102 pp. London: H.M.S.O.
RAMSAY, J. G. (1963). Structure and metamorphism of the Moine and Lewisian rocks of the north-west Caledonides. Pp. 143–75 in *The British Caledonides* (Johnson and Stewart, 1963).
RAMSAY, J. G. and SPRING, J. S. (1962). Moine stratigraphy in the western Highlands of Scotland. *Proc. Geol. Ass.* **73**, 295–326.
RAMSBOTTOM, W. H. C. (1965). A pictorial diagram of the Namurian rocks of the Pennines. *Trans. Leeds Geol. Ass.* **7**, 181–4.
RAMSBOTTOM, W. H. C., RHYS, G. H. and SMITH, E. G. (1962). Boreholes in the

Carboniferous rocks of the Ashover district, Derbyshire. *Bull. Geol. Surv. Gr. Br.* no. 19, 75–168.

Rast, N. (1963). Structure and metamorphism of the Dalradian rocks of Scotland. Pp. 123–41 in *The British Caledonides* (Johnson and Stewart, 1963).

Rast, N., Beavon, R. V. and Fitch, F. J. (1958). Sub-aerial volcanicity in Snowdonia. *Nature, Lond.* **181**, no. 4607, 508.

Rayner, D. H. (1953). The Lower Carboniferous rocks in the north of England: a review. *Proc. Yorks. Geol. Soc.* **28**, 231–315.

Read, H. H. (1958). A centenary lecture: stratigraphy in metamorphism. *Proc. Geol. Ass.* **69**, 83–102.

Read, H. H. (1961). Aspects of Caledonian magmatism in Britain. *L'pool Manchr Geol. J.* **2**, 653–83.

Read, H. H. and Watson, J. (1962). *Introduction to Geology*. Vol. 1. *Principles*, x+693 pp. London: Macmillan.

Reid, E. M. and Chandler, M. E. J. (1933). *The London Clay Flora*, viii+561 pp. London: British Museum (Natural History).

Richey, J. E. and Kennedy, W. Q. (1939). The Moine and Sub-Moine Series of Morar, Inverness-shire. *Bull. Geol. Surv. Gr. Br.* **1**, 26–45.

Roberts, J. L. and Treagus, J. E. (1964). A re-interpretation of the Ben Lui fold. *Geol. Mag.* **101**, 512–16.

Rolfe, W. D. I. (1960). The Silurian inlier of Carmichael, Lanarkshire. *Trans. R. Soc. Edinb.* **64**, 245–60.

Rolfe, W. D. I. (1961). The geology of the Hagshaw Hills Silurian inlier, Lanarkshire. *Trans. Edinb. Geol. Soc.* **18**, 240–69.

Rose, G. N. and Kent, P. E. (1955). A *Lingula*-bed in the Keuper of Nottinghamshire. *Geol. Mag.* **92**, 476–80.

Runcorn, S. K. (Ed.) (1962). *Continental Drift*, xii+338 pp. New York: Academic Press.

Sabine, P. A. and Watson, J. (1965). Isotopic age-determinations of rocks from the British Isles, 1955–64. *Q. Jl Geol. Soc. Lond.* **121**, 477–533.

Selley, R. C., Shearman, D. J., Sutton, J. and Watson, J. (1963). Some underwater disturbances in the Torridonian of Skye and Raasay. *Geol. Mag.* **100**, 224–43.

Shackleton, R. M. (1940). The succession of rocks in the Dingle peninsula, Co. Kerry. *Proc. R. Ir. Acad.* **46B**, 1–12.

Shackleton, R. M. (1954*a*). The structure and succession of Anglesey and the Lleyn peninsula. *Advmt Sci.* **11**, 166–8.

Shackleton, R. M. (1954*b*). The structural evolution of North Wales. *L'pool Manchr Geol. J.* **1**, 261–96.

Shackleton, R. M. (1958). Downward facing structures of the Highland Border. *Q. Jl Geol. Soc. Lond.* **113**, 361–92.

Sherlock, R. L. (1947). *The Permo-Triassic Formations*, 367 pp. London: Hutchinson.

Sherlock, R. L. (1960). *British Regional Geology:* London and the Thames Valley, 3rd edn. iv+62 pp. London: H.M.S.O.

Shotton, F. W. (1953). The Pleistocene deposits of the area between Coventry, Rugby and Leamington and their bearing upon the topographic development of the Midlands. *Phil. Trans. R. Soc.* B, **237**, 209–60.

Shotton, F. W. (1956). Some aspects of the New Red desert in Britain. *L'pool Manchr Geol. J.* **1**, 450–65.

Shotton, F. W. (1962). The physical background of Britain in the Pleistocene. *Advmt Sci.* **19**, 193–206.

SHROCK, R. R. (1948). *Sequence in Layered Rocks*, xiii + 507 pp. New York: McGraw-Hill.

SIMPSON, A. (1963). The stratigraphy and tectonics of the Manx Slate Series, Isle of Man. *Q. Jl Geol. Soc. Lond.* **119**, 367–400.

SIMPSON, S. (1951). Some solved and unsolved problems of the marine Devonian in Great Britain. *Abh. Senckenb. Naturforsch. Ges.* no. 485, 53–66.

SIMPSON, S. (1959*a*). Culm stratigraphy and the age of the main orogenic phase in Devon and Cornwall. *Geol. Mag.* **96**, 201–8.

SIMPSON, S. (1959*b*). Devonian. In *Lexique Stratigraphique International*, **1**, (3 a VI), 131 pp. Paris.

SIMPSON, S. (1962). Variscan orogenic phases. Pp. 65–73 in *Some aspects of the Variscan Fold Belt* (Coe, 1962).

SISSONS, J. B. (1965). Quaternary. Pp. 467–503 in *The Geology of Scotland* (Craig, 1965*a*).

SMITH, A. H. V. (1962). The palaeoecology of Carboniferous peats based on the miospores and petrography of bituminous coals. *Proc. Yorks. Geol. Soc.* **33**, 423–74.

SMITH, D. B. (1958). Observations on the Magnesian Limestone reefs of north-eastern Durham. *Bull. Geol. Surv. Gr. Br.* no. 15, 71–84.

SMYTH, L. B. (1950). The Carboniferous System in north County Dublin. *Q. Jl Geol. Soc. Lond.* **105**, 295–326.

SMYTH, L. B. *et al.* (1939). The geology of south-east Ireland together with parts of Limerick, Clare and Galway. *Proc. Geol. Ass.* **50**, 287–351.

SQUIRREL, H. C. (1958). New occurrences of fish remains in the Silurian of the Welsh Borderland. *Geol. Mag.* **95**, 328–32.

STANTON, W. I. (1960). The Lower Palaeozoic rocks of south-west Murrisk, Ireland. *Q. Jl Geol. Soc. Lond.* **116**, 269–96.

STEPHENS, J. V., MITCHELL, G. H. and EDWARDS, W. (1953). Geology of the country between Bradford and Skipton. *Mem. Geol. Surv. U.K.* vii + 180 pp. London: H.M.S.O.

STEVENSON, I. P. and MITCHELL, G. H. (1955). Geology of the country between Burton-upon-Trent, Rugeley and Uttoxeter. *Mem. Geol. Surv. U.K.* vii + 178 pp. London: H.M.S.O.

STEWART, A. D. (1962*a*). On the Torridonian sediments of Colonsay and their relationship to the main outcrop of northwest Scotland. *L'pool Manch Geol. J.* **3**, 121–56.

STEWART, A. D. (1962*b*). Greywacke sedimentation in the Torridonian of Colonsay and Oronsay. *Geol. Mag.* **99**, 399–419.

STEWART, F. H. (1954). Permian evaporites and associated rocks in Texas and New Mexico compared with those of northern England. *Proc. Yorks. Geol. Soc.* **29**, 185–235.

STEWART, F. H. (1963). The Permian Lower Evaporites of Fordon in Yorkshire. *Proc. Yorks. Geol. Soc.* **34**, 1–44.

STEWART, F. H. (1965). Tertiary igneous activity. Pp. 417–65 in *The Geology of Scotland* (Craig, 1965*a*).

STONE, M. (1957). The Aberfoyle anticline, Callander, Perthshire. *Geol. Mag.* **94**, 265–76.

STUBBLEFIELD, C. J. (1939). Some Devonian and supposed Ordovician fossils from south-west Cornwall. *Bull. Geol. Surv. Gr. Br.* no. 2, 63–71.

STUBBLEFIELD, C. J. (1956). Cambrian palaeogeography in Britain. *Int. Geol. Congr.* (20th, Mexico), Sect. 1, 1–43.

STUBBLEFIELD, C. J. (1959). Cambrian. In *Lexique Stratigraphique International*, **1** (3aIII), 95 pp. Paris.

STUBBLEFIELD, C. J. (1967). [Demonstration: Some results of a recent Geological Survey boring in Huntingdonshire.] *Proc. Geol. Soc. Lond.* no. 1637.

STUBBLEFIELD, C. J. and TROTTER, F. M. (1957). Divisions of the Coal Measures on Geological Survey maps of England and Wales. *Bull. Geol. Surv. Gr. Br.* no. 13, 1–5.

SUTTON, J. (1963). Some events in the evolution of the Caledonides. Pp. 249–69 in *The British Caledonides* (Johnson and Stewart, 1963).

SUTTON, J. and WATSON, J. (1951). The pre-Torridonian metamorphic history of the Loch Torridon and Scourie areas in the North-west Highlands, and its bearing on the chronological classification of the Lewisian. *Q. Jl Geol. Soc. Lond.* **106**, 241–307.

SUTTON, J. and WATSON, J. (1960). Sedimentary structures in the Epidotic Grits of Skye. *Geol. Mag.* **97**, 106–22.

SUTTON, J. and WATSON, J. (1964). Some aspects of Torridonian stratigraphy in Skye. *Proc. Geol. Ass.* **75**, 251–89.

SWINNERTON, H. H. and KENT, P. E. (1949). *The Geology of Lincolnshire*, 126 pp. Lincoln.

TAYLOR, B. J., PRICE, R. H. and TROTTER, F. M. (1963). Geology of the country around Stockport and Knutsford. *Mem. Geol. Surv. U.K.* viii + 183 pp. London: H.M.S.O.

TAYLOR, J. H. (1949). Petrology of the Northampton Sand Ironstone formation. *Mem. Geol. Surv. U.K.* vi + 111 pp. London: H.M.S.O.

TAYLOR, J. H. (1963). Geology of the country around Kettering, Corby and Oundle. *Mem. Geol. Surv. U.K.* ix + 149 pp. London: H.M.S.O.

THOMAS, G. E. and THOMAS, T. M. (1956). The volcanic rocks of the area between Fishguard and Strumble Head, Pembrokeshire. *Q. Jl Geol. Soc. Lond.* **112**, 291–314.

THURRELL, R. G. (1961). The sub-Cretaceous rocks of Norfolk. Pp. 271–9 in Larwood and Funnell (1961).

TRECHMANN, C. T. (1931). The Permian in Durham. *Proc. Geol. Ass.* **42**, 247–52.

TREMLETT, W. E. (1959). The Pre-Cambrian rocks of southern Co. Wicklow (Ireland). *Geol. Mag.* **96**, 58–68.

TROTTER, F. M. (1960). Upper Carboniferous. In *Lexique Stratigraphique International*, **1** (3aVIII), 365 pp. Paris.

TRUEMAN, A. E. (1946). Stratigraphical problems in the Coal Measures of Europe and North America. *Q. Jl Geol. Soc. Lond.* **102**, xlix–xciii.

TRUEMAN, A. E. (Ed.) (1954). The *Coalfields of Great Britain*, xi + 396 pp. London: Arnold.

TURNER, J. S. (1949). The deeper structure of central and northern England. *Proc. Yorks. Geol. Soc.* **27**, 280–97.

TURNER, J. S. (1952). The Lower Carboniferous rocks of Ireland. *L'pool Manchr Geol. J.* **1**, 113–47.

TWENHOFEL, W. H. (1950). *Principles of Sedimentation*, 2nd edn. xvi + 673 pp. New York: McGraw-Hill.

TYRRELL, G. W. (1928). The geology of Arran. *Mem. Geol. Surv. U.K.* viii + 292 pp. Edinburgh: H.M.S.O.

WALTON, E. K. (1963). Sedimentation and structure in the Southern Uplands. Pp. 71–97 in *The British Caledonides* (Johnson and Stewart, 1963).

WALTON, E. K. (1965*a*). Lower Palaeozoic rocks—stratigraphy. Pp. 161–200 in *The Geology of Scotland* (Craig, 1965*a*).

WALTON, E. K. (1965*b*). Lower Palaeozoic rocks—palaeogeography and structure. Pp. 201–27 in *The Geology of Scotland* (Craig, 1965*a*).
WATERSTON, C. D. (1965). Old Red Sandstone. Pp. 269–308 in *The Geology of Scotland* (Craig 1965*a*).
WATSON, J. V. (1963). Some problems concerning the evolution of the Caledonides of the Scottish Highlands. *Proc. Geol. Ass.* **74**, 213–58.
WATSON, J. V. (1965). Lewisian. Pp. 49–77 in *The Geology of Scotland* (Craig, 1965*a*).
WATTS, W. W. (1947). *Geology of the Ancient Rocks of Charnwood Forest, Leicestershire*, 160 pp. Leicester Lit. Phil. Soc.
WELCH, F. B. A. and TROTTER, F. M. (1961). Geology of the country around Monmouth and Chepstow. *Mem. Geol. Surv. U.K.* viii + 164 pp. London: H.M.S.O.
WELLER, J. M. (1960). *Stratigraphic Principles and Practice*, xvi + 725 pp. New York: Harper.
WELLS, A. J. (1960). Cyclic sedimentation: a review. *Geol. Mag.* **97**, 389–403.
WEST, R. G. (1961). The glacial and interglacial deposits of Norfolk. Pp. 365–75 in Larwood and Funnell (1961).
WEST, R. G. (1963). Problems of the British Quaternary. *Proc. Geol. Ass.* **74**, 147–86.
WEST, R. G. and DONNER, J. J. (1956). The glaciations of East Anglia and the East Midlands. *Q. Jl Geol. Soc. Lond.* **112**, 69–91.
WESTOLL, T. S. (1951). The vertebrate bearing strata of Scotland. *Int. Geol. Congr.* (18th, Gt Britain), part 11, 5–21.
WHITEHEAD, T. H., ANDERSON, W., WILSON, V. and WRAY, D. A. (1952). The Liassic ironstones. *Mem. Geol. Surv. U.K.* viii + 211 pp. London: H.M.S.O.
WHITAKER, J. H. MCD. (1962). Geology of the country around Leintwardine, Herefordshire. *Q. Jl Geol. Soc. Lond.* **118**, 319–51.
WHITTARD, W. F. (1928). The stratigraphy of the Valentian rocks of Shropshire. The main outcrop. *Q. Jl Geol. Soc. Lond.* **83**, 737–58.
WHITTARD, W. F. (1932). The stratigraphy of the Valentian rocks of Shropshire. The Longmynd–Shelve and Breidden outcrops. *Q. Jl Geol. Soc. Lond.* **88**, 859–902.
WHITTARD, W. F. (1952). A geology of south Shropshire. *Proc. Geol. Ass.* **63**, 143–97.
WHITTARD, W. F. (1955). The Ordovician trilobites of the Shelve inlier, west Shropshire. Part I. *Monogr. Palaeontogr. Soc.* pp. 1–40.
WHITTARD, W. F. (1960). Ordovician. In *Lexique Stratigraphique International*, **1** (3aIV), 296 pp. Paris.
WHITTARD, W. F. (1961). Silurian. In *Lexique Stratigraphique International*, **1** (3aV), 273 pp. Paris.
WHITTARD, W. F. (1962). Geology of the western approaches of the English Channel: a progress report. *Proc. R. Soc.* A, **265**, 395–406.
WILLIAMS, A. (1953). The geology of the Llandeilo district, Carmarthenshire. *Q. Jl Geol. Soc. Lond.* **108**, 177–207.
WILLIAMS, A. (1962). The Barr and Lower Ardmillan Series (Caradoc) of the Girvan district, south-west Ayrshire, with a description of the brachiopods. *Mem. Geol. Soc. Lond.* no. 3, 267 pp.
WILLS, L. J. (1929). *The Physiographical Evolution of Britain*, viii + 376 pp. London: Arnold.
WILLS, L. J. (1948). *The Palaeogeography of the Midlands*, vii + 144 pp. London: Hodder and Stoughton.

WILLS, L. J. (1951). *A Palaeogeographical Atlas*, 64 pp. London: Blackie.
WILLS, L. J. (1956). *Concealed Coalfields*, xiii+208 pp. London: Blackie.
WILSON, H. E. and ROBBIE, J. A. (1966). Geology of the country around Ballycastle. *Mem. Geol. Surv. U.K.* xv+370 pp. Belfast: H.M.S.O.
WILSON, V. (1948). *British Regional Geology:* East Yorkshire and Lincolnshire. iv+94 pp. London: H.M.S.O.
WILSON, V., BLACK, M. and HEMINGWAY, J. E. (1934). A synopsis of the Jurassic rocks of Yorkshire. *Proc. Geol. Ass.* **45**, 247–306.
WILSON, V., WELCH, F. B. A., ROBBIE, J. A. and GREEN, G. W. (1958). Geology of the country around Bridport and Yeovil. *Mem. Geol. Surv. U.K.* xii+239 pp. London: H.M.S.O.
WOOD, A. and SMITH, A. J. (1959). The sedimentation and sedimentary history of the Aberystwyth Grits (Upper Llandoverian). *Q. Jl Geol. Soc. Lond.* **114**, 163–95.
WOODLAND, A. W. and EVANS, W. B. (1964). The Geology of the South Wales Coalfield. Part IV. The country around Pontypridd and Maesteg. *Mem. Geol. Surv. U.K.* xiv+391 pp. London: H.M.S.O.
WOODWARD, H. B. (1907). *The History of the Geological Society of London*, xix+336 pp. The Geological Society.
WOOLDRIDGE, S. W. (1926). The structural evolution of the London basin. *Proc. Geol. Ass.* **37**, 162–96.
WOOLDRIDGE, S. W. (1960). The Pleistocene succession in the London basin. *Proc. Geol. Ass.* **71**, 113–29.
WOOLDRIDGE, S. W. and LINTON, D. L. (1955). *Structure, Surface and Drainage in South-east England*, viii+176 pp. London: George Philip.
WRIGHT, C. W. and WRIGHT, E. V. (1951). A survey of the fossil cephalopoda of the Chalk of Great Britain. *Monogr. Palaeontogr. Soc.* pp. 1–40.
WRIGLEY, A. and DAVIS, A. G. (1937). The occurrence of *Nummulites planulatus* in England, with a revised correlation of the strata containing it. *Proc. Geol. Ass.* **48**, 203–28.
ZEUNER, F. E. (1959). *The Pleistocene Period*, xviii+447 pp. London: Hutchinson.
ZITTEL, K. A. (1901). *History of Geology and Palaeontology* (translated by M. M. Ogilvie-Gordon); xiii+562 pp. London: Walter Scott.

INDEX

Authors and zonal index fossils are given in the list of References and the Appendix respectively, and do not appear in the Index. Page numbers in *italics* refer to illustrations. Subsidiary items under the main entries are arranged in page, not alphabetical, order.